SpringerWienNewYork

CISM COURSES AND LECTURES

The series presents lecture notes, monographs, edited works and proceedings in the field of Mechanics, Engineering, Computer Science and Applied Mathematics.
Purpose of the series is to make known in the international scientific and technical community results obtained in some of the activities organized by CISM, the International Centre for Mechanical Sciences.

INTERNATIONAL CENTRE FOR MECHANICAL SCIENCES

COURSES AND LECTURES - No. 489

WAVES IN GEOPHYSICAL FLUIDS

TSUNAMIS, ROGUE WAVES, INTERNAL WAVES AND INTERNAL TIDES

EDITED BY

JOHN GRUE AND KARSTEN TRULSEN
UNIVERSITY OF OSLO, NORWAY

SpringerWienNewYork

This volume contains 172 illustrations

Printed in Italy
SPIN 10817628

All contributions have been typeset by the authors.

ISBN-10 3-211-37460-4 SpringerWienNewYork
ISBN-13 978-3-211-37460-3 SpringerWienNewYork

Preface

This volume – *Waves in Geophysical Fluids* – contains extended versions of the lectures that were given at a summerschool in Centre International des Sciences Mécaniques (CISM), Udine, Italy, during the week of 11-15 July 2005. The subjects of the school covered: the formation of the very long *tsunamis*, the much shorter *rogue waves* taking place on the sea surface, very large *oceanic internal waves* and, finally, *internal tides and hot spots* in the ocean. The subjects were chosen because of their implications to the safety of human activity in and in relation to the ocean environment, and the impact of the processes to the ocean currents and climate. The physical phenomena and how to mathematically and numerically model them were described in the lectures.

The giant and destructive tsunami that occurred in the Indian Ocean on the 26th December 2004 urged scientists to communicate to the authorities and public the existing knowledge of tsunami wave modelling, and how the forecasts can be used for warning and evacuation of the people in the near shore region, corresponding to what is practice along the coasts of the Pacific Ocean. The first chapter of this book is precicsely describing the current practice of the forecasting and risk evaluation of tsunamis, and includes also an overview of the geograpical distribution of tsunami sources in the world's oceans. The mathematics and numerics behind the computations are described, including formulas for the tectonic generation, land slides and explosions, the typical length scales and magnitude of the resulting waves, and run-up on the coastline.

A prominent ship-owner said, during a visit to the hydrodynamic laboratory at the University of Oslo: "I have lost many men". He was speaking about how the *extreme – or rogue – waves* at sea had caused the loss of several of his ships. The extreme waves, like the Camille and Draupner waves, typically about 200 m long, are capable of lifting the sea surface almost 20 m above the level of equilibrium and causing a forward velocity of the water particles of more than 14 m/s, or 50 km/hour. The waves are caused by various physical mechanisms, which share a common factor, namely, focussing. A stochastic description based on Monte–Carlo simulations using deterministic models is given in chapter two. A deterministic description of focussing mechanisms is given in chapter three. The description is supplemented by a variety of nonlinear wave models, including analytical breather formulas. A novel fully nonlinear wave model in three dimensions is useful for rapid and accurate calculations of rogue waves, and compares favourably with PIV experiments (chapter four). The velocities and Lagrangian accelerations of the Draupner wave are estimated. The chapter culminates with numerical predictions of various crescent wave patterns in steep waves, and the competition between class I and class II instabilities.

Very large oceanic internal waves are a common phenomenon in all of the world's oceans. The full story about these waves is given in chapter five, including the discovery of the waves four decades ago, how the waves are visible from above through the signatures on the sea surface, and how to compute them. The strong interaction between the deep thermoclines in the ocean and the shelf slope causes strong near-bed currents. This is a factor that needs to be accounted for when it comes to new offshore installations in deep water, like the Ormen Lange gas field of Norsk Hydro at the slope of the Norwegian continental shelf. Weakly and fully nonlinear models of the waves are described including visualizations.

Energetic internal tides – hot spots – are formed by the barotropic tide interacting with submarine ridges and shelves, such as, *e.g.*, the Mid-Atlantic Ridge. Observations from several field campaigns in all parts of the world's oceans, and interpretation of the data including the description of spectra, are given in chapter six. The observation and the modeling of extreme internal tide generation at near critical latitudes in the Barents Sea as well as in, *e.g.*, the Strait of Gibraltar are described. The consequences of the internal tidal motion on the ocean currents and the climate, through the internal breaking and production of turbulent motion in the ocean, are elaborated.

We are grateful to Professor Manuel Velarde of Universidad Complutense de Madrid for suggesting us to submit a proposal for this summer school, and to the Rectors of CISM for approving it, giving us also the chance to prepare the material for this book. The school could not have been realized without the rather generous support from CISM and also by the economical support from the Strategic University Program: Modelling Waves and Currents for Sea Structures 2002-6 at the University of Oslo, funded by the Research Council of Norway.

Oslo and Marseille, May 2006

John Grue and Karsten Trulsen

CONTENTS

3

5

6

Hydrodynamics of Tsunami Waves

Efim Pelinovsky
Laboratory of Hydrophysics,
Institute of Applied Physics
Nizhny Novgorod, Russia

Abstract The giant tsunami that occurred in the Indian Ocean on 26th December 2004 draws attention to this natural phenomenon. The given course of lectures deals with the physics of the tsunami wave propagation from the source to the coast. Briefly, the geographical distribution of the tsunamis is described and physical mechanisms of their origin are discussed. Simplified robust formulas for the source parameters (dimension and height) are given for tsunamis of different origin. It is shown that the shallow-water theory is an adequate model to describe the tsunamis of the seismic origin; meanwhile for the tsunamis of the landslide or explosion origin (volcanoes, asteroid impact) various theories (from linear dispersive to nonlinear shallow-water equations) can be applied. The applicability of the existing theories to describe the tsunami wave propagation, refraction, transformation and climbing on the coast is demonstrated. Nonlinear-dispersive effects including the role of the solitons are discussed. The practical usage of the tsunami modeling for the tsunami forecasting and tsunami risk evaluation is described. The results of the numerical simulations of the two global tsunamis in the Indian Ocean induced by the catastrophic Krakatau eruption in 1883 and the strongest North Sumatra earthquake in 2004 are given.

1 Introduction

After the catastrophic tsunami in the Indian Ocean occurred on December 26, 2004 everybody knows the Japanese word "tsunami" meaning "harbor wave". Now tsunami waves are defined as surface gravity waves that occur in the ocean as the result of large-scale short-term perturbations: underwater earthquakes, eruption of underwater volcanoes, submarine landslides, underwater explosions, rock and asteroid falls, avalanche flows in the water from "land" mountains and volcanoes, and sometimes drastic changes of weather conditions (Murty, 1977; Pelinovsky, 1982, 1996; Bryant, 2001). The characteristic parameters of tsunamis are: duration, 5 to 150 min, length, 100 m to 1000 km, speed propagation, 1 to 200 m/s, and their heights can be up to tens of meters. Various data of observations in the world's oceans can be found in catalogues (Soloviev et al., 1988, 2000; Lander and Lockridge, 1989; O'Loughlin and Lander, 2003), and in sites of the National Geophysical Data Center - NGDC (http://www.ngdc.noaa.gov) and the Tsunami Laboratory (http://tsun.sscc.ru/htdbpac/). Most of them (more than 1000) occurred in the Pacific, and about 100 - in the Atlantic and Indian Oceans. Earthquakes

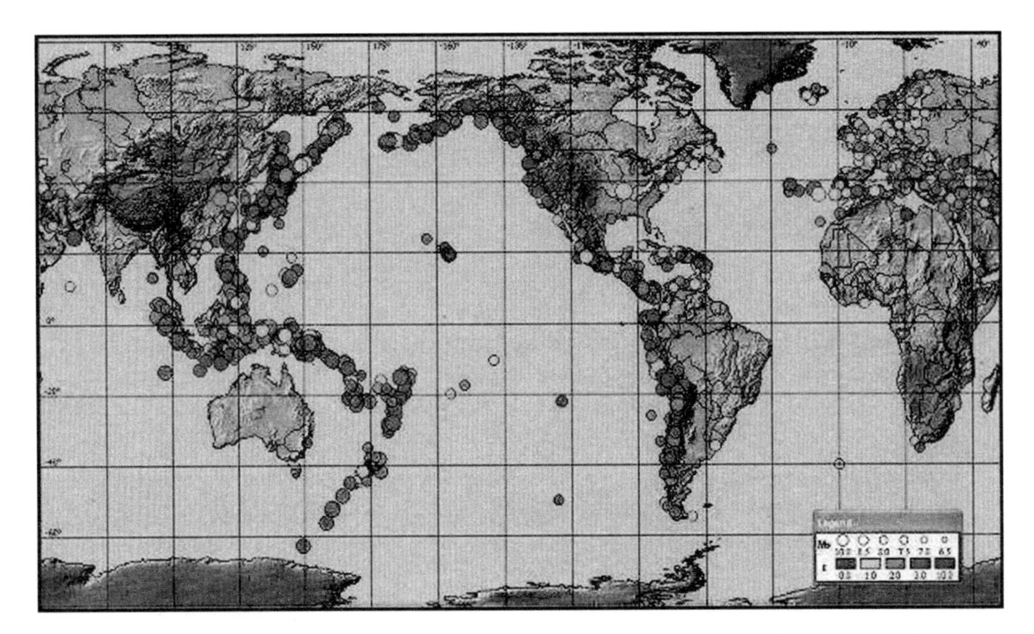

Figure 1. Geographical distribution of tsunami sources in the world's oceans. The size of circles is proportional to the earthquake magnitude, density of gray tone - to the tsunami intensity.

are responsible for 75% of all events, and their geographical distribution is shown in figure 1 (Gusiakov, 2005). A great attention is paid to tsunamis, which can occur due to the possible asteroid impact (diameter 100 m - 50 km) in the ocean, and the number of documented events exceeds 10 (figure 2).

By the number of deaths tsunami waves are in the fifth place after earthquakes, floods, typhoons and volcanic eruptions (very rough estimates). However, tsunamis have a high destructive potential for countries on large distances from the source, as it was demonstrated on December 26, 2004 when tsunami waves killed more than 250,000 people in Indonesia, Thailand, Sri Lanka, India, the Maldives, Kenya and Somali. Approximately 10 events occurred each year, and 10% of them are really damaging tsunamis (figure 3).

Short- and long-term tsunami predictions require complex studies of the mechanisms of tsunami excitation, the adequate modelling of tsunami propagation and the climbing on a beach, and the subsequent development of tsunami risk maps. This lecture deals with the hydrodynamics of the tsunami waves for all stages of tsunami propagation from the source to the coast.

2 Parameters of tsunami waves in the source

The determination of the tsunami wave characteristics in the source is an extremely difficult problem of the geophysics. Taking into account that the main goal of our study is

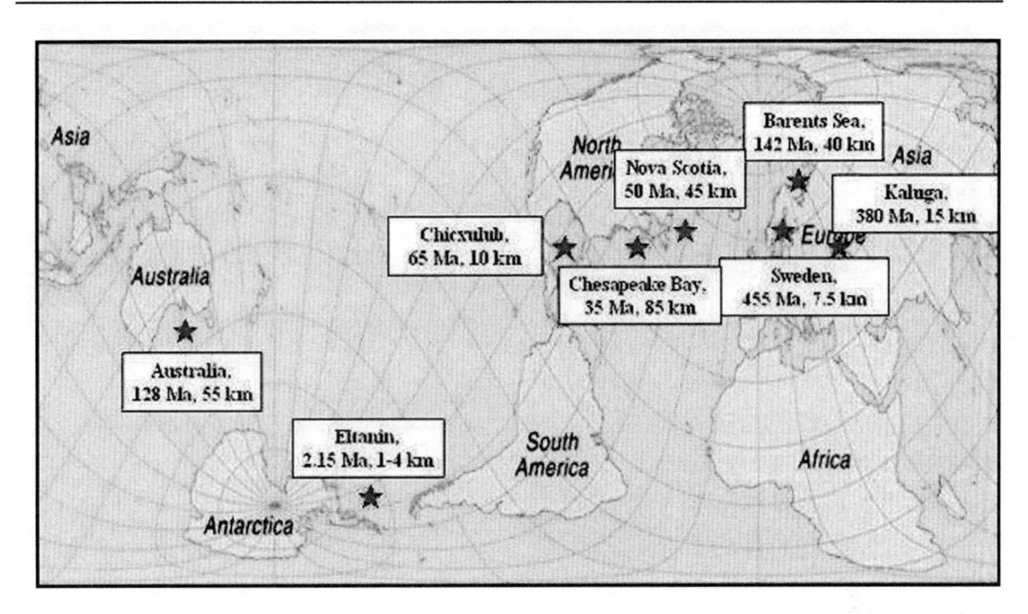

Figure 2. Map of known asteroid entries into the sea (Kharif and Pelinovsky, 2005).

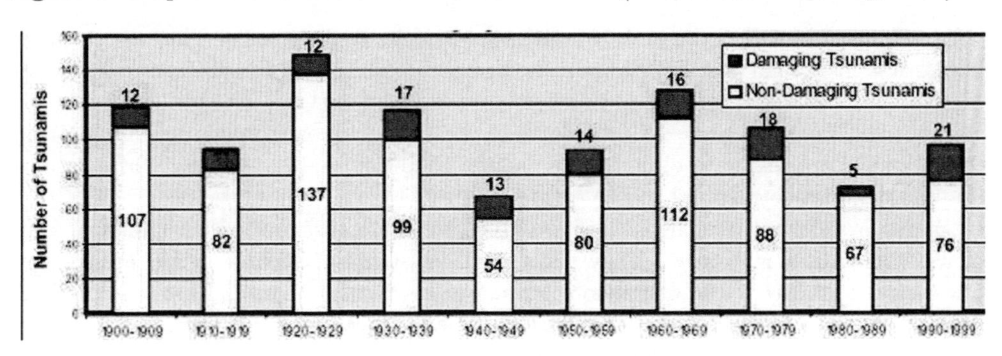

Figure 3. Tsunami statistics for the 20th century (NGDC).

the hydrodynamic theory, we briefly reproduce here some rough formulas to demonstrate the order of magnitudes of the height and dimensions of the initial water displacement in the tsunami source.

2.1 Tsunamis of seismic origin

The parameters of the source of tsunami waves of the seismic origin depend on many earthquake parameters: the fault length and width, the orientation in space and focal depth. These can be determined from the seismic models by Manshinha and Smylie (1971) or Okado (1985). Approximately, the tsunami source has an elliptical shape. Its effective radius, R_e and water displacement, H_e can very roughly be estimated through the earthquake magnitude, M which characterizes the earthquake energy (Hatori, 1966;

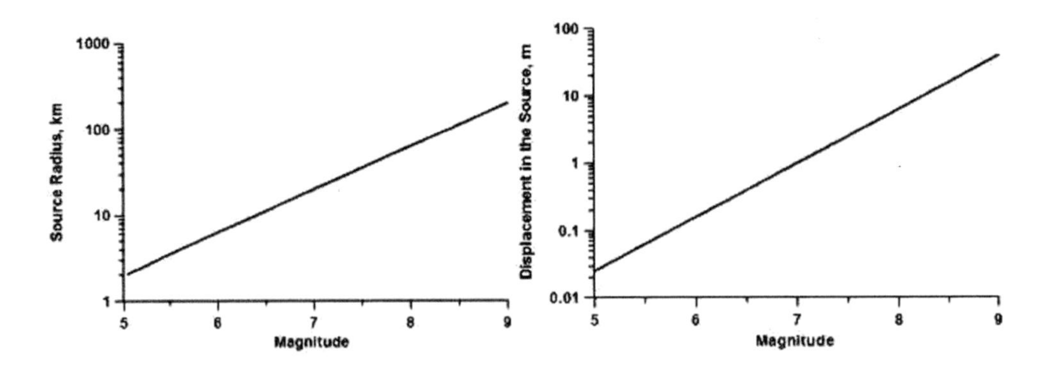

Figure 4. Rough estimates of the tsunami parameters in the source versus the earthquake magnitude.

Mirchina et al., 1982)

$$\log R_e = 0.5M - 2.2, \qquad \log H_e = 0.8M - 5.6 \tag{2.1}$$

where R_e is in km, H_e in m and M is measured in the Richter scale. They are displayed in figure 4. It is clearly seen, that the strong earthquakes with magnitudes 7-9 generate big tsunami waves several meters in height and one hundred km in length. For instance, the source of the last giant 2004 tsunami in the Indian Ocean (magnitude 9.0-9.3) had the following approximate dimensions: length, 1300 km; width, 200 km; and height, 15 m - these values correspond to the estimates in (2.1). Of course, the accuracy of these estimates is very low and here they are given to demonstrate the characteristic scales of the tsunami source only. They show that tsunamis of seismic origin are shallow-water waves, and that the dispersion parameter, $\mu = h/\lambda$ is small. Here h is the basin depth and λ the characteristic wavelength. The shape of tsunami waves in the source were in the early papers considered as a displacement of one polarity (like an ellipsoid). Nowadays components of opposite polarities appear in the shapes, in accordance with seismic solutions (Manshinha and Smylie, 1971; Okado, 1985); see also Synolakis et al. (1997).

2.2 Tsunamis from underwater explosions

Since the description of the mechanism of wave formation during an explosion, as such, is an extremely complicated problem, the main attention here is given to the choice of the equivalent source that gives the possibility to explain the wave field on long distances. The definition parameter for the deep-water explosion is the energy, and this characteristic can be used for tsunami waves generated by the explosion eruption of underwater volcanoes. Detailed experiments have been performed with underwater explosions with low energies (10^6 Joules), when the generated waves have small wavelengths compared with the water depth (deep-water waves). The best agreement of the experimental data from weak explosions with the results of the linear dispersive theory is achieved, when

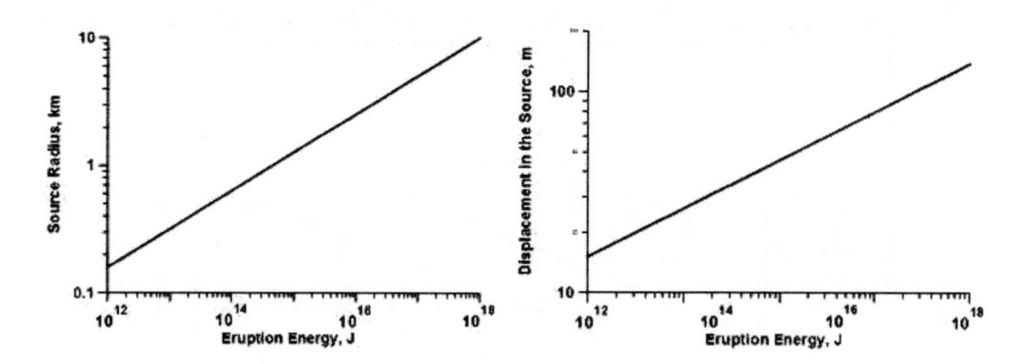

Figure 5. Parameters of the tsunami source of explosion origin.

the equivalent source has a shape of parabolic cavity with dimensions (Le Mehaute and Wang, 1996)

$$R_e = 0.04E^{0.3}, \qquad H_e = 0.02E^{0.24}, \tag{2.2}$$

where E is the explosion energy in Joules and R_e, H_e are in meters. These formulas are presented in figure 5. Formally, they should be used for deep water only but may also be applied for water of finite depth. For instance, the 1952-1953 Mijojin underwater volcano eruptions had an energy of 10^{16} J, and the parameters of the tsunami source were, $R_e \sim$ 2-3 km, $H_e \sim$ 100-200 m. In fact, due to cylindrical divergence the wave amplitude attenuates very quickly, and it is less than 10 m at distances of 30 km. For the 1883 Krakatau eruption ($8.4 \cdot 10^{17}$ Joules) the initial dimensions of volcano caldera are estimated as $R_e \sim 3.5$ km, and $H_e \sim 220$ m (Bryant, 2001). Such dimensions of the Krakatau tsunami source are chosen knowing the characteristics of volcano eruption without no using of (2.2). Thus, the tsunami waves from underwater volcano should be considered in general as the waves in the basin of finite depth.

The similar estimates are used for tsunami source of asteroid origin (Ward and Asphaug, 2000). Asteroids larger than 200 meters in diameter hit the Earth about every 3000-5000 years, so the probability of one impacting in a given human lifetime is about 2-3% (Hills and Gods, 1998). The cavity radius and depth depends from on the impactor radius and the "entry" velocity, and for typical conditions they are presented in figure 6 (Ward and Asphaug, 2000). Such waves also should also be considered as the waves in water of finite depth.

All classes of existing water theories may be used to explain the tsunami waves from underwater explosions: linear dispersive theory (Ward and Asphaug, 2000, 2003), shallow water simulations (Mader, 1998; Choi et al., 2003), and fully nonlinear computations (Crawford and Mader, 1998). For waves in an intermediate-depth basin, the shallow-water results overestimate the tsunami parameters to compare with simulations in the framework of incompressible Navier-Stokes equations (Crawford and Mader, 1998).

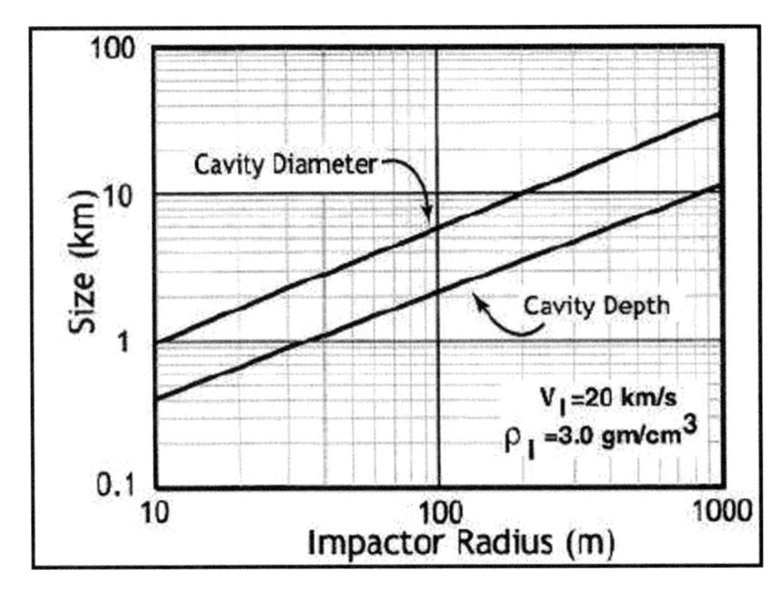

Figure 6. Characteristics of the equivalent source of asteroid tsunami.

2.3 Tsunamis generated by landslides

Parameters of submarine landslides as well as rock falls and the debris avalanches from "land" volcanoes vary within wide limits; they induce tsunami waves of different scales (Watts, 2000). As a result, all classes of theoretical models are applied here, they are, linear dispersive equations (Ward, 2001), nonlinear-dispersive Boussinesq system (Watts et al., 2003), nonlinear shallow-water equations (Assier-Rzadkiewicz et al., 2000) and fully nonlinear and dispersive computations (Grilli et al., 2002).

The data given above show the various scales of tsunami waves in the ocean. Earthquakes typically induce the large-scale waves (up to 1000 km) and the basic model for them is the shallow-water system. Tsunamis of the relatively weak explosion origin (underwater volcano eruptions, asteroid impact) are typically short-scales waves (0.1-1 km) and here the linear dispersive theory is the basic model. Many earthquakes initialize a submarine landslide motion, which then generates intense tsunami waves. This problem was the subject of the special discussion on the NATO Workshop (Yalciner et al., 2003). Tsunamis from landslides depending of characteristic scales can be described by one of these models. Due to strong variability of the ocean seafloor, the ratios depth/wavelength and amplitude/depth are changed along the wave path, and different theoretical models can be applied for the different stages of the wave propagation: in the source, in the open sea, in the coastal zone and on the beach. Typical models of tsunami waves will be discussed in the next sections.

3 Shallow water equations

The governing hydrodynamic model of tsunami generation by the underwater earthquake is based on the well-known Euler equations of ideal incompressible fluid

$$\frac{\partial \mathbf{u}}{\partial t} + \mathbf{u} \cdot \nabla_1 \mathbf{u} + w \frac{\partial \mathbf{u}}{\partial z} + \frac{\nabla_1 p}{\rho} = 0, \tag{3.1}$$

$$\frac{\partial w}{\partial t} + \mathbf{u} \cdot \nabla_1 w + w \frac{\partial w}{\partial z} + \frac{1}{\rho}\frac{\partial p}{\partial z} = -g, \tag{3.2}$$

$$\nabla_1 \cdot \mathbf{u} + \frac{\partial w}{\partial z} = 0. \tag{3.3}$$

with corresponding boundary conditions at the bottom and ocean surfaces (the geometry of the problem is shown in figure 7):

at the sea bottom ($z = -h(x, y, t)$),

$$w - \mathbf{u} \cdot \nabla_1 h = W_n(x, y, t), \tag{3.4}$$

at the free surface ($z = \eta(x, y, t)$), the kinematic condition is

$$w = \frac{\mathrm{d}\eta}{\mathrm{d}t} = \frac{\partial \eta}{\partial t} + \mathbf{u} \cdot \nabla_1 \eta, \tag{3.5}$$

and the dynamic equation,

$$p = p_{atm}. \tag{3.6}$$

Here $\eta(x, y, t)$ is the water displacement, $\mathbf{u} = (u, v)$ and w are horizontal and vertical components of the velocity field, x and y are coordinates in the horizontal plane, z-axis is directed upwards vertically, ρ is water density, p is pressure and p_{atm} is the atmospheric pressure, which is assumed constant, g is gravity acceleration, $h(x, y, t)$ is variable ocean depth due to the bottom displacement in the source area, and W_n is the velocity of the bottom motion on the perpendicular direction (when the bottom motion is vertical only, W coincides with $\mathrm{d}h/\mathrm{d}t$). The differential operator ∇_1 acts in the horizontal plane.

These equations can be applied to study tsunamis generated by bottom displacements in the source of underwater earthquake or submarine landslides. In the last case the function $h(x, y, t)$ can be given only for a solid-body landslide. In the opposite case, the equations of the slide motion should account the rheology of the landslide material (Kulikov et al., 1996; Heinrich et al., 1998).

As pointed out above, the tsunami waves from strong underwater earthquakes are long (as compared to the ocean depth). Therefore, it is natural first to consider the long-wave (or shallow-water) approximation as a model of tsunami generation, and then estimate the conditions of its applicability. The theory of long waves is based on the main assumption that the vertical velocity and acceleration are low as compared to horizontal ones and can be calculated from the initial system with the help of an asymptotic procedure. As small parameter it uses the ratio between the vertical velocity and the horizontal one or the ocean depth to the characteristic wavelength and will be given below.

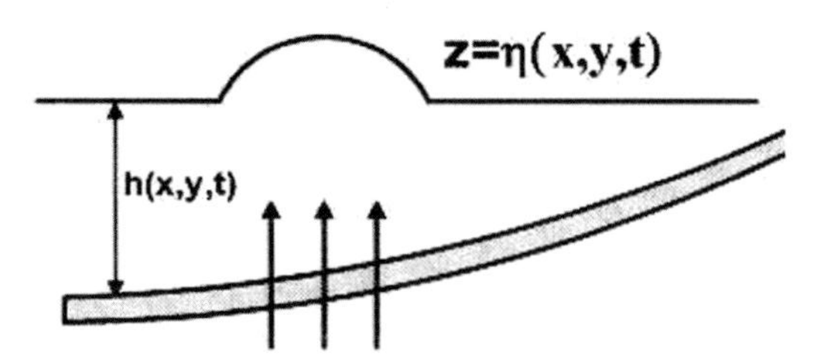

Figure 7. Problem geometry.

Here a simpler algorithm is used, which consists in neglecting vertical acceleration dw/dt in (3.2). In this case the equation (3.2) is integrated and, with the dynamic boundary condition (3.6) taken into account, determines the hydrostatic pressure:

$$p = p_{atm} + \rho g(\eta - z). \tag{3.7}$$

Substituting (3.7) into (3.1) and neglecting the vertical velocity once again, we obtain the first equation of the long-wave theory:

$$\frac{\partial \mathbf{u}}{\partial t} + \mathbf{u} \cdot \nabla_1 \mathbf{u} + g \nabla_1 \eta = 0. \tag{3.8}$$

The second equation is obtained by integration of (3.3) over the depth from the bottom ($z = h(x, y, t)$) to the surface ($z = \eta(x, y, t)$), taking into account boundary conditions (3.4) - (3.5), as well as the fact that the horizontal velocity does not depend on the vertical coordinate, z:

$$\frac{\partial \eta}{\partial t} + \nabla_1 \cdot [(h + \eta)\mathbf{u}] = W_n. \tag{3.9}$$

Equations (3.8) - (3.9) are closed as related to the functions η and $\mathbf{u}$. They are *nonlinear* (the so-called nonlinear shallow-water theory), *inhomogeneous* (the right-hand part is non-zero), and contain a preset function h that is variable in time and space.

Most generally used within the tsunami problem is the linear version of the shallow-water theory. In this case variations of depth are assumed weak, as well as the velocity of the bottom motion. As a result, we obtain a linear set of equations:

$$\frac{\partial \mathbf{u}}{\partial t} + g \nabla_1 \eta = 0, \qquad \frac{\partial \eta}{\partial t} + \nabla_1 \cdot (h\mathbf{u}) = W_n \tag{3.10}$$

with the right-hand part $W_n(x, y, t)$. It is convenient to exclude the velocity and pass over to the wave equation for the surface elevation:

$$\frac{\partial^2 \eta}{\partial t^2} - \nabla_1 \cdot (c^2 \nabla_1 \eta) = \frac{\partial W_n}{\partial t}, \tag{3.11}$$

where

$$c(x, y, t) = \sqrt{gh(x, y, t)} \tag{3.12}$$

is the long-wave speed (upper limit for the surface waves). The equation (3.11) is the basic one within the linear theory of tsunami generation and must be supplemented by initial conditions. It is natural to believe that at the initial moment the ocean is quiet, i.e.

$$\eta = 0, \quad \mathbf{u} = 0 \quad \text{or} \quad \partial\eta/\partial t = 0, \tag{3.13}$$

though due to linearity of (3.11) a more general case of the initial conditions can be also considered. From the point of view of mathematical physics the wave equation (3.11) is too well studied to discuss the details of its solution here. Note only that since the function W_n is "switched on"' at the moment $t = 0$, we formulate here the Cauchy problem for the wave equation in a generalized sense, i.e., we include into the consideration the generalized functions and do not require that the level and flow velocity should be differentiated twice.

The initial and long-wave systems of the equation presented here are basic hydrodynamic models of tsunami generation by underwater strong earthquakes and giant landslides.

4 Tsunami generation and propagation in the shallow sea of constant depth (linear approximation)

The simplest model here uses the linear wave equation (3.11) with the preset right-hand part $W_n(x, y, t)$. We can identify W_n with the vertical velocity of the bottom motion (dz_b/dt). In this case (3.11) is reduced to the classical wave equation

$$\frac{\partial^2\eta}{\partial t^2} - c^2\Delta\eta = \frac{\partial^2 z_b}{\partial t^2} \tag{4.1}$$

with constant long-wave speed, c and zeroth initial conditions.

Vertical bottom displacement. We will consider first a simple model of pulse perturbation when the bottom shifts momentarily to a finite value (the earthquake acts like a piston in this case)

$$h(x, y, t) = h_0[1 - \eta_0(x, y)\theta(t)], \tag{4.2}$$

where $\eta_0(x, y)$ describes the shape of the residual displacement of sea floor in the source and $\theta(t)$ is the Heaviside (unit) function with $\theta(t) = 1 (t > 0)$ and $\theta(t) = 0 (t < 0)$. Taking into consideration that $W_n = dh/dt$, the velocity of the bottom shift is the generalized Dirac function, $W_n = \eta_0\delta(t)$. It is easily shown in this case that the generalized Cauchy problem for the inhomogeneous wave equation is equivalent to the "usual" Cauchy problem for the homogeneous wave equation,

$$\frac{\partial^2\eta}{\partial t^2} - c^2\Delta\eta = 0 \tag{4.3}$$

with the following initial conditions

$$\eta(x, y, t = 0) = \eta_0(x, y), \quad \frac{\partial \eta}{\partial t}(x, y, t = 0) = 0. \tag{4.4}$$

We observe, when the piston shift is instantaneous, the ocean surface rises to the value of the bottom shift. This is evident due to the incompressibility of the fluid and the pressure being hydrostatic. This is the simplest and most comprehensive model of wave excitation (from the physical point of view), and has found the wide usage in the study of tsunamis. The shape of initial bottom displacement is determined from the solution of the seismic problem, see for instance, Manshinha and Smylie (1971) and Okada (1985). Roughly, the parameters of the tsunami source of the seismic origin are given in section 2 through the earthquake magnitude.

The radiation pattern depends on the source shape (the influence of the bottom relief will be analyzed later). If the tsunami source is an ellipse with large eccentricity like an 1D strip, the wave field is obtained simply by

$$\eta(x, t) = \frac{1}{2}[\eta_0(x - ct) + \eta_0(x + ct)], \tag{4.5}$$

and the shape of the tsunami wave repeats the shape of the initial displacement, with the height reduced by a factor of two. For a two-dimensional source, the solution of (4.3) is described by the Poisson formula

$$\eta(\mathbf{r}, t) = \frac{1}{2\pi c} \frac{\partial}{\partial t} \int \frac{\eta_0(\boldsymbol{\rho}) \mathrm{d}\boldsymbol{\rho}}{\sqrt{c^2 t^2 - |\mathbf{r} - \boldsymbol{\rho}|^2}}, \tag{4.6}$$

where $\mathbf{r}$ and $\boldsymbol{\rho}$ are two-dimensional vectors in the (x, y) plane, and integration is performed over the region of the circle

$$|\mathbf{r} - \boldsymbol{\rho}|^2 < c^2 t^2. \tag{4.7}$$

This integral is not calculated in the general case. One analytical example of the circular wave is displayed in figure 8. The simplest asymptotic is found for the wave field at the source for long times ($t >> r/c$)

$$\eta(r, t) \sim -\frac{V}{2\pi c^2 t^2}, \tag{4.8}$$

where V is the volume of the water displaced at the source

$$V = \int \eta_0(\mathbf{r}) \mathrm{d}\mathbf{r}. \tag{4.9}$$

Thus, unlike in the one-dimensional case, when the wave leaves the source completely, a motion remains in the source the water displacement that is damped with time. It is important to mention that if the water only uplifts in the source, the wave contains a leading elevation accompanied by a depression. The depth of the depression depends on the steepness of the leading wave and may exceed its height. In the opposite case, if the

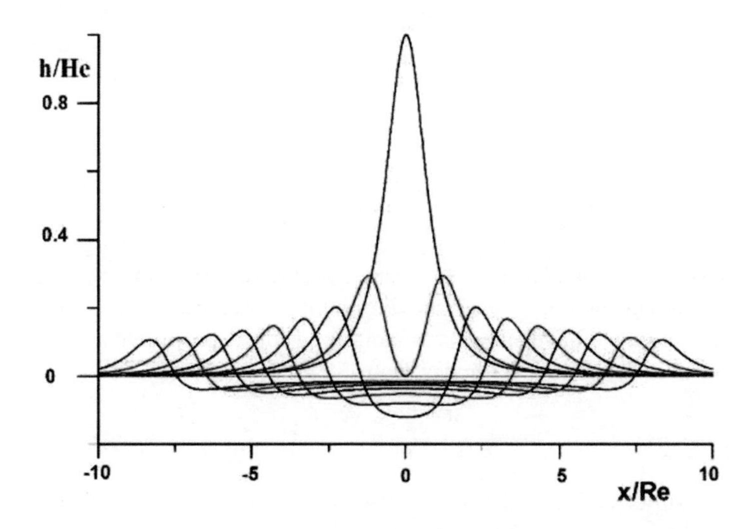

Figure 8. Cross-section of the circular wave for various times with unit R_e/c.

water falls in the source, the leading wave is negative (depression), but the next wave is positive with a higher amplitude. This explains why it is very dangerous to walk on the "open" seafloor when the water is receded; the next wave climbs on the beach.

Another asymptotic can be obtained for the amplitude of the leading wave, in particular for the elliptic uniform displacement in the source:

$$H = \frac{H_e}{2}\left(\frac{R}{d}\right)^{1/2}, \tag{4.10}$$

where R is radius of curvature of the ellipse at the point close to the observation, and d is the distance from the observation point to the center of the circle with curvature R^{-1}, which is r for large distance. Hence, the wave amplitude is attenuated as $r^{-1/2}$, but its value depends on the orientation of the source in relation to the observation point. Radiation directivity can be characterized by the ratio of the amplitude of the wave radiated towards the major axis H_a and the amplitude of the wave radiated towards the minor axis H_b. The latter is determined by the following simple equation obtained from (4.10)

$$\frac{H_a}{H_b} = \left(\frac{b}{a}\right)^{3/2}. \tag{4.11}$$

Thus, an elliptical source forms a radiation that is inhomogeneous in space. The strongest radiation is directed towards the minor axis of the ellipse. This position of the source, relative to the observation point, is the most dangerous as far as the formation of strong tsunami waves is considered. The sources of historical tsunami waves were mainly extended along the nearest coast, and, consequently, caused waves with larger amplitudes under other equal conditions.

The last conclusion concerns the arrival time. If the initial tsunami source is finite (η_0 is given only in the ellipse), the wave arrives at a given point r after a time $(r-R)/c$, i.e. it propagates with the speed, c. This is a conclusion that does not depend on the geometry of the problem, but is obtained by the shallow-water system being hyperbolic. This is of great importance for the prediction of the arrival time of tsunamis after an earthquake. More details are provided in section 8.

Landslide motion. Let us consider as source the simplest two-dimensional (x, z) bottom displacement as the model of a landslide moving horizontally with constant speed V. Actually, in this case we have the 1D problem: the solution, which satisfies the zeroth initial conditions, is easily found explicitly from (4.1)

$$\eta(x,t) = \frac{V^2}{V^2-c^2} z_b(x-Vt) - \frac{V^2}{2c(V-c)} z_b(x-ct) + \frac{V^2}{2c(V+c)} z_b(x+ct). \tag{4.12}$$

This solution is a superposition of three waves: one of them is bounded, and the two others are free. After some time they separate in space: the first wave remains attached to the source zone (moving with a landslide), while the two others leave. Let us discuss first the field in the source for sufficiently long times, when the waves become separated in space; it is described by the first term in (4.12). We see that in the case of a large speed ($V \to \infty$) the surface elevation responds to the bottom displacement ($\eta \approx z_b$), and when the motion is slow due to small vertical acceleration, the surface level is practically not perturbed ($\eta \approx V^2 z_b/c^2$). The free wave moving in the direction of the landslide is similarly amplified (but it is a trough, not a crest). The wave leaving the landslide zone in the opposite direction is limited with respect to the amplitude at any velocity (that is caused by great difference in motion velocities) and this wave is weak. Of a special interest is the case of synchronism between the landslide motion and the excited wave, when even a small bottom displacement causes the strong water motion; formally, bounded and free waves propagate together and the resulting wave is

$$\eta(x,t) = -\frac{ct}{2}\frac{\partial z_b(x-ct)}{\partial x}. \tag{4.13}$$

Let us consider now a model of the horizontally "growing" landslide, when its right end moves to the finite distance X, in such a way that the velocity of the bottom motion, V differs from zero only in the region $(0-X)$, see, figure 9

$$W_n = H_b \delta(t - x/V), \qquad 0 < x < X. \tag{4.14}$$

The solution of (4.1) is easily found in the form of the Duhamel integral

$$\eta(x,t) = \frac{1}{2c}\int_0^t \mathrm{d}\tau \int_{x-c(t-\tau)}^{x+c(t-\tau)} \mathrm{d}y \frac{\partial W_n}{\partial \tau}(y,\tau). \tag{4.15}$$

Out of the source region the wave is constituted of two rectangular pulses moving in opposite directions. The amplitude and duration of the pulse moving towards $(x > 0)$ are

$$H = \frac{H_b}{2}\frac{V}{|V-c|}, \qquad T = \frac{X}{c}\frac{|V-c|}{V}, \tag{4.16}$$

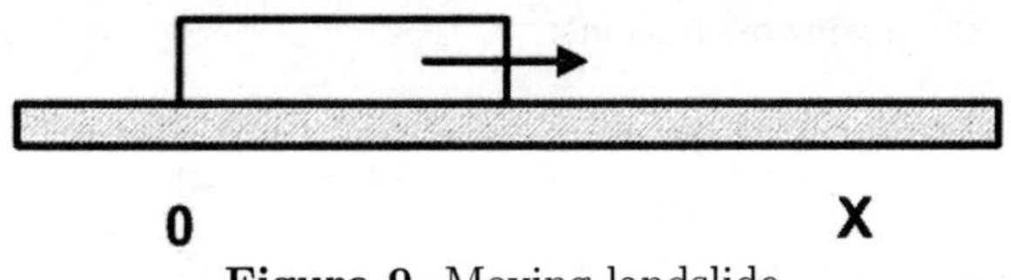

Figure 9. Moving landslide.

and in the opposite direction,

$$H = \frac{H_b}{2} \frac{V}{|V + c|}, \qquad T = \frac{X}{c} \frac{|V + c|}{V}. \tag{4.17}$$

It can be seen that here a resonance also takes place. When the speed of the bottom displacement motion, V approaches the velocity of wave propagation c, the amplitude of the resonance wave goes to infinity, and the pulse duration becomes very small. Such short pulses violate the applicability of the long-wave theory, and the resonance should be considered within more accurate models. Similar calculations can be made also for the plane (x, y) problem, when the bottom displacement runs within a rectangular region. It should be emphasized that the resonance is retained in the plane problem, wherein all the displacements with velocities higher than c are of a resonance character, and maximum radiation occurs along directions $\theta = \arccos(c/V)$, determined through the so-called Mach angle. Such relations are well-known in the theory of the wave radiation. Wave amplitude stays finite at $c \neq V$, and it is proportional to the factor $(1-c^2/V^2)^{-1/2}$ for the Mach direction.

Thus, horizontal motions of the bottom in the seismo-active zone or horizontally moving landslides cause intensification of the radiated waves. "Catastrophic" within the linear shallow-water theory is the case of resonance that occurs when $c = V$. The wave amplitude becomes then formally unbounded. This can explain how a weak earthquake sometimes generates a strong tsunami: at first the earthquake initiates a landslide motion which causes the water waves. The mechanism of the tsunami generation by submarine landslides is of great interest, and the special NATO workshop was recently organized (Yalciner et al., 2003).

5 Effects of finite depth for tsunami waves of seismic origin

Usually, the earthquake area has large dimensions (compared with water depth), and mainly shallow-water theory is used to describe the properties of tsunami waves. In principal., all results of the shallow-water theory can be re-examined using the finite-depth approximation for water waves. For instance, we may check the piston model of tsunami generation described in section 4. It is evident that the instantaneous bottom motion results in large values of the vertical velocity, in contrast to the assumptions of the shallow-water theory. Since the motion begins from the state of rest, the fluid motion can be considered to be potential. The equations for the fluid potential, $\Phi(x, y, z, t)$ can be easily derived from the Euler equations (3.1)-(3.6), and are well-known (for the fluid

of constant depth, linear approximation):
Laplace equation

$$\Delta\Phi + \frac{\partial^2\Phi}{\partial z^2} = 0, \qquad (-h < z < 0), \tag{5.1}$$

the boundary conditions on the sea surface ($z = 0$)

$$\frac{\partial^2\Phi}{\partial t^2} + g\frac{\partial\Phi}{\partial z} = 0, \tag{5.2}$$

and seafloor ($z = -h$)

$$\frac{\partial\Phi}{\partial z} = W_n(x, y, t). \tag{5.3}$$

The water displacement is expressed through potential as

$$\eta(x, y, t) = -\frac{1}{g}\frac{\partial\Phi}{\partial t} \quad \text{at} \quad z = 0. \tag{5.4}$$

The solution of this boundary problem can be found using the time-dependent Green function, G. The latter is a harmonic function in space with a peculiarity of the source type, and it is found analytically (Stoker, 1955)

$$G(\mathbf{r}, z, t|\mathbf{r}_0, z_0, \tau) = \frac{1}{h}\int_0^{+\infty} \frac{J_0(m|\mathbf{r}-\mathbf{r}_0|/h)}{\cosh(m)} \left\{\sinh\left[m\left(1 - \frac{z - z_0}{h}\right)\right] - \right.$$
$$- \sinh\left[m\left(1 + \frac{z + z_0}{h}\right)\right] + \frac{2m}{\gamma^2\cosh(m)}\left[1 - \right.$$
$$\left.\left. - \cos\left(\gamma\sqrt{g/h}(t-\tau)\right)\right] \cosh\left[m\left(1 + \frac{z}{h}\right)\right] \cosh\left[m\left(1 + \frac{z_0}{h}\right)\right]\right\} dm, \tag{5.5}$$

where $\gamma = [m\tanh(m)]^{1/2}$. As a result, the water displacement induced by the bottom motion is determined by the integral

$$\eta(\mathbf{r}, t) = \frac{1}{4\pi g}\int\int d\mathbf{r}_0 \int_0^t W_n(\mathbf{r}_0, t)\frac{\partial^2 G(\mathbf{r}, 0, t|\mathbf{r}_0, -h, \tau)}{\partial t\partial\tau} d\tau, \tag{5.6}$$

where

$$G(\mathbf{r}, 0, t|\mathbf{r}_0, -h, \tau) = \frac{2}{h}\int_0^{+\infty} \frac{mJ_0(m|\mathbf{r}-\mathbf{r}_0|/h)}{\gamma^2\cosh(m)} dm \left[1 - \cos\left(\gamma\sqrt{g/h}(t-\tau)\right)\right]. \tag{5.7}$$

Assuming a piston motion, $W_n(x, y, t) = \eta_b(x, y)\delta(t)$, the final expression for the water displacement is (Kajiura, 1963)

$$\eta(\mathbf{r}, t) = \frac{1}{2\pi h^2}\int\int \eta_b(\mathbf{r}_0) d\mathbf{r}_0 \int_0^{+\infty} dm \frac{mJ_0(m|\mathbf{r}-\mathbf{r}_0|/h)}{\cosh(m)} \cos\left(\gamma\sqrt{g/h}t\right). \tag{5.8}$$

Hence, at $t = 0$ we find the elevation of the water level caused by the piston motion of the basin bottom. The latter is used as initial value when solving the shallow-water wave equation (4.3)

$$\eta_0(x, y) = \frac{1}{2\pi h^2} \int\int \eta_b(x_0, y_0) \mathrm{d}x_0 \mathrm{d}y_0 \int_0^{+\infty} \mathrm{d}m \frac{mJ_0(m|\mathbf{r} - \mathbf{r}_0|/h)}{\cosh(m)}. \tag{5.9}$$

We can see that the initial water displacement, generally speaking, does not repeat the bottom shift. Kajiura (1963) has performed specific calculations using (5.9). Particularly, for a homogeneous source of a circular shape the water displacement above the source center answers the bottom elevation at $R_e > (2 - 4)h$, i.e., in the framework of the usual assumptions of the shallow-water theory. The water displacement at the ends of the source is smoother than the bottom elevation and does not contain jumps as in the shallow-water approximation. Meanwhile, this effect is not essential for applications, and the tsunami source of the seismic origin can be determined in the shallow-water theory.

However, the finite-depth approximation changes radically the wave field far from the source if even if it is well described well in the theory by the shallow-water theory. Considering the 1D wave field out of the source area the Fourier transformation can be applied

$$\eta(x, t) = \int \eta(\omega) \exp\left[\mathrm{i}(\omega t - kx)\right] \mathrm{d}\omega, \tag{5.10}$$

where $\eta(\omega)$ is the frequency spectrum of the radiated wave (it can be found within the shallow-water theory), and the wavenumber, k and wave frequency, ω satisfy to the dispersion relation

$$\omega(k) = \sqrt{gk \tanh(kh)}. \tag{5.11}$$

In the shallow-water approximation, $\omega = ck$, where as early $c = (gh)^{1/2}$, and (5.10) gives the trivial answer, $\eta(x, t) = \eta(t - x/c)$, and the wave does not change its shape in the process of the wave propagation, as it is expected from (4.5). In the finite-depth approximation, water waves have dispersion, and the spectral components propagate with the different velocities resulting in a wave shape deformation. Considering almost long waves ($kh << 1$), the relation (5.11) can be simplified

$$k \approx \frac{\omega}{c}\left(1 + \frac{\omega^2 h^2}{6c^2}\right). \tag{5.12}$$

Asymptotic expression of the integral (5.10) with (5.12) for large distances from the source is expressed through the Airy function

$$\eta(x, t) = cH_0T_0 \left(\frac{2}{h^2 x}\right)^{1/3} Ai\left[\left(\frac{2}{h^2 x}\right)^{1/3} (x - ct)\right], \tag{5.13}$$

which is plotted in figure 10 in normalized variables. Here H_0 and T_0 are the amplitude and duration of the wave radiated from the source. In fact, the value, cH_0T_0 can be replaced by V, the volume of the initial water displacement, and the solution (5.13) is valid for $V \neq 0$. So, the time record of the tsunami wave far from the source begins with the

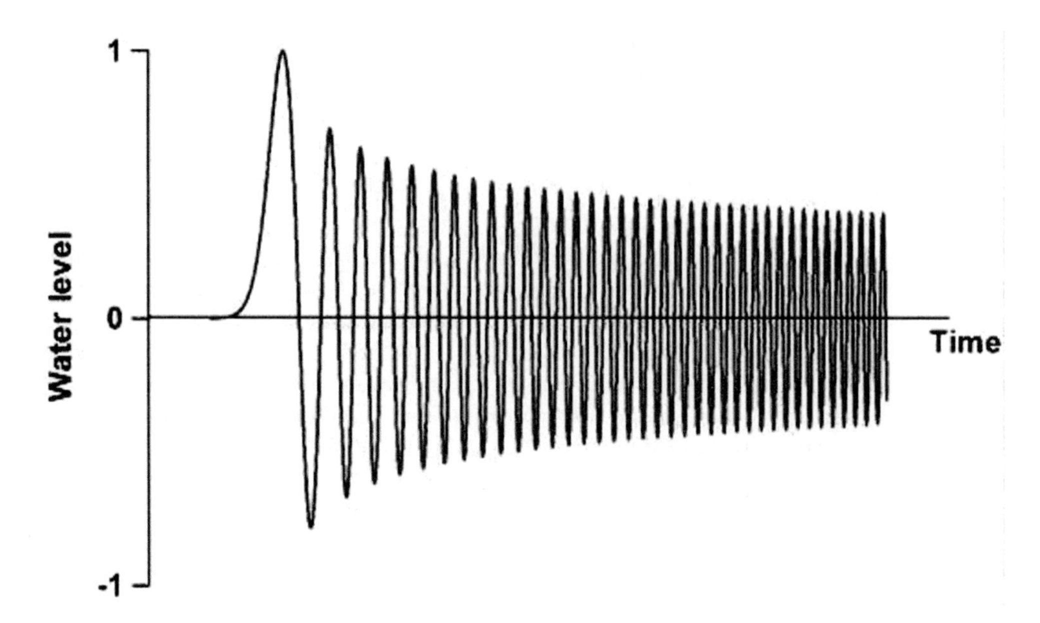

Figure 10. Time series of the dispersive wave (5.13) far from the source.

long wave accompanied by a train of smaller waves with decaying period. This is simply explained by the effect of dispersion, where the longer waves have larger propagation speed than the shorter. It is important to mention that the height of the leading wave is attenuated with distance as $x^{-1/3}$, and its duration grows as $x^{1/3}$. In the 2D case, far from the source, due to cylindrical divergence, the leading wave height is proportional to $r^{-5/6}$, and its duration - to $r^{1/3}$. If the earthquake generates the sign-variable disturbance with zeroth volume, V, the leading wave height is attenuated as $x^{-2/3}$ in the 1D case, and $r^{-7/6}$ in the 2D case. But in both cases the leading wave is not the wave of maximal amplitude. The asymptotic for the height of maximal amplitude is $x^{-1/2}$ (1D) and r^{-1} (2D), and for duration, $x^{1/2}$ or $r^{1/2}$; these formulas can be obtained by using the method of stationary phase.

We discussed above impulse (instantaneous) bottom displacements during the earthquake. Very often during the earthquake, the short and steep waves are exited in the epicentral zone. Seaquakes are followed by strong hydro-acoustic and cavitation effects, and they are dangerous for navigation in the open sea, unlike tsunami waves that are dangerous in coastal zones only. Amplitudes of the seismic oscillations are practically confined within the range 10^{-6} through 10^{-3} m, frequencies of such oscillations are 0.1 to 100 Hz, oscillation velocities, 10^{-3} through 0.1 m/s, acceleration, 0.1 through 10 m/s^2 (Aki and Richardson, 1980). Witnesses have reported that there appears on the sea surface a grid of waves with lengths 10-20 m and frequency 1 Hz during 10-60 s. The simplest model of a seaquake can be considered again with the use of (4.1) in the shallow-water approximation, or (5.6) taking into account the effects of the finite depth, if we assume that the bottom elevation is a homogeneous rise in the band $|x| < L/2$ oscillating in time with frequency, ω . In particular, the wave field directly over the epicenter of the underwater earthquake in the shallow-water approximation is easily found from the Duhamel integral (4.15)

$$\eta(t, x = 0) = -\frac{2H_e}{\pi} \sin[\omega(t - L/4c)]. \tag{5.14}$$

It follows from (5.14) that at frequencies $\omega L/4c = \pi/2 + \pi n(n = 0, 1, ...)$ the amplitude of the water level oscillations practically coincides with the amplitude of bottom shifts. This resonance condition can be written in a more traditional form

$$L = cT_n(2n + 1), \tag{5.15}$$

and is fulfilled well for typical conditions of seaquakes. The finite-depth approximation leads to the decrease of the seaquake amplitudes.

Earlier we took into account the bottom motion only as "forcing" in the wave equation. However, the water depth varies also, and, therefore, we have equations with time-variable coefficients. The joint account of variability of the coefficients and external forcing is rather complicated. Therefore, we simplify the problem and assume the bottom shift to be homogeneous in space, $h = h(t)$. In this case we may in the governing Euler equations pass over to the system of coordinates connected with an oscillating bottom; and only equations (3.2) and (3.4) are changed by

$$\frac{\partial w}{\partial t} + \mathbf{u} \cdot \nabla_1 w + w\frac{\partial w}{\partial z} + \frac{1}{\rho}\frac{\partial p}{\partial z} = -g + \frac{\mathrm{d}^2 z_b}{\mathrm{d}t^2}, \tag{5.16}$$

$$w = 0 \quad \text{at} \quad z = -h, \tag{5.17}$$

where z_b is the ordinate of the bottom elevation. Thus, all the changes manifest themselves in re-normalization of the gravity acceleration. In the linear approximation for spatial Fourier harmonics the following ordinary differential equation can be derived (Rabinovich et al., 2000)

$$\frac{\mathrm{d}^2 Q}{\mathrm{d}t^2} + \left[g - \frac{\mathrm{d}^2 z_b}{\mathrm{d}t^2}\right] k \tanh(kh) Q = 0, \tag{5.18}$$

where $Q = \partial\eta(k, t)/\partial t$. In the specific case of the periodic bottom motions $z_b(t) = H_e \cos(\omega t)$, equation (5.18) reduces to the Mathieu equation and its general solution is expressed by the Mathieu functions. Under certain conditions between coefficients the solution grows exponentially, and parametric resonance occurs. This is possible under the following conditions

$$\sqrt{gk_n \tanh(k_n h)} = n\omega/2, \qquad (n = 1, 2, ...) \tag{5.19}$$

and determines wavenumbers of unstable waves. The wave increment, γ at the basic frequency $(n = 1)$ is

$$\gamma = \frac{\omega^3 H_e}{2g}. \tag{5.20}$$

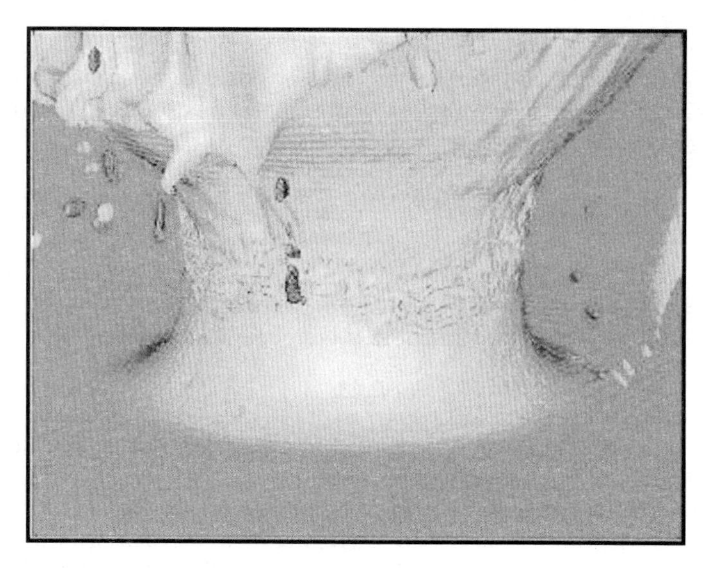

Figure 11. Entry of 1-km comet into the ocean (Numerical simulations by Crawford, http://sherpa.sandia.gov/planet-impact/comet/).

The instability of the water waves above the oscillating bottom is a well known Faraday effect for capillary ripples (Rabinovich et al., 2000), and it is manifested for gravity waves due to the earthquake. It is important to mention that the value of the increment does not depend on the depth, and may manifest in deep water, where the "usual" tsunami waves will not be generated. The characteristic time of the increase of the oscillation amplitude at frequency 0.1 Hz, and bottom displacement 1 m is about 1 min; this value agrees well with the data of the observations. The wavelengths of unstable waves are 2-200 m, and this also coincides well with the observations of seaquakes.

6 Explosive generated tsunamis (deep-water approximation)

Tsunami waves that are caused by explosions (underwater volcanoes, asteroid impact, etc) have shorter scales than the tsunamis of seismic origin. For the sake of simplicity we will analyze here the waves with lengths of some hundreds meters when the effects of the seafloor can be ignored. First of all, we should say that the wave generation due to explosions is extremely difficult to compute even if powerful computers are employed. This is due to the large number of physical effects: strong nonlinearity, wave breaking, heat processes and transfer to the gas fraction, deformation (and particularly vanishing) of the volcanic or asteroid body and so on. An example of such calculations is given in figure 11. The relatively complicated initial flow structure is the reason why an empirical parametrization of the tsunami source is popular for the simulation of propagation and estimation of the wave impact on the coasts. With little difference, the tsunami source of the explosion origin is parametrized by the parabolic cavity

$$\eta_e(r) = \begin{cases} H_e[2(r/R)^2 - 1] & r \leq R, \\ 0 & r > R, \end{cases} \tag{6.1}$$

$$\mathbf{u}_e(r) = 0,$$

where the equivalent height, H_e and radius, R of the source are connected to the explosion energy through expressions (2.2). The formal solution of the Euler equations in the linear deep-water approximation can be written through the Fourier-Hankel transformation

$$\eta(r,t) = \int_0^\infty k\eta(k)J_0(kr)\cos(\omega t)\mathrm{d}k, \qquad \omega = \sqrt{gk}, \tag{6.2}$$

where

$$\eta(k) = \int_0^R r\eta_e(r)J_0(kr)\mathrm{d}r. \tag{6.3}$$

The generated wave packet is very dispersive (as any wave train in deep water), and the stationary phase approximation is effective to obtain the analytical expression of explicit form far from the source (Le Mehaute and Wang, 1996; Mirchina and Pelinovsky, 1988; Ward and Ashpang, 2000). For large r and t (but constant ratio r/t), the stationary phase approximation gives

$$\eta(r,t) \approx \sqrt{2}\frac{H_e R}{r} J_3\left(\frac{gRt^2}{4r^2}\right)\cos\left(\frac{gt^2}{4r}\right), \tag{6.4}$$

where $J_3(z)$ is the third-order Bessel function. At a fixed time, the wave train consists of a leading group of long waves followed by groups shorter waves with lengths that decays with increasing distance from the leading group. Their amplitudes are decreased slightly within the group. The shape in dimensionless variables, η/H_e and $t\sqrt{(g/R)}$ for $r/R=10$ and 20 is presented in figure 12.

Maximal amplitude of the leading packet is achieved when the Bessel function has a maximum; it is

$$H \approx 0.6H_e\frac{R}{r}, \tag{6.4}$$

and attenuates as r^{-1}. The wave of maximal amplitude has period T given by

$$T \approx \pi\sqrt{R/g}. \tag{6.5}$$

The leading wave train propagates with constant speed

$$c_{gr} \approx \frac{\sqrt{gR}}{4}. \tag{6.6}$$

It is important to stress that the number of maximal amplitude wave increases linearly with distance. This is a feature that can be exploited in a Tsunami Warning System intended for evacuation. As can be seen, the parameters of the wave are expressed through

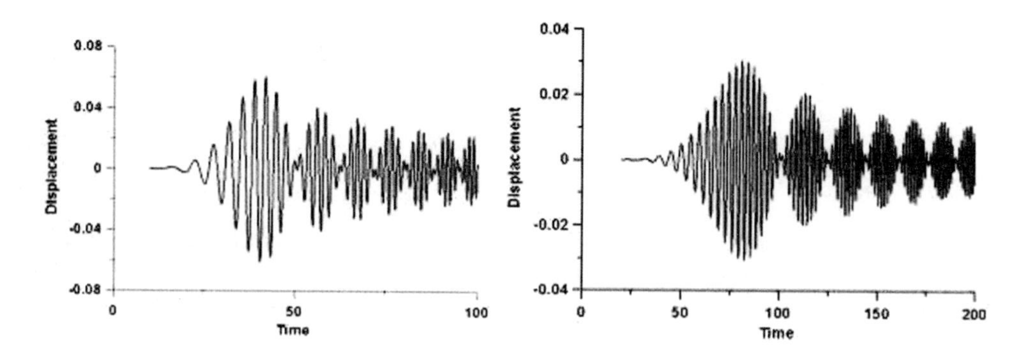

Figure 12. Time series of tsunami wave of the explosion origin at distances $r/R = 10$ (left) and 20 (right).

the initial equivalent radius (explosion energy), which can be estimated from the observation data. Mirchina and Pelinovsky (1988) using the observation data, estimated the energy of the 1952-1953 Mijojin underwater volcano (near the Japanese coast) eruptions as 10^{16} Joules.

Ward and Asphaug (2000, 2003) used this approach to estimate the parameters of past and possible asteroid tsunamis. In fact, they used the finite-depth dispersion relation for water waves (not deep-water), and also some corrections taking into account the variability of the bottom relief. In particular, they developed the scenario of the possible collision of a 1.1 km diameter asteroid (1950 DA) with the Earth on 2880, March 16 (Ward and Asphaug, 2003). This collision is predicted with a probability of 0.33%. Traveling at 17.8 km/s the asteroid may strike the ocean 600 km east of the United States coast blowing a cavity of 19 km in diameter. Predictions show that waves may strike the coasts of Europe and Africa within 12 hours, with heights that reach 23 m in Ireland, 16 m in England, 17-21 m in France, and 15 m in Portugal.

Usually only the main eruption is analyzed. Meanwhile, the volcanic eruption lasts for a long time, and not a single one, but several explosions being powerful. These explosions generate independent tsunami waves, which can intersect and interact between each other. For instance, there were at least four strong eruptions of the Krakatau volcano during less than one day: the first explosion took place on August 26, 1883 at 17:00 and induced tsunami waves up to 2 m, the second one on the 27th of August at 5:44 generated tsunami waves up to 10 m, and the strongest one, which took place on the 27th of August at 10:00 induced the giant waves up to 42 m (Simskin and Fiske, 1983). During the subaquatic volcano eruption in Karymskoye Lake (Kamchatka, Russia) on January 2, 1996 tsunami waves with the maximal runup height up to 30 m were generated; "explosions occurred every 4 to 12 min. Six explosions were observed with an average interval of 6 min" (Belousov et al., 2000). The interaction of the tsunami waves with different speeds and directions can induce the appearance of short-lived large-amplitude tsunami waves (*freak waves*), as it is pointed for wind waves (Kharif and Pelinovsky, 2003).

7 Nonlinear-dispersive theory of tsunami waves

For tsunami waves with intermediate scales (100 m - 10 km) the nonlinear-dispersive effects become important. Briefly, the derivation of the simplified model can be demonstrated for the potential fluid motion. In this case, the exact Laplace equation (5.1) should be considered together with the kinematic and dynamics boundary conditions on the sea surface, $z = \eta(x, y, t)$

$$\frac{\partial \Phi}{\partial z} = \frac{\partial \eta}{\partial t} + \nabla_1 \Phi \cdot \nabla_1 \eta, \qquad \frac{\partial \Phi}{\partial t} + \frac{1}{2}\left(\frac{\partial \Phi}{\partial z}\right)^2 + g\eta = 0, \tag{7.1}$$

and the sea bottom, $z = -h(x, y)$

$$\frac{\partial \Phi}{\partial z} - \nabla_1 \cdot \Phi \nabla_1 h = 0, \tag{7.2}$$

where we neglect for simplicity terms responsible for the tsunami generation. The "shallowness" of the tsunami waves is equivalent to the a variation of the fluid potential with depth, and we may seek the solution in the form of a power series with respect to the basin depth

$$\Phi(x, y, z, t) = \sum_{n=0} \Psi_n(x, y, t)(z + h)^n. \tag{7.3}$$

Substitution of (7.3) into the Laplace equation (5.1) yields the recurrent correlations for the unknown functions, Ψ_n

$$\Psi_{n+2} = -\frac{\triangle \Psi_n}{(n+2)(n+1)}, \tag{7.4}$$

and only two of them (namely, Ψ_0 and Ψ_1) are independent. Now, having substituted series (7.3) to the bottom boundary condition (7.2), we find Ψ_1

$$\Psi_1 = -\frac{\nabla_1 \Psi_0 \cdot \nabla_1 h}{1 + (\nabla_1 h)^2}, \tag{7.5}$$

and the series (7.3) becomes fully determined by one function $\Psi_0(x, y, t)$. Having substituted the series (7.3) into the kinematic and dynamic boundary conditions at the sea surface (7.2) we obtain the system of two partial differential equations for water level, η and velocity, $\nabla_1 \Psi_0$. The use of these equations is complicated by the absence of physically apparent treatment of value Ψ_0, since it does neither determine the depth-averaged flow rate, nor the surface velocity. The choice of the corresponding value, which have the dimensional representation of the velocity on the fixed depth, z_0 leads to a difference in models of the nonlinear dispersive theory that are used by various authors (Chen and Liu, 1995; Wei et al., 1995; Shi et al., 2001; Madsen et al., 2002, 2003; Lynett et al., 2002; Cheung et al., 2003). This depth, z_0 can be used to fit experimental data. For simplicity (or: for the sake of simplicity), we use here the depth-averaged velocity

$$\mathbf{u} = \frac{1}{h + \eta} \int_{-h}^{\eta} \nabla_1 \Phi \mathrm{d}z, \tag{7.6}$$

and re-calculate $\nabla_1 \Psi_0$ through depth-averaged velocity. As a result, the following equations can be derived

$$\frac{\partial \eta}{\partial t} + \nabla_1 \cdot [(h+\eta)\mathbf{u}] = 0, \tag{7.7}$$

$$\frac{\partial \mathbf{u}}{\partial t} + \mathbf{u} \cdot \nabla_1 \mathbf{u} + g \nabla_1 \eta = \mathbf{D}, \tag{7.8}$$

$$\mathbf{D} = \frac{1}{h+\eta} \nabla_1 \left[\frac{(h+\eta)^3}{3} R + \frac{(h+\eta)^2}{2} Q \right] - \left[\frac{h+\eta}{2} R + Q \right] \nabla_1 h, \tag{7.9}$$

$$R = \frac{\partial \nabla_1 \cdot \mathbf{u}}{\partial t} + (\mathbf{u} \cdot \nabla_1) \nabla_1 \cdot \mathbf{u} - (\nabla_1 \cdot \mathbf{u})^2, \qquad Q = \frac{\partial \mathbf{u}}{\partial t} \cdot \nabla_1 h + \mathbf{u} \cdot \nabla_1 (\mathbf{u} \cdot \nabla_1 h).$$

We give here also the expression for the pressure within the nonlinear-dispersive theory

$$p(x,y,z,t) = p_{atm} + \rho g(\eta - z) + \frac{\rho}{2} \left[z^2 + 2h(z-\eta) - \eta^2 \right] R - \rho(\eta - z) Q. \tag{7.10}$$

These equations (Boussinesq-like system) differ from the shallow-water system (3.8)-(3.9) by the presence of the dispersion, D. We would like to stress that nonlinearity here stays arbitrary, not weak, but the dispersion is weak (in the opposite case, we should take into account more terms in series). Taking into account the weakness of the dispersion, we may change temporal derivatives on spatial ones (and vice versa) in the term, D in accordance with the connections yielded by the shallow-water equations. As a result, we have a series of the equivalent Boussinesq systems. In the linear long-wave approximation all of them yield the same dispersion relation (5.12), but numerically, when the Boussinesq system is considered as the "exact" one, the dispersion properties of such systems are different, and some of them are unstable in the short-wave limit. In conclusion, we would like to mention the fact that the extended Boussinesq systems contained almost the full dispersion of water waves and full nonlinearity (see, for instance, Madsen et al., 2002, 2003), they are now beginning to be applied in tsunami practice.

If the nonlinearity and dispersion are weak, and basin depth is constant, the Boussinesq system for unidirectional tsunami wave can be reduced to the famous Korteweg-de Vries equation

$$\frac{\partial \eta}{\partial t} + \sqrt{gh} \left(1 + \frac{3\eta}{2h} \right) \frac{\partial \eta}{\partial x} + \frac{\sqrt{gh} h^2}{6} \frac{\partial^3 \eta}{\partial x^3} = 0. \tag{7.11}$$

Equation (7.11) is derived in the first order of the asymptotic procedure in the small parameters: nonlinearity, $\epsilon \sim \eta/h$, and dispersion, $\mu \sim (h/\lambda)^2$, assuming $\epsilon \sim \mu$. This equation plays an important role in the theory of nonlinear waves due to its integrability; see for instance, Drazin and Johnson (1992). The existence of solitons (solitary waves propagating with no change in its shape) was proved first for the Korteweg-de Vries equation. It is widely discussed in the theory of tsunami waves (Hammack, 1973; Murty, 1977; Pelinovsky, 1982). The water soliton is an elevation wave described by

$$\eta(x,t) = H \operatorname{sech}^2 \left[\sqrt{\frac{3H}{4h}} \frac{x - \sqrt{gh}(1 + H/2h)t}{h} \right], \tag{7.12}$$

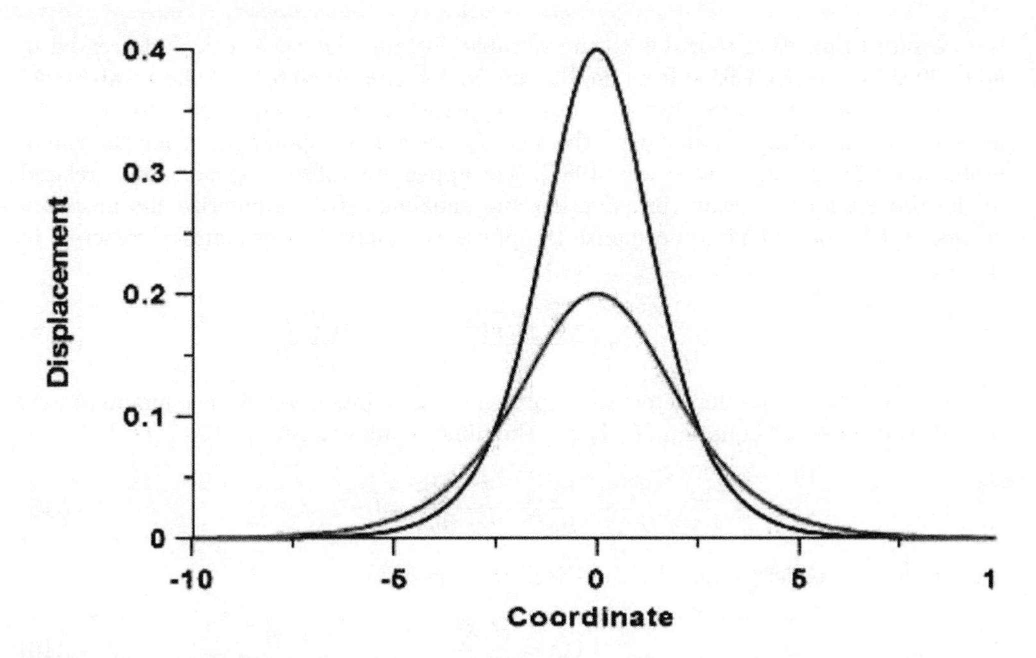

Figure 13. Solitary wave - soliton for various amplitudes (all variables are normalized on the water depth).

where H is the soliton height (figure 13). The characteristic wavelength determined by the level $0.5H$ is

$$\lambda = 2h\sqrt{\frac{4h}{3H}}\ln(1+\sqrt{2}) \cong 2h\sqrt{\frac{h}{H}}. \tag{7.13}$$

The values of the soliton length for typical values of depth and wave height are given in Table 7.1. The soliton length is typically of the order of 10-100 km in deep oceans (1-5 km) and less than 1 km in shallow water (50-100 m).

Table 7.1 Characteristic soliton length in km

depth (m)\ Height (m)	1	5	10	20
50	0.72	0.32	0.23	0.16
100	2	0.91	0.64	0.46
500	22.8	10.2	7.2	5.1
1000	64.4	28.8	20.4	14.4
5000	720	322	227	161

The source radius of the "seismic" tsunamis has an order of magnitude 10-100 km, and corresponds to the soliton length in deep oceans (1-5 km). In shallow water (50-100 m), the soliton length corresponds to the scales of tsunamis of the "explosion" origin. It is necessary to mention that due to the variable bottom, the wave scale is decreased in up to 30 times (for transition from depth 4 km in open sea to 40 m in shallow water). As a result, the tsunami waves of seismic origin approaching to shallow water also may also have a "soliton" scale. That is why the solitary wave is a popular input for the runup problem; see for instance, Synolakis (1987). The appearance of solitons can also be related to the processes of the wave propagation and shoaling. To characterize the unsteady processes it is convenient to normalize the physical variables on tsunami parameters in the sources

$$\xi = \frac{\eta}{H_e}, \qquad y = \frac{x - \sqrt{gh}t}{\lambda_e}, \qquad \tau = \frac{3\sqrt{gh}t}{2\lambda_e}, \tag{7.14}$$

where H_e and λ_e are the characteristic displacement and dimension of the tsunami source. In such variables, the equation (7.11) has the dimensionless form

$$\frac{\partial \xi}{\partial \tau} + \xi \frac{\partial \xi}{\partial y} + \frac{1}{9Ur} \frac{\partial^3 \xi}{\partial y^3} = 0 \tag{7.15}$$

and its solution is determined by the Ursell parameter

$$Ur = \frac{H_e \lambda_e^2}{h^3}. \tag{7.16}$$

For the solitary wave (soliton) using its length definition (7.13) the Ursell parameter equals to $Ur_s = 4$. This value separates two different regimes of the wave dynamics: linear-dispersive ($Ur << 1$) and nonlinear-nondispersive ($Ur >> 1$). In fact, the estimates of the Ursell parameter can be done by comparison of the tsunami source dimension and the soliton length. Taking into account the characteristic dimension of the "seismic" tsunami source of 10-100 km it is simple to conclude that in shallow seas tsunami waves are non-dispersive ($Ur >> 1$). This is why the nonlinear-dispersive theory is the basic model for tsunami waves in the coastal zone. Waves of the explosion origin have characteristics scales less ocean depth, and such waves can be considered as linear and dispersive ($Ur << 1$).

In fact, to characterize the nonlinear and dispersive properties of the wave field requires the additional estimates of the spatial scales of nonlinearity and dispersion. If, for instance, $Ur >> 1$ and dispersion can be neglected, the equation (7.15) reduces to the quasi-linear equation

$$\frac{\partial \xi}{\partial \tau} + \xi \frac{\partial \xi}{\partial y} = 0, \tag{7.17}$$

having the solution

$$\xi(y, \tau) = \xi_0(y - \xi\tau). \tag{7.18}$$

This solution is smooth for finite time only and then the wave breaks ("gradient catastrophe"). The characteristic time of the wave breaking is easily found from (7.18) and in the dimension variables can be re-written for the length of the wave breaking

$$L_n \sim \lambda_e \frac{h}{H_e}, \tag{7.19}$$

where the numerical coefficient in (7.19) is determined by the initial wave shape. Taking into account that nonlinearity is weak in the deep ocean ($H_e/h \sim 10^{-3}$), the characteristic nonlinear scale exceeds the ocean dimension, and, therefore, the nonlinear effects in average are not essential for tsunami waves of seismic origin in the deep water. For shallow water the role of nonlinear effects growths. These arguments demonstrate why the nonlinear shallow-water equations are popular to simulate wave tsunamis in the coastal zone.

For the case of small values of the Ursell parameter ($Ur << 1$) the nonlinear term in (7.15) can be neglected and the equation reduces to

$$\frac{\partial \xi}{\partial \tau} + \frac{1}{9Ur}\frac{\partial^3 \xi}{\partial y^3} = 0 \tag{7.20}$$

Far from the source, its solution tends to the self-similar solution having the shape of the Airy-function (5.13). As a result, if the water level in the source is uplifted only and the generated wave is of elevation, the wave shape deforms with time, and the oscillations of both polarities are appear behind the leading wave. The characteristic scale of dispersion is

$$L_d \sim \frac{\lambda_e^3}{h^2}. \qquad (7.21))$$

For tsunamis of the explosion origin the dispersion parameter (h/λ_e) is not weak, and the characteristic dispersion scale in deep ocean (1-5 km) is too small, and the wave should be considered as a linear-dispersive one.

The estimates given above characterize the tsunami wave field on average. It is important to mention that the characteristic scales of nonlinearity and dispersion are comparable with the scales of the sea bottom inhomogeneities. As a result, nonlinear and dispersive properties (for instance, soliton) may be "hidden" in the real wave field. Most of tsunami registrations have been done near the coastal locations where dispersion vanishes. This process will be demonstrated in the solution of the runup problem when the initial elevation of soliton-like shape transforms into a sign-variable pulse on the beach; see section 9. Nonlinear bottom friction also changes the ratio nonlinearity/dispersion. Really complicated geometry of the coastal zone leads to the generation of the long-time resonance oscillations in the bays, and solitons can be "hidden" in such oscillations. That is why the soliton-like disturbances in the tsunami wave field have not been recorded so far.

8 Tsunami waves in the ocean of variable depth

The bottom relief of the ocean is rather complicated (figure 14), and now the bathymetric charts are available in the digitized format with the grid resolution of 1 km for the world's oceans. In the Pacific, about 90% of the seafloor are occupied by hills with heights 100 to 300 m and slopes of the order of 0.01. For such smoothed variation of the seafloor (on tsunami wave scales), the ray method taking into account nonlinearity and dispersion of the tsunami waves can be developed (Pelinovsky, 1982, 1996; Dingemans, 1996). Moreover, the hydrodynamic equations in the shallow-water approximation are

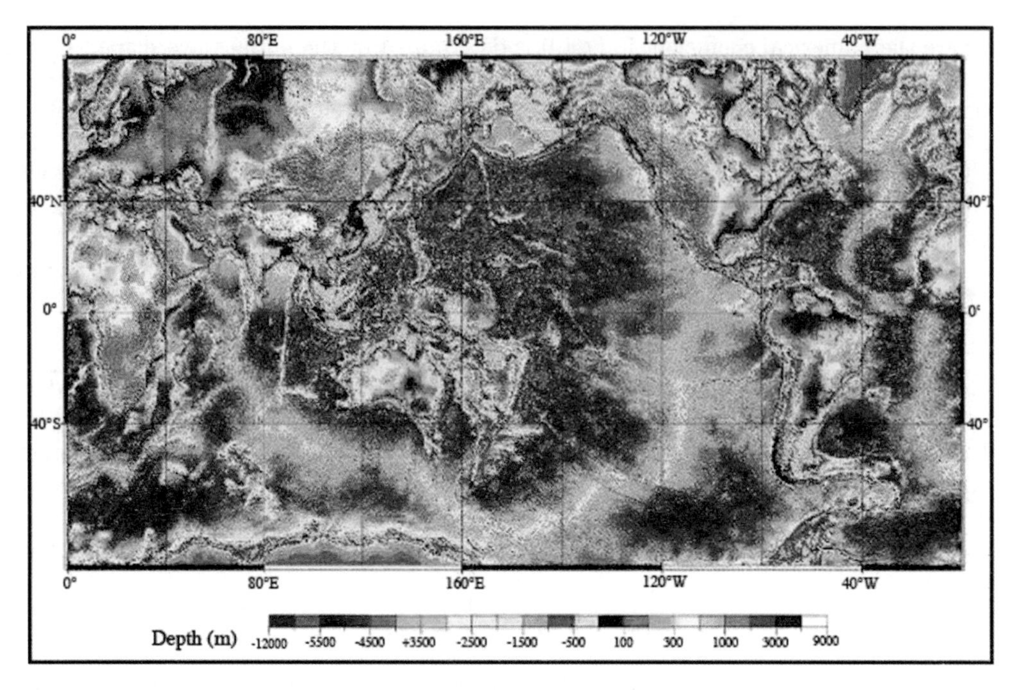

Figure 14. Bathymetry chart of the world's oceans.

hyperbolic, and, therefore, the wave front propagates along rays with the fixed speed, $c = \sqrt{gh}$ for any relation between the depth and wave scales. Basic ray equations for the waves of any physical nature, as it is known, can be written in the Hamilton form

$$\frac{\mathrm{d}\mathbf{r}}{\mathrm{d}t} = \frac{\partial \omega}{\partial \mathbf{k}}, \qquad \frac{\mathrm{d}\mathbf{k}}{\mathrm{d}t} = -\frac{\partial \omega}{\partial \mathbf{r}}, \tag{8.1}$$

and the specificity of the tsunami waves is the dispersion relation

$$\omega = c|\mathbf{k}| = \sqrt{gh(x,y)}k, \qquad k = |\mathbf{k}| = \sqrt{k_x^2 + k_y^2}. \tag{8.2}$$

Let us consider the simplest geometry of the seafloor with parallel isobaths, $h = h(x)$, and the plane wave approached to the coast under angle, ϑ_0 to isobaths. In this case from (8.1) it is immediately follows that $k_y = const$ and the well-known Snell formula for the refraction angle is derived

$$\cos\vartheta(x) = \cos\vartheta_0 \frac{c(x)}{x_0} = \cos\vartheta_0 \sqrt{\frac{h(x)}{h_0}}. \tag{8.3}$$

In particular, if the wave propagates in shallow water, $\vartheta \to \pi/2$, and the wave turns to the coast. The ray trajectory can be determined in the integral form

$$y = y_0 + \int_{x_0}^{x} \left(\frac{h_0}{h(x)} - \cos^2 \vartheta_0 \right)^{-1/2} \mathrm{d}x \tag{8.4}$$

and for the specific case of the parabolic bottom ($h \sim x^2$), the rays are arcs of circle

$$(y - y_0 - x_0 \tan \vartheta_0)^2 + x^2 = \frac{x_0^2}{\cos^2 \vartheta_0} \tag{8.5}$$

with its center at the coastal line and its radius depending on the initial sliding angle. Even by that simple analytical example one can see a number of general properties of rays in a basin with a varying depth: straightening of the rays in the shallow water, turning of the rays and their reflection from the zone of great depth, formation of wave-guides over underwater mountain ridge, etc. Really the ray pattern can be very complicated, and figure 15 demonstrates the ray pattern of tsunami waves in the Japan (East) Sea (Choi et al., 2002).

For simulation of the tsunami waves on long distances, the ray equation should be modified taking into account the Earth radius (Satake, 1988)

$$\frac{\mathrm{d}\theta}{\mathrm{d}t} = \frac{\cos\zeta}{nR}, \qquad \frac{\mathrm{d}\varphi}{\mathrm{d}t} = \frac{\sin\zeta}{nR\sin\theta},$$

$$\frac{\mathrm{d}\zeta}{\mathrm{d}t} = -\frac{\sin\zeta}{n^2R}\frac{\partial n}{\partial\theta} + \frac{\cos\zeta}{n^2R\sin\theta}\frac{\partial n}{\partial\varphi} - \frac{\sin\zeta\cot\theta}{nR}, \tag{8.6}$$

where θ and φ are latitude and longitude of the ray, $n = \sqrt{gh}$ is the slowness, R is the radius of the earth, and ζ is the ray direction measured counter-clockwise from the south. Figure 16 displays the computed rays for the 1883 Krakatau volcanic tsunami (Choi et al., 2003) demonstrated the passage of the tsunami waves from the Indian Ocean to the Atlantic and Pacific Oceans. This tsunami had the global character and was recorded world-wide. The last giant 2004 tsunami in the Indian Ocean also was recorded world-wide. Figure 17 shows the ray pattern in the Indian Ocean computed by Prof. Choi (unpublished). It is clearly seen that tsunami waves affect coasts of Thailand, India, Sri Lanka, Somali and Kenya. Bangladesh and Malaysia are in the "shadow" zone, and tsunamis here were weak in agreement with the observations.

The orthogonals to the rays determine the location of the wave front with time, and can be used to calculate tsunami travel time

$$\tau = \int \frac{\mathrm{d}l}{c(l)} = \int \frac{\mathrm{d}l}{\sqrt{gh(l)}}, \tag{8.7}$$

where the integral is calculated along the ray. That is why that the orthogonals are called as isochrones, and they are shown in Figs. 8.3 and 8.4 with the interval of 1 hr. Tsunami waves crossed the Indian Ocean for approximately 12 hrs and reached the coasts of France and the UK for about 30 hrs. Unfortunately, manifestations of the Krakatau tsunami

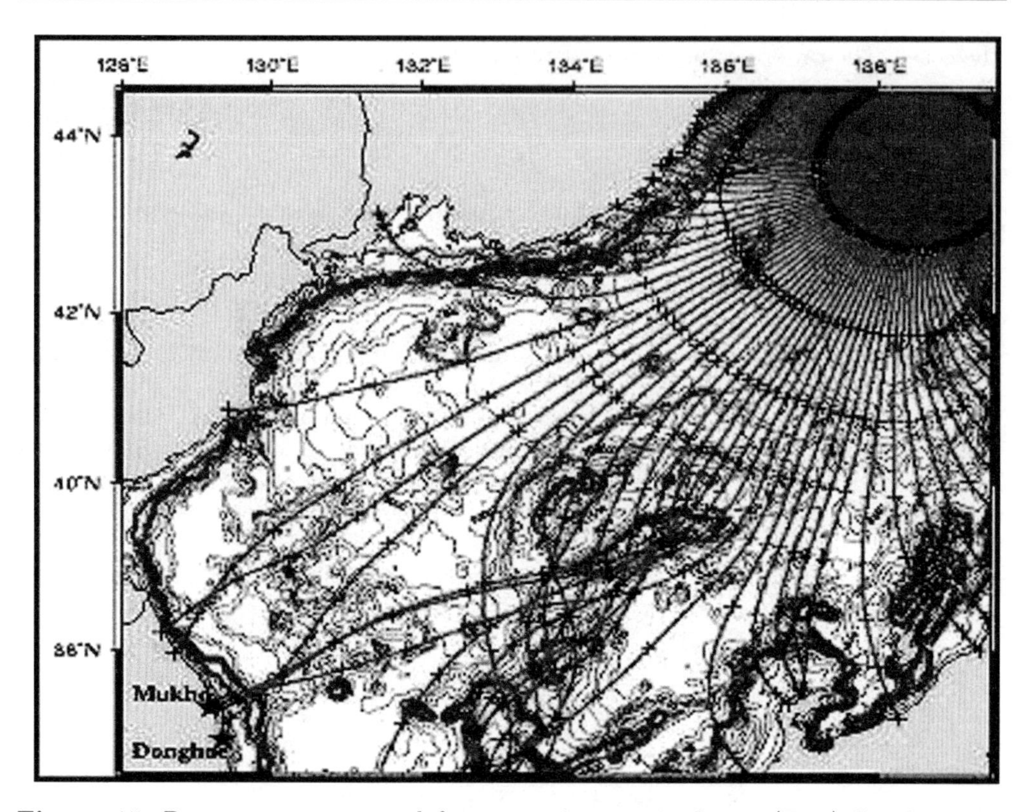

Figure 15. Ray pattern computed for tsunami waves in Japan (East) Sea from the isotropic source.

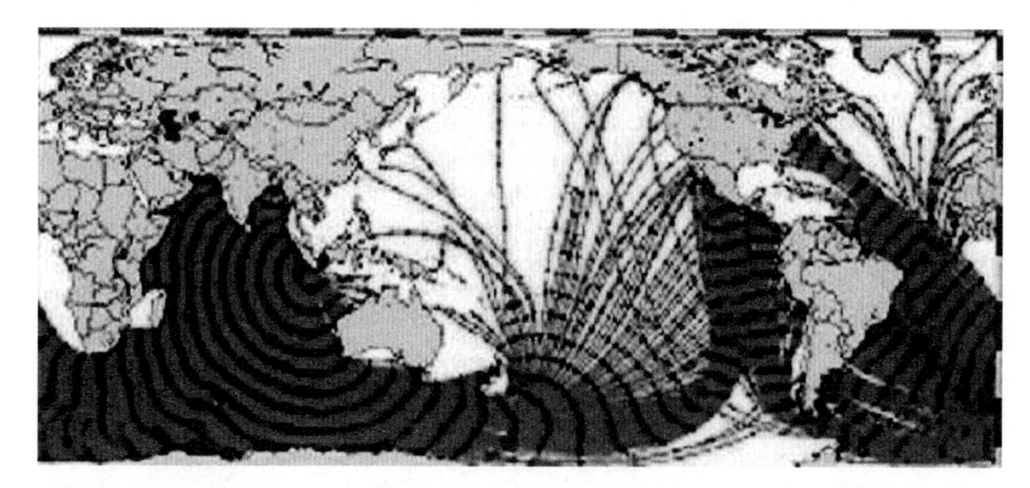

Figure 16. Ray pattern for the 1883 Krakatau volcanic tsunami.

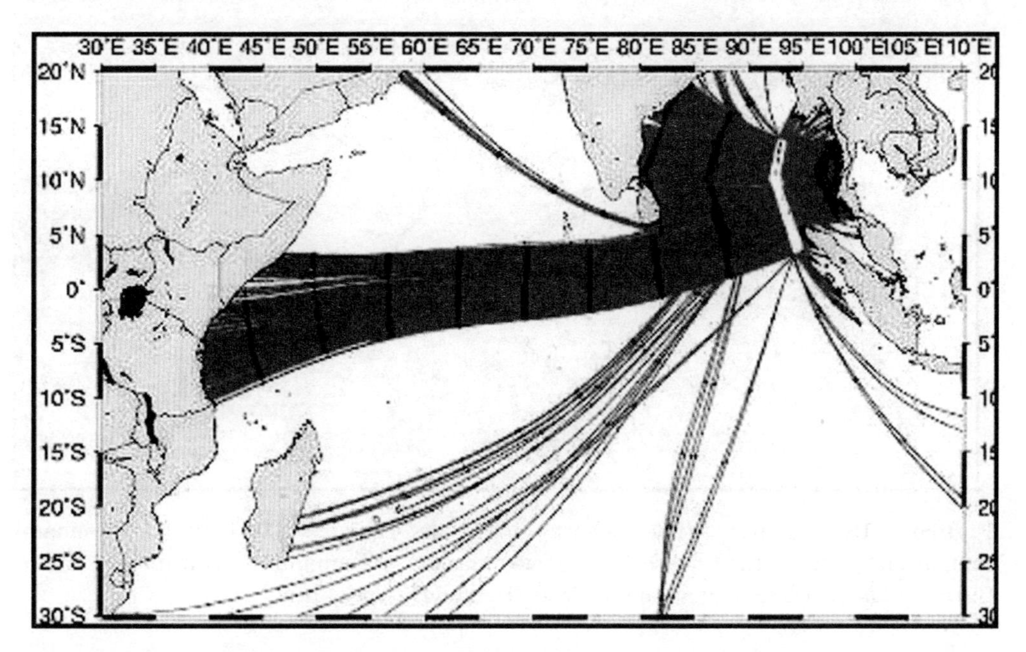

Figure 17. Ray pattern for the 2004 Indonesian earthquake tsunami (B.H.Choi, 2005, unpublished.)

on large distances are not clearly detected on tide-gauge records in Europe (Choi et al., 2003). For comparison, the travel time chart for the giant 2004 Sumatra tsunami in the Indian Ocean is shown in figure 18. Penetration of the tsunami waves in the Pacific and Atlantic Oceans was documented for the last event; in particular, the wave amplitude in Canada (on both, the Atlantic and Pacific Coasts) was about 10 cm after the long trip through all the oceans. The 2004 Sumatra tsunami of the seismic origin has larger dimensions than the 1883 Krakatau volcanic tsunami, and as a result, more visible on large distances.

Computed travel time charts are extremely important for warning the inhabitants and organization of their evacuation in the safety places, and now it is the "routine" work of the Tsunami Warning Centers. For instance, the Pacific Tsunami Warning Center detected the strongest 2004 Sumatra earthquake (but determined its magnitude as 8) and after 15 minutes evaluated that "based on location and magnitude the earthquake was not sufficient to generate a tsunami damaging to California - Oregon, Washington - British Columbia or Alaska. Some areas may experience small sea level changes. In areas of intense shaking locally generated tsunamis can be triggered by slumping" (http://www.prh.noaa.gov/ptwc/olderwmsg). It also advised that there should be a tsunami watch going on but did not non an alert about the coming danger. 65 minutes later, the Pacific Warning Center issued the next bulletin revising the magnitude on

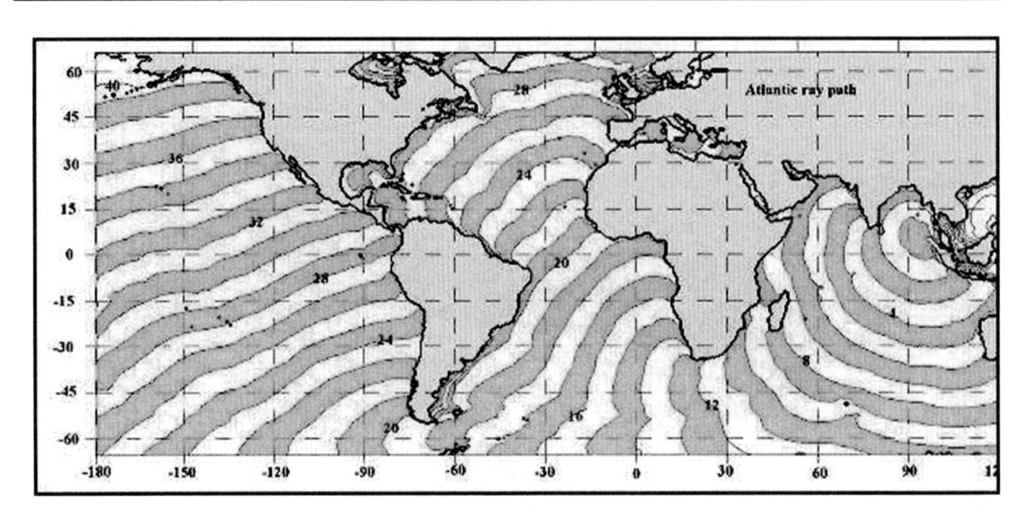

Figure 18. Travel time chart for the giant 2004 Indian tsunami, simulation of the West Coast/Alaska Tsunami Warning Center (http://wcatwc.arh.noaa.gov/IndianOSite/IndianO12-26-04.htm).

8.5 and informed everyone that, "this earthquake is located outside of the Pacific. No destructive tsunami threat exists for the Pacific basin based on historical earthquake and tsunami data". As we can see, the action of the Pacific Tsunami Warning Center was absolutely correct for the Pacific. The tragedy of the last event is that there is no Tsunami Warning System for the Indian Ocean region. As a result, the officers of the Pacific Tsunami Warning Center did not have the addresses of the appropriate officials in the Indian Ocean.

The computing of the wave heights in the ocean of variable depth is a very complicated task to compare with the travel time charts. If the main assumption of the ray theory (smoothness of the bottom relief in scales of tsunami wavelengths) will be used, it will leads to the energy flux conservation in the channel formed by the two neighboring rays

$$cbE = \sqrt{gh}b\rho g \int \eta^2 \mathrm{d}t = const, \tag{8.8}$$

where b is a differential ray width (the distance between two neighboring rays) determined by the travel time. In fact, the differential width can be calculated as the Jacobian between the Cartesian coordinate system and the curvilinear system formed by the rays and their orthogonals. The equation (8.8) is simple integrated for the linear waves

$$H(l) = H_0 \left[\frac{h_0}{h(l)}\right]^{1/4} \left[\frac{b_0}{b(l)}\right]^{1/2}. \tag{8.9}$$

In particular, the amplitude of the waves approaching the beach grows as $h^{-1/4}$ (in the 1D case), and if the depth varies from 4 km (open sea) to 10 m (shallow water),

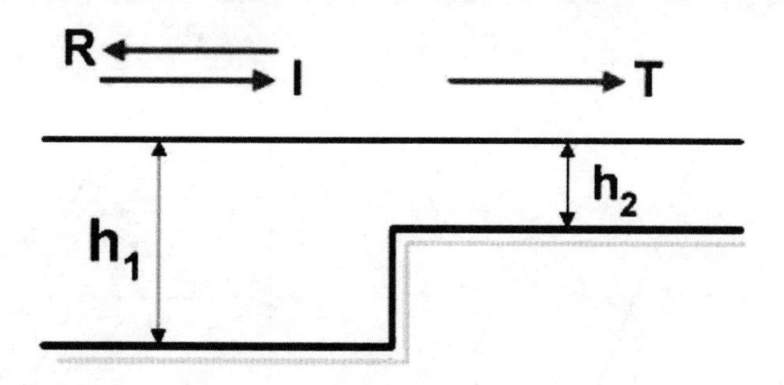

Figure 19. "Bottom step" geometry of the coastal zone.

the tsunami wave height amplifies 4.5 times. If the incident wave is a soliton (7.12), it follows from (8.8)

$$H(l) = H_0 \left[\frac{h_0}{h(l)}\right] \left[\frac{b_0}{b(l)}\right]^{1/3}. \tag{8.10}$$

and the soliton height grows more significantly than the height of the linear wave.

Unfortunately, the bottom relief is not smoothed enough in the scales of the tsunami waves of the seismic origin. As a result, the tsunami waves reflect and scatter by the bottom topography in the coastal zone, and the formulas (8.9) and (8.10) overestimate the wave heights.

To demonstrate the reflection of the tsunami wave let us consider the simplest geometry of the coastal zone containing a bottom step (figure 19). Here **I** is the incident wave with amplitude A, **R** is the reflected wave with amplitude A_r and **T** is the transmitted wave with amplitude A_t. The relation between all amplitudes can be found from two boundary conditions at the bottom step; they are continuity of the pressure (water level in the hydrostatic approximation) and discharge (hu)

$$\eta_1 = \eta_2, \qquad (hu)_1 = (hu)_2. \tag{8.11}$$

These two conditions determines the amplitudes of the reflected and transmitted waves

$$\frac{A_t}{A} = \frac{2c_1}{c_2 + c_1}, \qquad \frac{A_r}{A} = \frac{c_1 - c_2}{c_2 + c_1}. \tag{8.12}$$

It is important to note that the energy posses in this process,

$$A_r^2 + A_t^2 = A^2. \tag{8.13}$$

The relation between the wave amplitudes is displayed in figure 20. In the case when there is no bottom step $(h_2 = h_1)$, reflection is absent, and the wave energy is transmitted with no loss of energy. If the wave moves in the shallowest region $(h_1 > h_2)$, the reflected wave has the same polarity as the incident wave, and the amplitude of the transmitted wave is increased. The maximal amplification is achieved for $h_1 >> h_2$, and in this case the transmitted wave becomes twice the incident wave, in contrast to the smoothed relief, see (8.9). If the wave moves in the deepest depth $(h_1 < h_2)$, the reflected wave has the

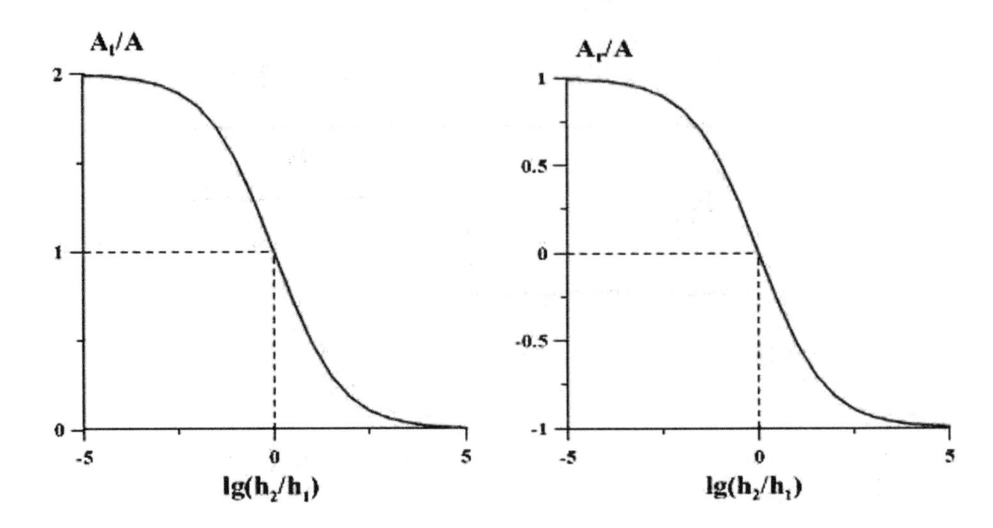

Figure 20. Amplitudes of the transmitted and reflected waves versus the bottom step.

opposite polarity, and the amplitude of the transmitted wave is decreased. In the case of multiple reflections the wave shape can be very complicated, even if the incident wave contains only one crest; the wave contains the different pulses of various polarities and amplitudes and its time record looks like the wave train.

The two-dimensional bottom relief leads to the existing of surface waves captured by the bottom inhomogeneities. For instance, if the beach has a constant slope ($h = -\alpha y$), the solution of the linear equations of the shallow-water theory can be expressed through the ***edge waves***

$$\eta_n(x, y, t) = A_n \exp(-ky) L_n(2ky) \exp\left[\mathrm{i}(\omega_n t - kx)\right], \tag{8.14}$$

where L_n is the Laguerre polynomial describing the offshore modal structure. The frequencies, ω_n and the along wavenumber, k are related by the dispersion relation

$$\omega_n = \sqrt{(2n+1)\alpha g k}, \tag{8.15}$$

n is the integer indicated the modal number, and A_n is the wave amplitude (in general., complex value). The offshore structure of the edge waves is shown in figure 21, and the dispersion relation (8.15) in figure 22 (for beach slope, $\alpha = 10^{-3}$).

It is important to mention that the edge waves are highly dispersive (like wind waves) and the wave pattern in space varies with time. There is a lot of tsunami observations where strong intensity can be explained by the theory of trapped waves only. For instance, the 25 April, 1992 Cape Mendocino earthquake generated a tsunami characterized by edge waves that propagated north and south along the U.S. West Coast (Gonzalez et al., 1995). Due to frequency dispersion of the edge waves, such waves approach significantly later than the leading wave, and their amplitudes are significantly higher. The existence

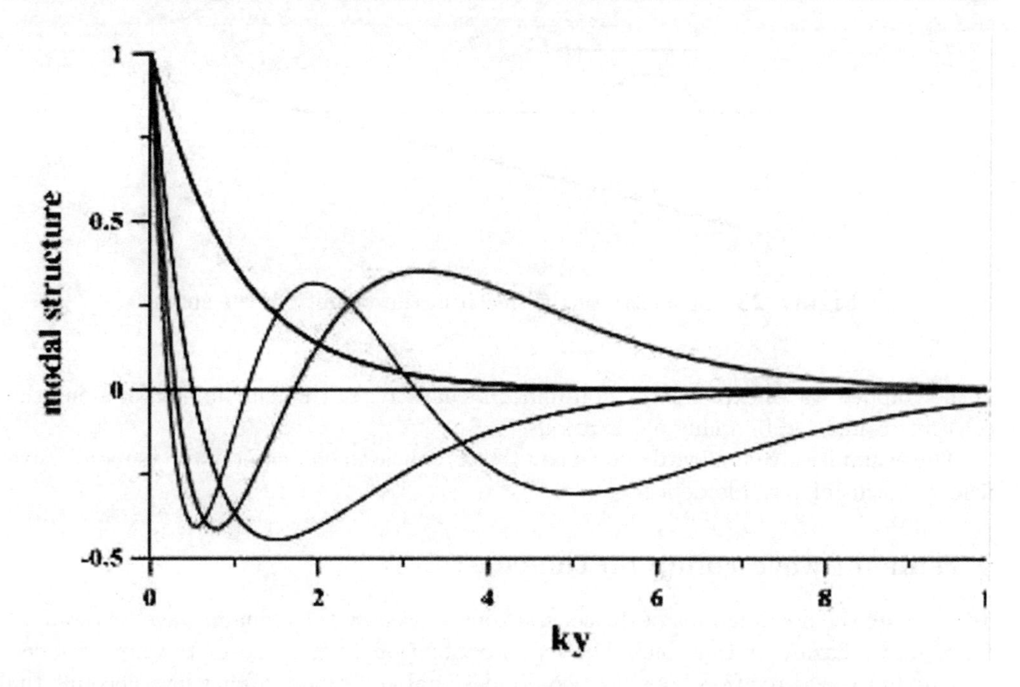

Figure 21. Offshore structure of the edge waves above the beach of constant slope.

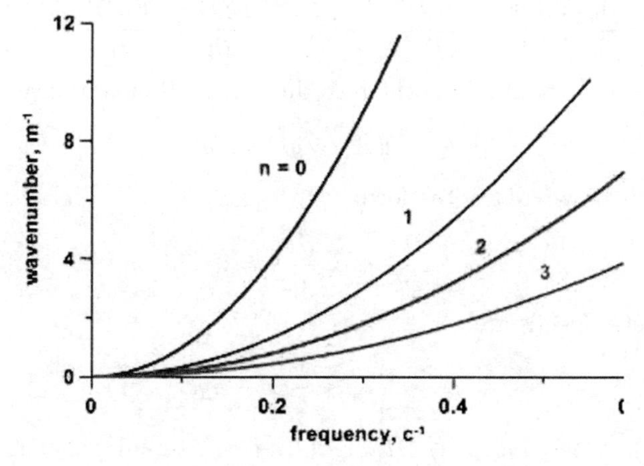

Figure 22. Dispersion relation of the edge waves for the beach slope, $\alpha = 10^{-3}$.

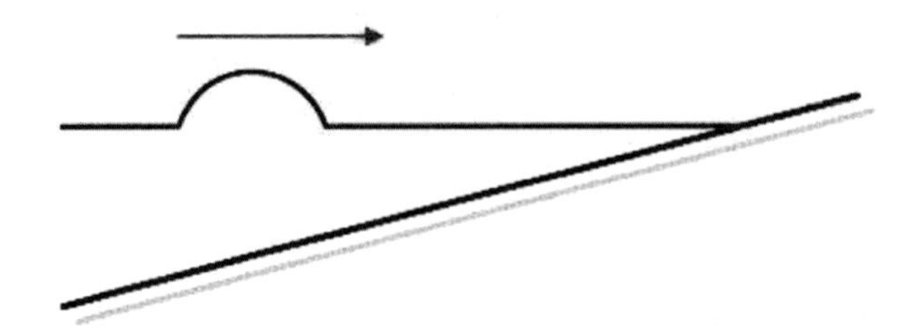

Figure 23. Geometry of the beach in the vicinity of the shore.

of the trapped waves explains the non-uniform character of the tsunami height along the coastline confirmed by many observations.

The examples given above demonstrate the complicated character of the tsunami wave field in basins of variable depth.

9 Tsunami wave runup on the coast

To estimate the flooding area of the coastal zone caused by the tsunami waves is essential for tsunami hazard mitigation. The exact estimation of the size of the area flooded by tsunami waves requires to solve two-dimensional equations, taking into account the complex geometry of the coastal line containing bays, straits, estuaries, etc. Here we will analyze the simplest geometry of the beach of the constant slope (figure 23)

$$h(x) = -\alpha x. \tag{9.1}$$

The nonlinear shallow-water system (3.8) and (3.9) in the 1D case reduces to

$$\frac{\partial u}{\partial t} + u\frac{\partial u}{\partial x} + g\frac{\partial D}{\partial x} = -\alpha g, \qquad \frac{\partial D}{\partial t} + \frac{\partial (uD)}{\partial x} = 0, \tag{9.2}$$

where $D = h(x) + \eta$ is the total depth. Introducing the Riemann invariants

$$I_\pm = u \pm 2\sqrt{gD} + \alpha g t, \tag{9.3}$$

the system (9.2) is re-written in the form

$$\frac{\partial I_\pm}{\partial t} + c_\pm \frac{\partial I_\pm}{\partial x} = 0, \tag{9.4}$$

where characteristic speeds are

$$c_\pm = \frac{3}{4}I_+ + \frac{1}{4}I_- - \alpha g t. \tag{9.5}$$

Multiplying (9.4) by the Jacobian $\partial(t,x)/\partial(I_+, I_-)$, assuming that it is not zero (this important question will be discussed later), the equation can be transformed to

$$\frac{\partial^2 t}{\partial I_+ \partial I_-} + \frac{3}{2(I_+ - I_-)}\left(\frac{\partial t}{\partial I_-} - \frac{\partial t}{\partial I_+}\right) = 0. \tag{9.6}$$

Let us introduce new arguments:

$$\lambda = \frac{I_+ + I_-}{2} = u + \alpha g t, \qquad \sigma = \frac{I_+ - I_-}{2} = 2\sqrt{gD}. \tag{9.7}$$

Then, equation (9.6) takes the form

$$\frac{\partial^2 t}{\partial \lambda^2} - \frac{\partial^2 t}{\partial \sigma^2} - 3\frac{\partial t}{\partial \sigma} = 0. \tag{9.8}$$

Extracting time from (9.7) and substitute

$$u = \frac{1}{\sigma}\frac{\partial \Phi}{\partial \sigma}, \tag{9.9}$$

equation (9.8) is re-written in the final form

$$\frac{\partial^2 \Phi}{\partial \lambda^2} - \frac{\partial^2 \Phi}{\partial \sigma^2} - \frac{1}{\sigma}\frac{\partial \Phi}{\partial \sigma} = 0. \tag{9.10}$$

It is convenient to give formulas to determine all physical variables

$$\eta = \frac{1}{2g}\left[\frac{\partial \Phi}{\partial \lambda} - u^2\right], \quad u = \frac{1}{\sigma}\frac{\partial \phi}{\partial \sigma}, \quad x = \frac{\eta}{\alpha} - \frac{\sigma^2}{4g\alpha}, \quad t = \frac{\lambda - u}{g\alpha}. \tag{9.11}$$

So, the initial set of nonlinear shallow water equations has reduced to the linear wave equation (9.10) and all physical variables can be found via Φ using simple operations. The main advantage of this form is that the moving (unknown) shoreline corresponds now to $\sigma = 0$ (since the total depth $D = 0$) and, therefore, equation (9.10) is solved in the half-space $-\infty < \sigma < 0$ with the fixed boundary. Such transformation was suggested by Carrier and Greenspan (1958) and then generalized for more complicated geometry of the coastal zone.

The formulas (9.11) are implicit and this is the main difficulty in finding the wave field analytically. Meanwhile, an important problem for practice - the dynamics of the moving shoreline (the boundary of the flooded zone) - can be found explicitly for waves generated far from the coast. At large distances from the shore, the wave field is linear, and formulas (9.11) give the explicit relations between all variables:

$$\eta_l = \frac{1}{2g}\frac{\partial \Phi_l}{\partial \lambda_l}, \quad u_l = \frac{1}{\sigma_l}\frac{\partial \phi_l}{\partial \sigma_l}, \quad x_l = -\frac{\sigma_l^2}{4g\alpha}, \quad t_l = \frac{\lambda_l}{g\alpha}, \tag{9.12}$$

where we particularly mark all variables as linear variables. Having initial conditions for the zone of the tsunami generation, or knowing the characteristics of the approaching tsunami wave, it is easy to find the function $\Phi_l(t_l, x_l)$ or $\Phi_l(\lambda_l, \sigma_l)$. Because the functions $\Phi(\lambda, \sigma)$ and $\Phi_l(\lambda_l, \sigma_l)$ are described by the same equation (9.10), the solution of the wave equation is determined fully, in the linear approximation, as well as in the nonlinear exact formulation. Considering now $\sigma_l = 0$, we may obtain the wave record at the non-moving shoreline ($x_l = 0$) in the linear approximation (water elevation and fluid velocity), $R_l(t_l) = \eta_l(t_l, 0)$ and $U_l(t_l) = u_l(t_l, 0)$. In the nonlinear formulation, the point $\sigma = 0$

corresponds to the moving shoreline, and $R(t) = \eta(\lambda, 0)$ and $U(t) = u(\lambda, 0)$ describe the "real" runup characteristics. Formally, both functions, $R_l(\lambda_l)$ and $R(\lambda)$ are the same, but the argument of the nonlinear runup depends on the velocity. At the moment $U = 0$ (maximum runup or rundown), arguments coincide, and, therefore, maximum runup characteristics can be found in the linear formulation. This is an important conclusion for tsunami practice, which is used sometimes with no strong mathematical proof. Moreover, the solution of the linear problem can be used to describe the dynamics of the moving shoreline in a simple form. According to the last formula in (9.11) for the time, the following relation is obtained (Pelinovsky and Mazova, 1992)

$$U(t) = U_l \left(t + \frac{U}{g\alpha} \right), \tag{9.13}$$

and the "nonlinear" velocity of the moving shoreline is obtained from the "linear" velocity with the variable deformation of time. Such deformation is well-known for the Riemann (simple) wave in gas dynamics deformed in space and time, but here it is valid for the moving boundary only. Using (9.13), the runup height can be calculated

$$R(t) = \alpha \int U(t)\mathrm{d}t = R_l \left(t + \frac{U}{g\alpha} \right) - \frac{1}{2g} U^2 \left(t + \frac{U}{g\alpha} \right). \tag{9.14}$$

All theoretical conclusions can be illustrated by the example of the sine wave runup. The particular bounded solution of the wave equation (9.10) has the following form

$$\Phi(\lambda, \sigma) = Q J_0(p\sigma) \sin(p\lambda), \tag{9.15}$$

where Q and p are constants which to be determined. We assume that the characteristics of the incident tsunami waves are known far from the shore, where we may use the linear approximation (9.12). Using standard asymptotics of the Bessel functions for small and large values of the argument, we may determine all the parameters and find the relation between the maximum runup amplitude, R_{max} and the amplitude A_0 of the incident wave at depth h_0

$$\frac{R_{max}}{A_0} = 2\sqrt{\pi} \left(\frac{\omega^2 h_0}{g\alpha^2} \right)^{1/4} = 2\pi \sqrt{\frac{2L}{\lambda_0}}, \tag{9.16}$$

where L is the distance from the isobath h_0 to the shoreline, and λ_0 is the wavelength of the incident wave. In fact, the same formula will be obtained if the linear approximation is used everywhere, from the open sea to the beach, and is obtained in the framework of the nonlinear shallow-water theory. For more complicated geometry of the coastal zone, we could formally not obtain the proof of the applicability of the linear theory. But if the beach is varied significantly in the zone of the linear wave and tends to the beach of constant depth in the zone of the nonlinear wave, the applicability of the linear approach is evident. We will give here the generalization of (9.16) for the geometry presented in figure 24,

$$\frac{R}{A_0} = \frac{2}{\sqrt{J_0^2(2kL) + J_1^2(2kL)}}, \tag{9.17}$$

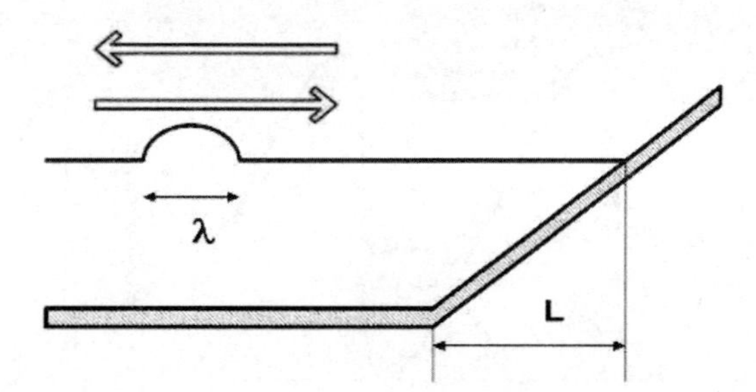

Figure 24. Beach geometry.

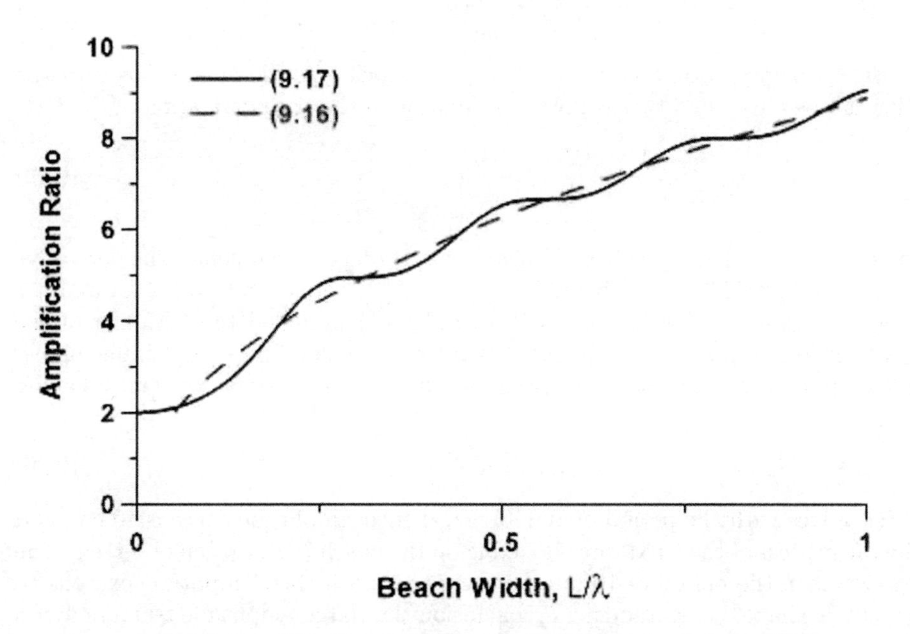

Figure 25. Runup wave height versus the beach width.

and (9.17) transforms to (9.16) for large values of the argument, kL (k is the wavelength on depth h_0); see, figure 25. As it can be seen the asymptotic nonlinear formula (9.16) almost coincides with the linear formula (9.17). Meanwhile, if the beach width tends to zero, the runup amplitude exceeds twice the incident wave amplitude; in fact the nonlinear theory predicts a little more.

So, the linear theory can predict correctly the runup height. Nonlinearity influences on the wave profile according to (9.13) and (9.14), and this is displayed in figure 26

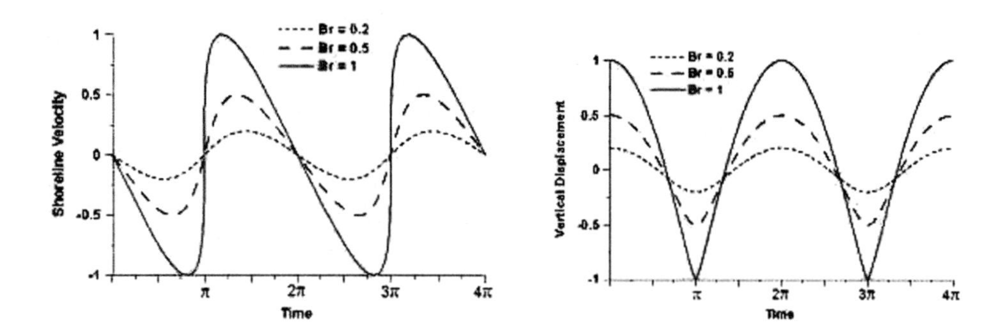

Figure 26. Velocity and vertical displacement of the moving shoreline.

(dimensionless variables, ωt, R/R_{max}, U/U_{max}) for various values of the breaking parameter

$$Br = \frac{\omega^2 R_{max}}{g\alpha^2}. \tag{9.18}$$

When the breaking parameter tends to 1, the velocity profile transforms into the shock. This can be illustrated by the direct calculation of the time-derivation of the $U(t)$

$$\frac{\mathrm{d}U}{\mathrm{d}t} = \frac{U_l'}{1 - \frac{U_l'}{g\alpha}}, \tag{9.19}$$

where prime means the derivative from "linear" velocity on its argument. The derivative $\mathrm{d}U/\mathrm{d}t$ tends to infinity (the wave breaks) when $Br = 1$. This condition is an another form of the zero value of the Jacobian $\partial(t,x)/\partial(I_+, I_-)$ provided the breaking of the hodograph transformation. The form (9.18) is more convenient because it immediately gives the maximum possible value of the runup height of tsunami waves without breaking (figure 27),

$$R_{breaking} = \frac{g\alpha^2}{\omega^2}. \tag{9.20}$$

The wind waves with the period of 10 s break, if their amplitude exceeds 10 cm, and this is why the process of the wind wave breaking on the beach is very often observed. The tsunami waves with the period of 10 min and more break, if their amplitude exceeds 10 m. As a result, we have the "paradoxical" conclusion that large-amplitude tsunami waves may climb on the beach without breaking meanwhile small-amplitude wind waves break near the shore. According to the observations, approximately 70% of tsunamis climbed on the beach without breaking. Breaking waves lost the energy and penetrated shorter distances inland compare to the non-breaking waves. This explains why the tsunami waves penetrate longer distances inland than the storm waves resulting in flooding of the coastal area. One of the photos of the flooding area during the giant 2004 Indian Ocean tsunami is given in figure 28.

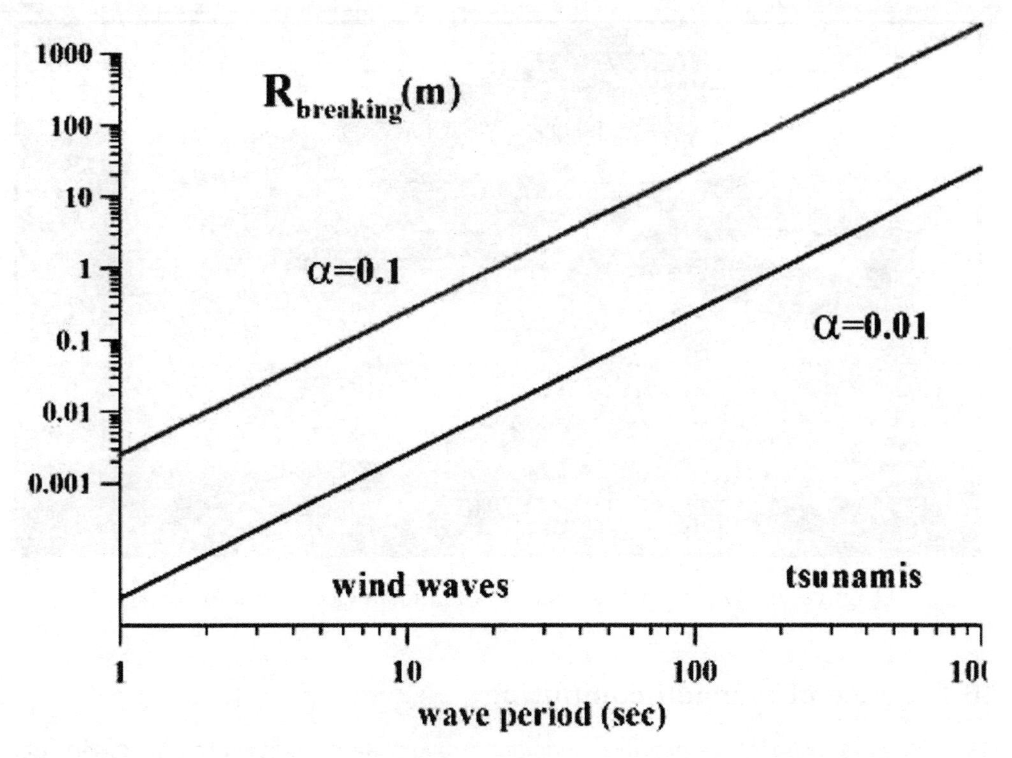

Figure 27. Breaking criterion for the long waves on the beach of constant slope.

Using the Fourier solution (9.15), we may analyze the runup of the tsunami wave of any shape and obtain (9.16) for the runup amplitude, but this formula will contain a numerical coefficient depending on the wave shape. In particular, if the incident wave is the solitary wave (7.12), it immediately follows from (9.16) that $R \sim H^{5/4}$. This case has been studied by Synolakis (1987), and we give here his formula for the runup amplitude

$$\frac{R_{max}}{h} = 2.83\sqrt{\cot\alpha}\left(\frac{H}{h}\right)^{5/4}. \tag{9.21}$$

In conclusion, we may say, that very often the tsunami computing is performed with non-reflected boundary conditions on the land (approximation of the vertical wall in the last "sea point" of numerical scheme on depth 5-10 m), because the computations of the tsunami wave run up and inundation require large computer resources. In such cases, the analytical formulas given above can be used (and they are used) for the rough estimates of the tsunami flooding zone (tsunami inundation maps).

Figure 28. Giant 2004 tsunami in the Indian Ocean (photo AFP).

10 Practice of tsunami computing

As we described above, the popular model for tsunami waves induced by the underwater earthquakes is the shallow-water system derived in section 3. In practice of the tsunami computing, two additional factors should be taken into account: the sphericity and rotation of the Earth, and the wave dissipation in the bottom turbulent boundary layer. As we have pointed out above, the nonlinear effects are not essential for tsunami waves in the open ocean, and here the basic shallow-water system becomes

$$\frac{\partial M}{\partial t} + \frac{gD}{R\cos\theta}\frac{\partial\eta}{\partial\varphi} = fN, \qquad \frac{\partial N}{\partial t} + \frac{gD}{R}\frac{\partial\eta}{\partial\theta} = -fM,$$

$$\frac{\partial\eta}{\partial t} + \frac{1}{R\cos\theta}\left[\frac{\partial M}{\partial\varphi} + \frac{\partial}{\partial\theta}(N\cos\theta)\right] = 0, \tag{10.1}$$

where θ and φ are latitude and longitude, R is the radius of the earth, $f = 2\omega_e\sin(\theta)$ is the Coriolis parameter, ω_e is the Earth rotation frequency, M and N are discharge fluxes along the latitude and longitude, $D = h(x,y) + \eta$ is the total water depth, $h(x,y)$ is unperturbed basin depth. This model can be applied to compute the tsunami waves on long distances (1000 - 10000 km), In particular, the numerical simulation of two global tsunamis in the Indian Ocean (1883 Krakatau volcano eruption, and 2004 Indonesian earthquake), when the waves crossed the Indian Ocean and penetrated in the Pacific and Atlantic Oceans, have been performed in the framework of the linear spherical model (10.1).

In the coastal zone and seas, having relatively small scales (up to 1000 km), the model of plane Earth with no rotation is applied

$$\frac{\partial M}{\partial t} + \frac{\partial}{\partial x}\left(\frac{M^2}{D}\right) + \frac{\partial}{\partial y}\left(\frac{MN}{D}\right) + gD\frac{\partial \eta}{\partial x} + \frac{k}{2D^2}M\sqrt{M^2+N^2} = 0,$$

$$\frac{\partial N}{\partial t} + \frac{\partial}{\partial x}\left(\frac{MN}{D}\right) + \frac{\partial}{\partial y}\left(\frac{N^2}{D}\right) + gD\frac{\partial \eta}{\partial y} + \frac{k}{2D^2}N\sqrt{M^2+N^2} = 0,$$

$$\frac{\partial \eta}{\partial t} + \frac{\partial M}{\partial x} + \frac{\partial N}{\partial y} = 0, \qquad (10.2)$$

where M and N are discharge fluxes along the horizontal coordinates, x and y, and $k = 0.025$ is the bottom friction coefficient (the same value is used in the theory of tidal waves). The use of discharges instead of velocities can provide automatically the stability of the numerical scheme when the wave breaks and transforms into the shock wave (bore).

On the open boundaries of the computed domain (for instance, straits) radiation condition is applied

$$\frac{\partial \eta}{\partial t} + \sqrt{gh}\frac{\partial \eta}{\partial n} = 0, \qquad (10.3)$$

where n is the external normal to the boundary of the domain. Formally, the condition (10.3) is exact only in the framework of (10.2) with no friction and rectangular boundary. On the "land" boundary the total reflection condition

$$\frac{\partial \eta}{\partial n} = 0 \qquad (10.4)$$

is applied (usually, on depth 5-10 m), or the runup problem is solved, when the equations (10.2) are solved in the domain with a moving boundary,

$$D = h(x,y) + \eta(x,y,t) \geq 0. \qquad (10.5)$$

As an initial condition, the conception of the equivalent tsunami source is applied

$$\eta(x,y,t=0) = \eta_0(x,y), \qquad u(x,y,t=0) = v(x,y,t=0) = 0, \qquad (10.6)$$

and the initial water displacement in the source is determined using the existing seismological models (Manshinha and Smylie, 1971; Okada, 1985), or in the simple "hydrodynamic" form with the empirically determined scales through the earthquake magnitude.

The models described above are realized numerically by using various numerical algorithms. One of the models, TUNAMI is developed in Tohoku University (Japan) and provided free through the Tsunami Inundation Modeling Exchange (TIME) program funded by UNESCO; see Goto el al., (1997). The model solves the governing equations by the finite difference technique with the leap-frog scheme.

The number of the cases studied is now big, and such models are applied to simulate the historical tsunamis and to evaluate the characteristics of the prognostic tsunamis in

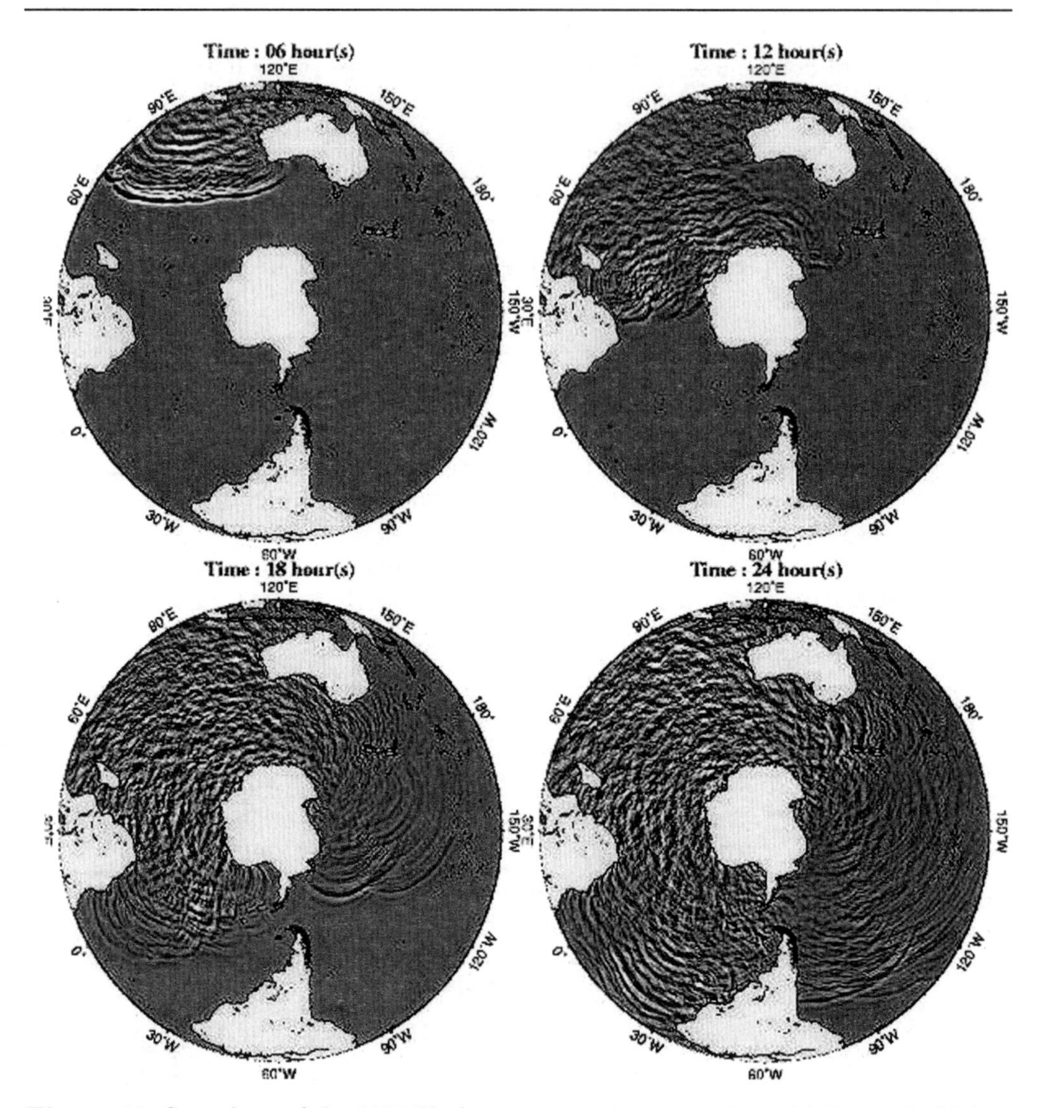

Figure 29. Snapshots of the 1883 Krakatau tsunami propagation with interval of 6 hrs.

all the areas of the world's oceans. Figure 29 displays the snapshots of the 1883 Krakatau volcanic tsunami with the interval 6 hrs (Choi et al., 2003). It is clearly seen that, the waves are scattered by the bottom inhomogeneous relief and reflected from the land, forming the complicated wave pattern.

A similar compution has been done for the last giant tsunami in the Indian Ocean practically immediately after the 26th of December, 2004 (most of the results were available during the first two weeks after the occurred event). Figure 30 presents the results of the calculations of the maximum wave amplitude made by Dr. Titov (PMEL, USA). Computing of the wave amplitudes confirms the results of the ray theory giving that tsunami waves were intense in Thailand, India, Sri Lanka, Maldives, Somali and Kenya.

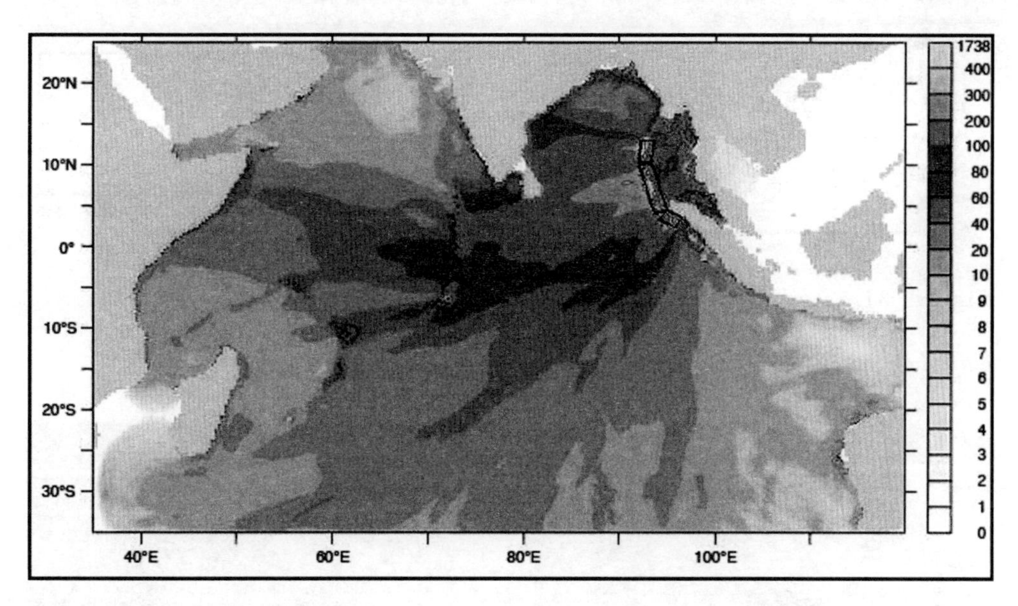

Figure 30. Directivity of wave amplitudes (in cm) in the Indian Ocean during the 2004 tsunami (*http : //pmel.noaa.gov/tsunami/indo*_01204*.html*).

The difference in the results of various research groups is related to the difference in the description of the tsunami source. The analysis of the seismic information leads to the increase of the magnitude of the earthquake, the last estimation is 9.3, and this changes the dimensions of the tsunami source. The fault length is estimated from 450 to 1200 km. The tsunami source had an elliptical shape in the first simulations, and then - in the form of several blocks (they are seen in figure 30). All variations in the tsunami source influences on the wave pattern relatively weakly, but dramatically - on the wave records in the selected coastal locations. Figure 31 shows the computed tide-gauge records of tsunami in Thailand (near Phuket) computed in the Tohoku and Kyoto Universities; they demonstrate the influence of the initial conditions on the computed results. The measured runup height at Phuket is 4-5 m and this correlates with computed heights. Now the field surveys are conducted in many countries of the Indian Oceans and their process will help to choice the adequate model of the tsunami source.

Numerical simulations of tsunami wave propagation is an effective tool to evaluate the tsunami risk for areas where the number of tsunami observations is very limited. For instance, 91 tsunamis have been reported in the Caribbean Sea in the past 500 years (O'Loughlin and Lander, 2003). Of these, 27 are judged to be true, verified tsunamis and the additional nine are considered to be very likely true tsunamis. Many of documented events have no quantitative data of wave heights and earthquakes. Such basins are now analyzed using the concept of tsunami potential, where the waves are computed from all the sources which can generate tsunamis. Figure 32 displays the location of the

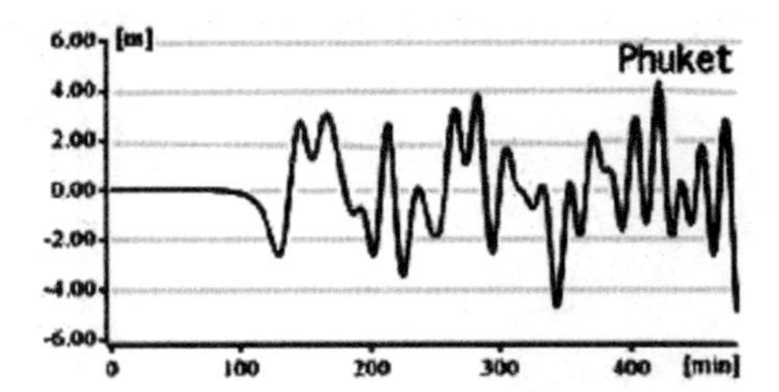

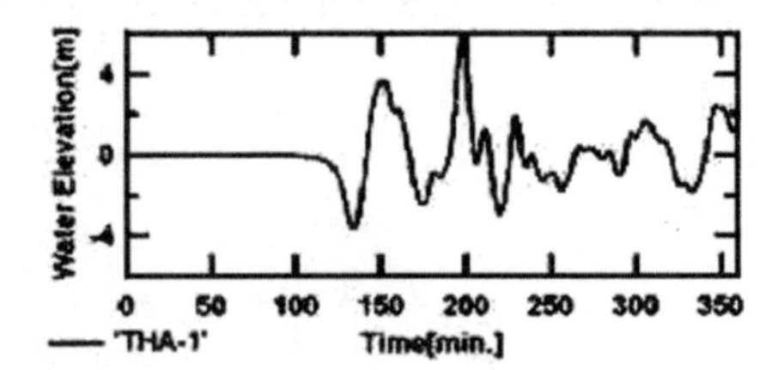

Figure 31. Computed tsunami records for Thailand in Tohoku (left) and Kyoto (right) universities (from *http : //www.drs.dpri.kyoto − u.ac.jp/sumatra/index − e.html#model*).

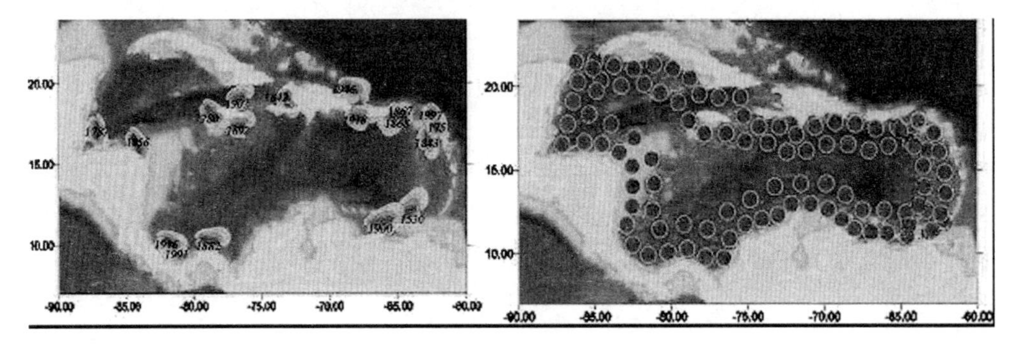

Figure 32. Location of historic (left) and hypothetic (right) tsunamis in the Caribbean Sea.

historical tsunamis in the Caribbean Sea, and also the location of hypothetic tsunamis (121 prognostic events) used for simulations (Zahibo et al., 2003). As a result, we have a lot of prognostic wave heights for each coastal location and may compare possible wave heights in different locations. The wave heights may have weak amplitudes in some areas (compared with others), and we may conclude that such zones have low tsunami risk. Computed zones with low tsunami risk in the Caribbean Sea are shown in Figure 33.

As we have mentioned above, dispersion can be important for the transoceanic tsunami propagation. This effect can be taken into account even without an explicit use of dispersive terms in the governing equations. The first correction to the dispersion relation (5.2) can be modeled by exploiting the numerical dispersion inherent in finite-difference algorithms. This modification is used in the numerical model MOST (Method of Splitting Tsunami) developed by PMEL (Titov and Gonzalez, 1997). In fact, the results of simulation of the giant 2004 tsunami in the Indian Ocean displayed in figure 30 are obtained in the framework of the MOST model. Another variant of nonlinear - dispersive model (GEOWAVE) is developed using the fully nonlinear extended Boussinesq system (FUNWAVE); it is available through the Internet (http://www.tsunamicommunity.org).

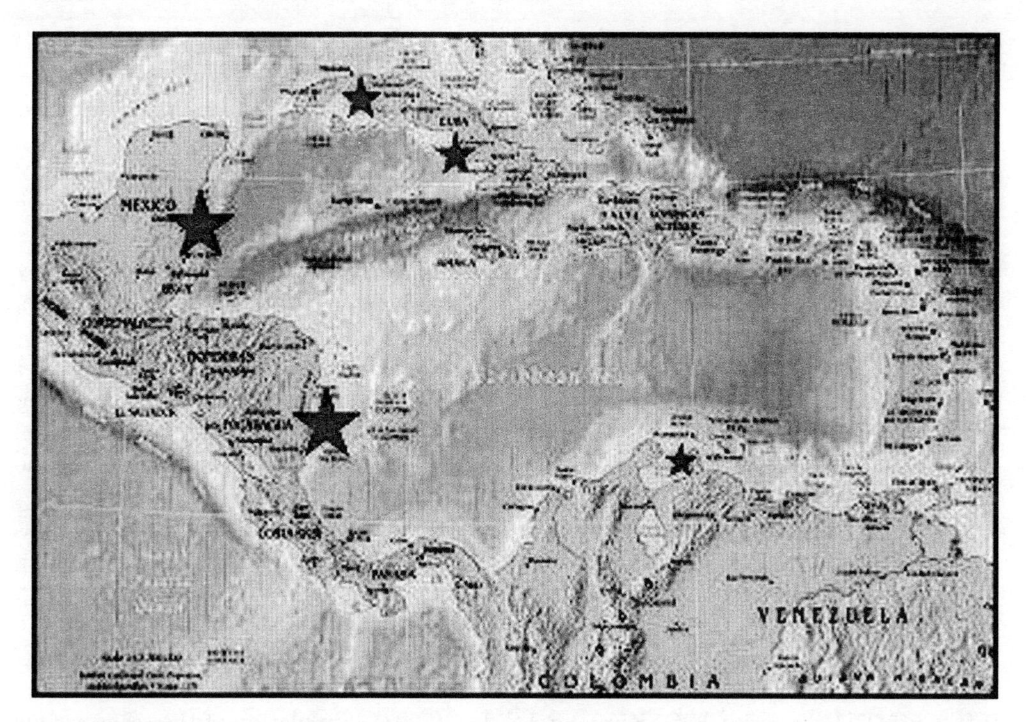

Figure 33. Geographical distribution of the zones with low tsunami risk (stars) in the Caribbean Sea.

This model includes special blocks to generate initial water displacement as for seismic, as well as for landslide processes.

11 Conclusion

During a long time, tsunami practice has been operating using a few of hydrodynamic approaches: the ray tracing method to estimate the tsunami travel time, and the nonlinear shallow-water theory to calculate the wave amplitudes. All other results of the hydrodynamic theory were only little used, and mainly then to demonstrate the physics of the tsunami generation, propagation and runup. The last giant 2004 Indian Ocean tsunami was very well documented by observations, and it is evident that this tsunami gives a strong impulse to develop and apply the theoretical results of modern hydrodynamics realized in the computer codes. The author hopes that the given course of lectures will be useful for readers to understand the key ideas existing in the hydrodynamics of tsunami and "entry" in the tsunami society.

Support from grants from INTAS (03-51-4286) and RFBR (05-05-64265) is acknowledged. I would like to thank Prof. J. Grue for great assistance with the preparation of this text.

Bibliography

K. Aki, K. and P. Richards (1980). Quantitative Seismology, Theory and Methods. Freeman, San Francisco.

S. Assier-Rzadkiewicz, P. Heinrich, P.C. Sabatier, B. Savoye and J.F. Bourillet (2000). Numerical modelling of a landslide-generated tsunami: the 1979 Nice event. Pure and Applied Geophysics, Vol. 157, 1707-27.

A. Belousov, B. Voight, M. Belousiva and Y. Muravyev (2000). Tsunamis generated by subaquatic volcanic explosions: unique data from 1996 eruption in Karymskoye Lake, Kamchatka, Russia. Pure and Applied Geophysics, Vol. 157, 1135-43.

E. Bryant (2001). Tsunamis. The underrated hazard. Cambridge University Press.

G.F. Carrier and H.P. Greenspan (1958). Water waves of finite amplitude on a sloping beach. J. Fluid Mech., Vol. 4, 97 - 109.

Y. Chen and P.L.-F. Liu (1995). Modified Boussinesq equations and associated parabolic models for water wave propagation. J. Fluid Mech., Vol. 228, 351-81.

K.F. Cheung, A.C. Phadke, Y. Wei, R. Rojas, Y.J.-M. Douyere, C.D. Martino, S.H. Houston, P.L.-F. Liu, P.J. Lynett, N. Dodd, S. Liao and E. Nakazaki, E (2003). Modeling of storm-induced coastal flooding for emergency management. Ocean Engineering, Vol. 30, 1353-86.

B.H. Choi, E. Pelinovsky, I. Ryabov and S.J. Hong (2002). Distribution functions of tsunami wave heights, Natural Hazards Vol. 25, 1-21.

B.H. Choi, E. Pelinovsky, K.O. Kim and J.S. Lee (2003). Simulation of the transoceanic tsunami propagation due to the 1883 Krakatau volcanic eruption. Natural Hazards and Earth System Sciences, Vol. 3, 321 - 32.

D.A. Crawford and C.L. Mader (1998). Modeling asteroid impact and tsunami. Science Tsunami Hazards, Vol. 16, 21-30.

M.W. Dingemans (1996). Water Wave Propagation over Uneven Bottom. World Sci., Singapore.

P.G. Drazin and R.S. Johnson (1992). Solitons: An Introduction. Cambridge University Press.

F.I. Gonzalez, K. Satake, E.F. Boss and H.O. Mofjeld (1995). Edge wave and non-trapped modes of the 25 April 1992 Cape Mendocino tsunami. Pure and Applied Geophysics, Vol. 144, 409-26.

C. Goto, Y. Ogawa, N. Shuto and F. Imamura (1997). Numerical method of tsunami simulation with the leap-frog scheme (IUGG/IOC Time Project), IOC Manual, UNESCO, No. 35.

S.T. Grilli, S. Vogelmann and P. Watts (2002). Development of a 3D numerical wave tank for modelling tsunami generation by underwater landslides, Engineering Analysis with Boundary Elements, Vol. 26, 301-13.

V.K. Gusiakov (2005). Tsunamis as destructive aftermath of oceanic impacts. Comet/-Asteroid Impacts and Human Society (Eds, P. Bobrowsky and H. Rickman), Springer.

J.L. Hammack (1973). A note on tsunamis: their generation and propagation in an ocean of uniform depth. J. Fluid Mech., Vol. 60, 769-99.

T. Hatori (1966). Vertical displacement in a tsunami source area and the topography of the sea bottom. Bull. Earthquake Research Institute, Vol. 44, 1449-64.

F. Heinrich, A. Mangeney, S. Guibourg and R. Roche (1998). Simulation of water waves generated by a potential debris avalanche in Montserrat, Lesser Antilles. Geophys. Research Letters, Vol. 25, 3697-700.

J.G. Hills and M.P. Gods (1998). Tsunami from asteroid and comet impacts: the vulnerability of Europe. Science Tsunami Hazards, Vol. 16, 3-10.

K. Kajiura (1963). The leading wave of a tsunami. Bull. Earthquake Research Institute, Vol. 41, 535-71.

Ch. Kharif and E. Pelinovsky (2003). Physical mechanisms of the rogue wave phenomenon. European J Mechanics / B - Fluid, Vol. 22, 603-34.

Ch. Kharif and E. Pelinovsky (2005). Asteroid Impact Tsunamis. Comptes Rendus Physique, Vol. 6, 361-6.

E.A. Kulikov, A.B. Rabinovich, R.E. Thomson and B.D. Bornhold (1996). The landslide tsunami of November 3, 1994, Skagway Harbor, Alaska. J. Geophys. Research, Vol. 101C, 6609-15.

J.F. Lander and P.A. Lockridge (1989). United States Tsunamis 1690-1988. NOAA, National Geophysical Data Center, Boulder.

B. Le Mehaute and S. Wang (1996). Water waves generated by underwater explosion. World Sci., Singapore.

P.J. Lynett, T.R. Wu and P.L.-F. Liu (2002). Modeling wave runup with depth-integrated equations. Coastal Engineering, Vol.46, 89-108.

C.L. Mader (1998). Modeling the Eltanin asteroid tsunami. Science Tsunami Hazards, Vol. 16, 17-20.

P.A. Madsen, H.B. Bingham and H. Liu (2002). A new Boussinesq method for fully nonlinear waves from shallow to deep water. J. Fluid Mech., Vol. 462, 1-30.

P.A. Madsen, H.B. Bingham and H.A. Schäffer (2003). Boussinesq-type formulations for fully nonlinear and extremely dispersive water waves: derivation and analysis. Proc. Roy. Soc. London, Vol. A459, 1075-104.

L. Manshinha and D.E. Smylie (1971). The displacement fields of inclined faults. Bull. Seism. Soc. America, Vol. 61, 1433-40.

N. Mirchina, E. Pelinovsky and S. Shavratsky (1982). Parameters of tsunami waves in the source. Science Tsunami Hazards, Vol. 2, B1 - B7.

N.R. Mirchina and E.N. Pelinovsky (1988). Estimation of underwater eruption energy based on tsunami wave data, Natural Hazards, Vol. 1, 277-83.

T. Murty (1977). Seismic Sea Waves - Tsunamis, Bull. Dep. Fisheries, Canada.

Y. Okada (1985). Surface deformation due to shear and tensile faults in a half-space, Bull. Seism. Soc. America, Vol. 75, 1135-54.

K.F. O'Loughlin and J.F. Lander (2003). Caribbean Tsunamis: A 500-year history from 1498-1998. Advances in Natural and Technological Hazards Research, 20. Kluwer, Dordrecht.

E.N. Pelinovsky (1982). Nonlinear Dynamics of Tsunami Waves. Gorky: Inst. Applied Phys. Press.

E.N. Pelinovsky (1996). Tsunami Wave Hydrodynamics. Nizhny Novgorod: Inst. Applied Physics Press.

E. Pelinovsky and R. Mazova (1992). Exact analytical solutions of nonlinear problems of tsunami wave run-up on slopes with different profiles. Natural Hazards, Vol. 6, 227-49.

M.I. Rabinovich, A.B. Ezersky and P. Weidman (2000). The Dynamics of Patterns. World Sci., Singapore.

F. Shi, R.A. Dalrymple, J.T. Kirby, Q. Chen and A. Kennedy (2001). A fully nonlinear Boussinesq model in generalized curvilinear coordinates. Coastal Engineering, Vol. 42, 337-58.

T. Simskin and R.S. Fiske (1983). Krakatau 1883 - the volcanic eruption and its effects, Smithsonian Institution Press, Washington, D.C.

S.L. Soloviev, Ch, N. Go and Kh. S. Kim (1988). A Catalog of tsunamis in the Pacific, 1969-1982. Amerind Publishing Company, New Delhi, 1988.

S.L. Soloviev, O.N. Solovieva, Ch.N. Go, Kh.S. Dim and N.A. Schetnikov (2000). Tsunamis in the Mediterranean Sea 2000 B.C.-2000 A.D. Advances in Natural and Technological Hazards Research, 13. Kluwer, Dordrecht.

J.J. Stoker (1955). Water Waves, John Willey.

C.E. Synolakis (1987). The runup of solitary waves. J. Fluid Mech., Vol. 185, 523-45.

C.E. Synolakis, P.L.-F. Liu, H. Yeh and G.F. Carrier (1997). Tsunamigenic seafloor deformations. Science, Vol. 278, 598-600.

V.V. Titov and F.I. Gonzalez (1997). Implementation and testing of the method of splitting tsunami (MOST) model. NOAA Technical Memorandum ERL PMEL-112; http://www.pmel.noaa.gov/tsunami/

S.N. Ward (2001). Landslide tsunami. J. Geophys. Research, Vol. 106, 11,201-215.

S.N. Ward and E. Asphaug (2000). Asteroid impact tsunami: a probabilistic hazard assessment. Icarus, Vol. 145, 64-78.

S.N. Ward and E. Asphaug (2003). Asteroid impact tsunami of 2880 March 16. Geophys. J. Int., Vol. 153, F6-F10.

P. Watts (2000). Tsunami features of solid block underwater landslides. J. Waterway, Port, Coastal and Ocean Engineering, Vol. 126, 144-52.

P. Watts, S.T. Grilli, J.T. Kirby, G.J. Fryer and D.R. Tappin (2003) Landslide tsunami case studies using a Boussinesq model and a fully nonlinear tsunami generation model. Natural Hazards and Earth System Sciences, Vol. 3, 391-402.

G. Wei, J.T. Kirby, S.T. Grilli and R. Subramanya (1995). A fully nonlinear Boussinesq model for surface waves. I. Highly nonlinear, unsteady waves. J. Fluid Mech., Vol. 294, 71-92.

N. Zahibo, E. Pelinovsky, A. Kurkin and A. Kozelkov (2003). Estimation of far-field tsunami potential for the Caribbean Coast based on numerical simulation. Science Tsunami Hazards. Vol. 21, 202 - 22.

A.C. Yalciner, E.N. Pelinovsky, E. Okal and C.E. Synolakis (2003). Submarine landslides and tsunamis. NATO Science Series: IV. Earth and Environmental Sciences, Kluwer, Vol. 21.

Weakly nonlinear and stochastic properties of ocean wave fields. Application to an extreme wave event

Karsten Trulsen

Mechanics Division, Department of Mathematics, University of Oslo, Norway

Abstract There has been much interest in freak or rogue waves in recent years, especially after the Draupner "New Year Wave" that occurred in the central North Sea on January 1st 1995. From the beginning there have been two main research directions, deterministic and statistical. The deterministic approach has concentrated on focusing mechanisms and modulational instabilities, these are explained in Chap. 3 and some examples are also given in Chap. 4. A problem with many of these deterministic theories is that they require initial conditions that are just as unlikely as the freak wave itself, or they require idealized instabilities such as Benjamin–Feir instability to act over unrealistically long distances or times. For this reason the deterministic theories alone are not very useful for understanding how exceptional the freak waves are. On the other hand, a purely statistical approach based on data analysis is difficult due to the unusual character of these waves. Recently a third research direction has proved promising, stochastic analysis based on Monte–Carlo simulations with phase-resolving deterministic models. This approach accounts for all possible mechanisms for generating freak waves, within a sea state that is hopefully as realistic as possible. This chapter presents several different modified nonlinear Schrödinger (MNLS) equations as candidates for simplified phase-resolving models, followed by an introduction to some essential elements of stochastic analysis. The material is aimed at readers with some background in nonlinear wave modeling, but little background in stochastic modeling. Despite their simplicity, the MNLS equations capture remarkably non-trivial physics of the sea surface such as the establishment of a quasi-stationary spectrum with ω^{-4} power law for the high-frequency tail, and nonlinear probability distributions for extreme waves. In the end we will suggest how often one should expect a "New Year Wave" within the sea state in which it occurred.

1 Introduction

The material contained here is to a large extent motivated by the so-called Draupner "New Year Wave", an extreme wave event that was recorded at the Draupner E platform in the central North Sea on January 1st 1995 (Haver 2004; Karunakaran et al. 1997). This location has an essentially uniform depth of 70 m. The platform is of jacket type and is not expected to modify the wave field in any significant way. The platform had a foundation of a novel type, and for this reason was instrumented with a large number of sensors measuring environmental data, structural and foundation response. We are particularly interested in measurements taken by a down looking laser-based wave sensor, recording surface elevation at a speed of 2.1333 Hz during 20 minutes of every hour. The full 20 minute time series recorded starting at 1520 GMT is shown in figure 1. To remove any doubt that the measurements are of good quality, figure 2 shows a close-up view with the individual measurements indicated. It is clear that the extreme wave is not an isolated erroneous measurement. The minimum distance between the sensor and the water surface was 7.4 m.

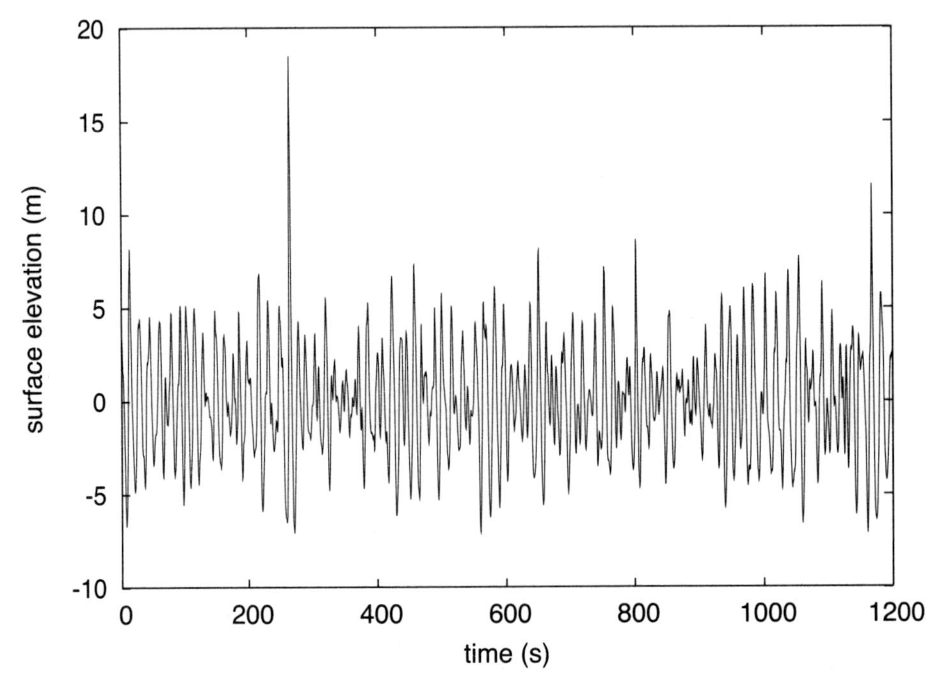

Figure 1. Draupner 20 minute time series starting at 1520 GMT.

The significant wave height was $H_s = 11.9$ m while the maximum wave height was 25.6 m with a crest height of 18.5 m (see section 2). Haver (2004) states that the wave itself was not beyond design parameters for the platform, but basic engineering approaches did not suggest that such a large wave should occur in a sea state with such a small value of H_s. Some damage was reported to equipment on a temporary deck. The wave has often been referred to as a freak or rogue wave.

During the past decade, a considerable effort has been undertaken by many people to understand the Draupner wave, and rogue waves in general. Statoil should be commended for their policy of releasing this wave data to the public, thus igniting an avalanche of exciting research.

The goal of the present chapter is to understand how unusual the Draupner "New Year Wave" is. More precisely, given an overall sea state like that shown in figure 1, how often should we expect an extreme wave of the observed magnitude to occur. Sev-

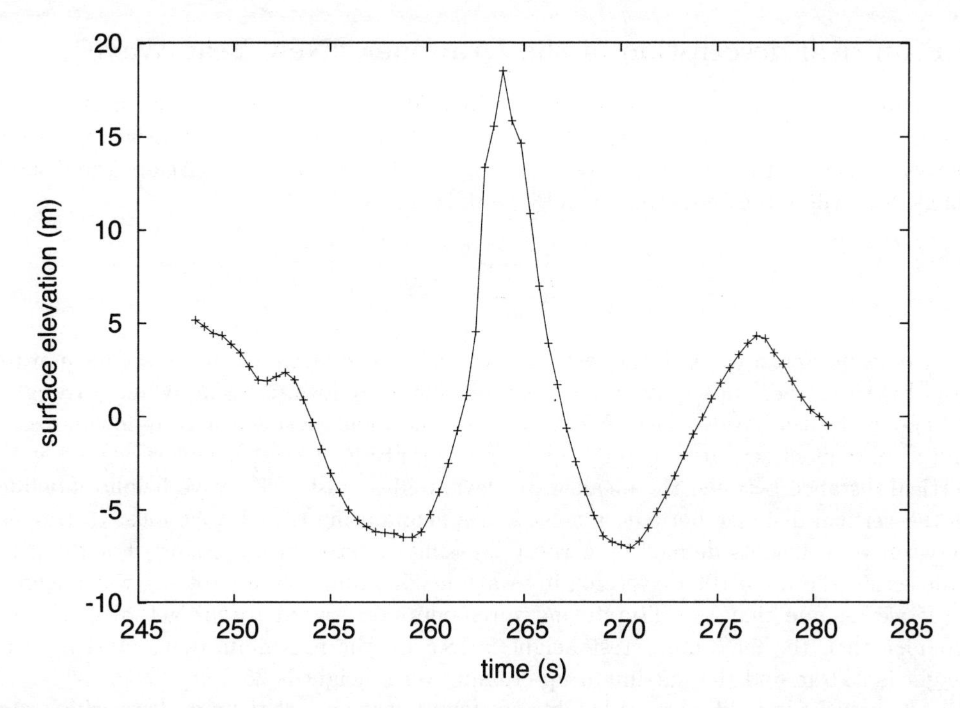

Figure 2. Extract of Draupner time series close to the extreme wave, with discrete measurements indicated.

eral idealized deterministic mechanisms are known to produce extreme waves, they are reviewed in Chap. 3 with further examples given in Chap. 4, however they do not suggest the probability of occurrence of extreme waves. The approach taken in the present chapter is to start out with a collection of random realizations of hopefully realistic sea states, and to find out how the deterministic evolution of these realizations may produce extreme waves due to any mechanism. The probability of a wave of given magnitude may then be deduced from an ensemble of such simulations. The philosophy of this chapter is thus to divert attention away from the immediate neighborhood of a particular extreme wave, and rather understand general properties of the entire wave field.

Three important steps must be taken to carry out the proposed line of investigation: First we seek some understanding of what is a realistic sea state in section 2, then we develop some appropriate computationally efficient deterministic evolution models (modified nonlinear Schrödinger equations) in sections 3–7, and then we develop necessary stochastic theory in sections 8–12. Finally, we are able to say something about how unusual the Draupner "New Year Wave" might be in section 13.

2 Empirical description of the Draupner "New Year Wave"

We define the mean water level as the arithmetic mean position during the $N = 2560$ discrete measurements of duration 20 minutes. Thus figures 1 and 2 show the surface elevation relative to the mean water level, $\eta_n \equiv \eta(t_n)$, where $t_n = n\Delta t$ are the discrete times, and where the time average is by definition

$$\frac{1}{N}\sum_{n=0}^{N-1}\eta_n = 0.$$

We define a crest to be the largest positive surface displacement above the mean water level between a zero-up-crossing and a subsequent zero-down-crossing, while a trough is defined to be the largest negative surface displacement between a zero-down-crossing and a subsequent zero-up-crossing. The wave amplitude or crest height is defined as the vertical distance between the mean water level and a crest. The wave height is defined as the vertical distance between a crest and a neighboring trough. We must distinguish between wave heights defined by zero-up-crossing or zero-down-crossing. For the given time series, there are 106 down-crossing wave heights and 105 up-crossing wave heights.

If we assume that the discrete measurements correspond to the actual crests and troughs, then the maximum crest height is 18.5 m, the maximum down-crossing wave height is 25.6 m and the maximum up-crossing wave height is 25.0 m.

After all the individual wave heights are found, we can sort them in decreasing order. For any positive number α, the average of the $1/\alpha$ highest waves is denoted as $H_{1/\alpha}$. Particularly common is the average of the 1/3 highest waves; the down-crossing $H_{1/3}$ is 11.6 m and the up-crossing $H_{1/3}$ is 11.4 m.

Given a discrete time series, we likely do not have measurements of the exact crests and troughs. More robust estimates for the overall amplitude and wave height are achieved through the standard deviation, or root-mean-square, of the discrete measurements, which can be computed as

$$\sigma = \sqrt{\overline{\eta^2}} = \sqrt{\frac{1}{N}\sum_{n=0}^{N-1}\eta_n^2} = 2.98 \text{ m}.$$

We define the *characteristic amplitude* as

$$\bar{a} = \sqrt{2}\sigma = 4.2 \text{ m}. \tag{2.1}$$

The *significant wave height* is defined as four times the standard deviation

$$H_s = 4\sigma = 11.9 \text{ m}. \tag{2.2}$$

The reason why $H_{1/3}$ is almost identical to H_s will be explained in section 12.2.

The maximum crest height of 18.5 m is equal to 6.2 standard deviations. The maximum wave height of 25.6 m is equal to 8.6 standard deviations, or equivalently, 2.1 significant wave heights. A common definition of a freak wave is that the maximum wave height should be larger than twice the significant wave height. The Draupner "New Year

Wave" is a freak wave according to this definition, however the crest height appears to be more "freak" than the wave height.

The frequency spectrum $S(\omega)$ can be estimated by the square magnitude of the Fourier transform of the time series $2|\hat{\eta}(\omega)|^2$. Continuous and discrete Fourier transforms are reviewed in Appendix A. We employ the discrete Fourier transform of the time series

$$\hat{\eta}_j = \frac{1}{N} \sum_{n=0}^{N-1} \eta_n \mathrm{e}^{\mathrm{i}\omega_j t_n}, \tag{2.3}$$

and use only non-negative frequencies $\omega_j = 2\pi j/T$ for $j = 0, 1, \ldots, N/2$ for the representation of the frequency spectrum. Figure 3 shows the estimated spectrum with linear axes, while figure 4 shows the same with logarithmic axes, in both cases without any smoothing. Figure 4 suggests how ω^{-5} and ω^{-4} power laws fit the high-frequency tail of the observed spectrum; the first law is often used for engineering design spectra (Goda 2000; Ochi 1998) while the second law is often found to be a better model near the spectral peak (Dysthe et al. 2003).

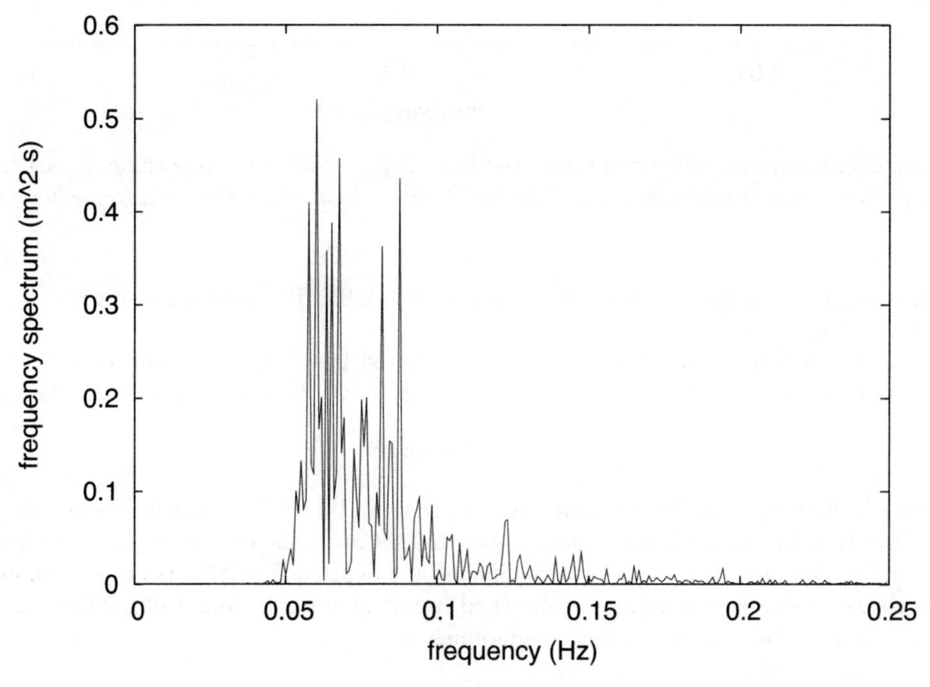

Figure 3. Frequency spectrum estimated from $2|\hat{\eta}(\omega)|^2$ without smoothing, linear axes.

Based on the Fourier transform of the time series for surface elevation, we may estimate a characteristic angular frequency as the expected value of the spectral distribution

$$\omega_c = 2\pi f_c = \frac{\sum_j |\omega_j||\hat{\eta}_j|^2}{\sum_j |\hat{\eta}_j|^2} = 0.52 \text{ s}^{-1} \tag{2.4}$$

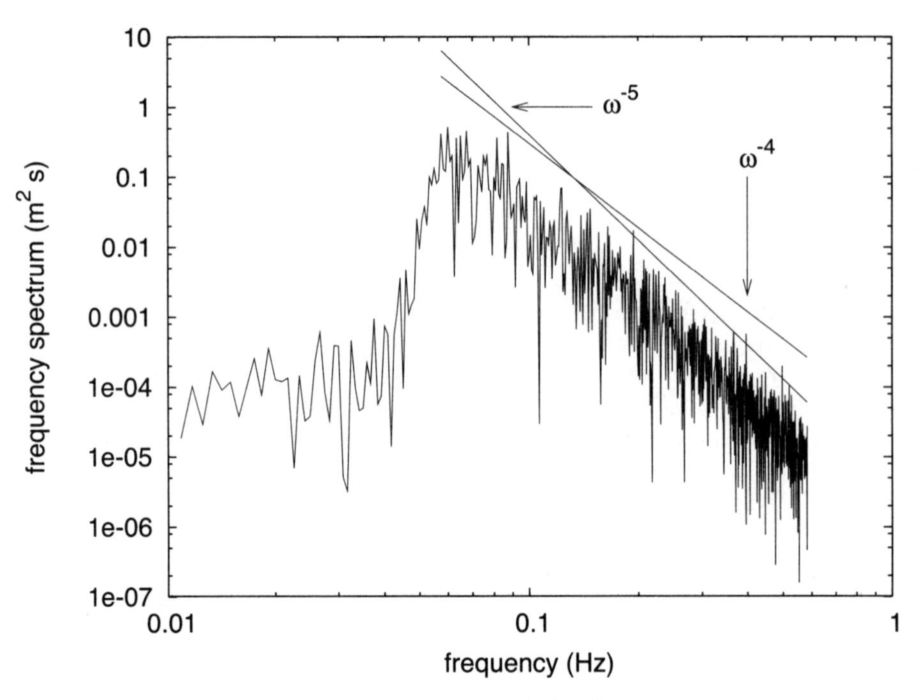

Figure 4. Frequency spectrum estimated from $2|\hat{\eta}(\omega)|^2$ without smoothing, logarithmic axes. Two possible power laws, ω^{-4} and ω^{-5}, are indicated for the high frequency tail.

corresponding to a characteristic frequency of $f_c = 0.083$ Hz and a characteristic period of $T_c = 12$ s.

The characteristic wavenumber can be estimated based on the assumption that the waves are linear to leading order. Then we simply solve the linear dispersion relation

$$\omega^2 = gk \tanh kh$$

were $g = 9.81$ m/s^2 is the acceleration of gravity and $h = 70$ m is the relevant depth. We find the characteristic wavenumber $k_c = 0.0289$ m^{-1}, the characteristic wave length $\lambda_c = 217$ m and the characteristic non-dimensional depth $k_c h = 2.0$. As far as estimating the characteristic wavenumber is concerned, no great error is done using infinite depth in which case the dispersion relation simplifies to

$$\omega^2 = gk$$

and we get $k_c = 0.0279$ m^{-1} and $\lambda_c = 225$ m.

Knowing the characteristic wavenumber, we may proceed to compute the characteristic steepness $\epsilon = k_c \bar{a} = 0.12$ which is a measure of the strength of nonlinearity.

It will be useful to have some measure of the characteristic deviation $\Delta\omega$ of the frequency spectrum around the characteristic frequency ω_c. We define the dimensionless bandwidth to be the ratio between the deviation and the mean

$$\delta_\omega = \frac{\Delta\omega}{\omega_c}.$$

The most simple-minded approach to estimate $\Delta\omega$ is to stare at figure 3. This leads to the conclusion that $\Delta\omega/2\pi \approx 0.025$ Hz, and thus $\delta_\omega \approx 0.32$.

Some other approaches to compute $\Delta\omega$, for example computing the standard deviation of the frequency spectrum, typically yield unreasonably large and less useful values of $\Delta\omega$. This is because the high-frequency tail of the spectrum is largely due to nonlinearly bound waves (waves that are not free, i.e. waves that do not satisfy the linear dispersion relation) and measurement noise. Our estimate of bandwidth will only be applied to the derivation of the simplified models to be described in the following. The most relevant measure of spectral width should then only account for linear free waves, and therefore the simple-minded approach turns out to be quite recommendable.

In conclusion of these empirical considerations, the important observations are given by three non-dimensional numbers: The characteristic steepness is $\epsilon = k_c\bar{a} \approx 0.12$, the characteristic bandwidth is $\delta_\omega = \Delta\omega/\omega_c \approx 0.32$ and the characteristic non-dimensional depth is $k_c h \approx 2.0$. We shall say that the wave field is weakly nonlinear, has small bandwidth and is essentially on deep water. Notice also that the magnitudes of the bandwidth and the steepness have the approximate relationship

$$\delta_\omega \sim \sqrt{\epsilon} > \epsilon.$$

In the next sections we show how to take advantage of the magnitudes of these non-dimensional numbers to construct simplified mathematical models that can describe this wave field.

It is instructive to see how the sea state of the Draupner wave compares to typical sea states in the northern North Sea. Figure 5 contains approximately 70 000 data points, each representing a 20 minute wave record from the northern North Sea. T_p denotes the wave period corresponding to the frequency at the spectral peak. From figure 4 we deduce the value $T_p \approx 15$ s for the Draupner time series, greater than the estimated value of T_c (in general we typical find $T_p \gtrsim T_c$). It appears that the sea state of the Draupner "New Year Wave" may be considered slightly unusual due to its large H_s, but the sea state is definitely not unusual due to its overall steepness.

The above discussion reveals the need to distinguish two questions: How unusual is the sea state in which the Draupner "New Year Wave" occurred? And how unusual is the Draupner "New Year Wave" within the sea state in which it occurred? The first question will not be dealt with here. The second question is the main motivation for the rest of this chapter.

3 The governing equations

As our starting point we take the equations for the velocity potential $\phi(\mathbf{r}, z, t)$ and surface displacement $\eta(\mathbf{r}, t)$ of an inviscid, irrotational and incompressible fluid with uniform depth h,

$$\nabla^2\phi = 0 \qquad \text{for} \quad -h < z < \eta, \tag{3.1}$$

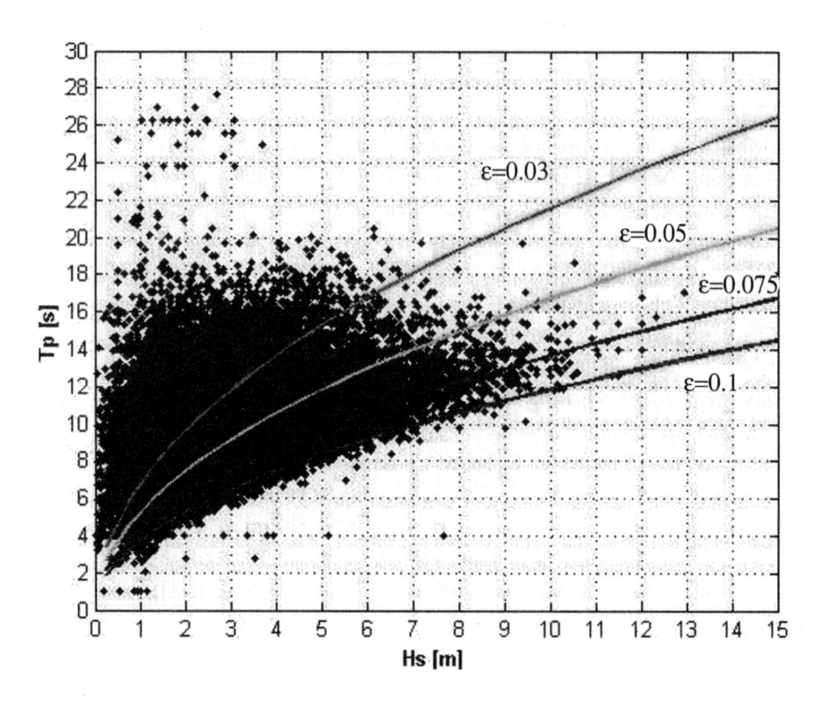

Figure 5. Scatter diagram for H_s and T_p from the northern North Sea. Pooled data 1973–2001, from the platforms Brent, Statfjord, Gullfax and Troll. Curves of constant steepness are also shown, using a definition of steepness $\epsilon = \sqrt{2}\pi^2 H_s/(gT_p^2)$ slightly different from that used in the text above since $T_p \gtrapprox T_c$. The figure was prepared by K. Johannessen, Statoil.

$$\frac{\partial \eta}{\partial t} + \nabla\phi \cdot \nabla\eta = \frac{\partial \phi}{\partial z} \qquad \text{at} \quad z = \eta, \tag{3.2}$$

$$\frac{\partial \phi}{\partial t} + g\eta + \frac{1}{2}(\nabla\phi)^2 = 0 \qquad \text{at} \quad z = \eta, \tag{3.3}$$

$$\frac{\partial \phi}{\partial z} = 0 \qquad \text{at} \quad z = -h. \tag{3.4}$$

Here g is the acceleration of gravity, the horizontal position vector is $\mathbf{r} = (x, y)$, the vertical coordinate is z, $\nabla = (\partial/\partial x, \partial/\partial y, \partial/\partial z)$ and t is time.

The solution of the linearization of equations (3.1)–(3.4) is a linear superposition of simple harmonic waves such as

$$\eta(\mathbf{r}, t) = a\cos(\mathbf{k}\cdot\mathbf{r} - \omega t)$$

$$\phi(\mathbf{r}, z, t) = \frac{\omega a}{k}\frac{\cosh k(z+h)}{\sinh kh}\sin(\mathbf{k}\cdot\mathbf{r} - \omega t)$$

subject to the dispersion relation for gravity waves on finite depth

$$\omega^2 = gk \tanh kh, \tag{3.5}$$

where a is the amplitude, ω is the angular frequency, $\mathbf{k} = (k_x, k_y)$ is the wave vector and $k = \sqrt{k_x^2 + k_y^2}$ is the wavenumber.

In section 2 we computed the values of ω_c, k_c and $\bar{a}$, which can be used to introduce properly normalized dimensionless quantities

$$\bar{a}\eta' = \eta, \qquad \frac{\omega_c \bar{a}}{k_c}\phi' = \phi, \qquad (\mathbf{r}', z') = k_c(\mathbf{r}, z), \qquad t' = \omega_c t, \qquad h' = k_c h.$$

Dropping the primes, the normalized equations become

$$\nabla^2 \phi = 0 \qquad \text{for} \quad -h < z < \epsilon\eta, \tag{3.6}$$

$$\frac{\partial \eta}{\partial t} + \epsilon \nabla\phi \cdot \nabla\eta = \frac{\partial \phi}{\partial z} \qquad \text{at} \quad z = \epsilon\eta, \tag{3.7}$$

$$\frac{\partial \phi}{\partial t} + \frac{\eta}{s} + \frac{\epsilon}{2}(\nabla\phi)^2 = 0 \qquad \text{at} \quad z = \epsilon\eta, \tag{3.8}$$

$$\frac{\partial \phi}{\partial z} = 0 \qquad \text{at} \quad z = -h, \tag{3.9}$$

where we use the notation $\epsilon = k_c \bar{a}$ and $s = \tanh h$.

With small steepness $\epsilon \ll 1$ it is reasonable to assume perturbation expansions

$$\left.\begin{array}{rcl} \eta & = & \eta_1 + \epsilon\eta_2 + \epsilon^2\eta_3 + \ldots \\ \phi & = & \phi_1 + \epsilon\phi_2 + \epsilon^2\phi_3 + \ldots \end{array}\right\} \tag{3.10}$$

4 Weakly nonlinear narrowbanded equations

In Chap. 3 the classical nonlinear Schrödinger (NLS) equation is reviewed. For the classical NLS equation it is assumed that the magnitudes of the steepness and bandwidth are similar, and that the depth is infinite. For finite depth, the corresponding NLS equation is known as the Davey–Stewartson system. There are several reasons why we here desire more accurate modified nonlinear Schrödinger (MNLS) equations. Primarily, the MNLS equation captures essential physics of the sea surface that the NLS equation does not account for. The MNLS equation has nicer properties for numerical simulation of short-crested waves. Finally, the empirical considerations in section 2 reveal that the steepness is small, but the bandwidth is not quite as small. With no constraint on bandwidth, a straightforward application of the perturbation expansions (3.10) leads to the much more complicated Zakharov integral equation (Zakharov 1968; Krasitskii 1994). The non-dimensional numbers derived in section 2 suggest that some intermediate model may be optimal in terms of mathematical and computational complexity (Trulsen and Dysthe 1997), and here the MNLS equation is proposed as a promising candidate.

4.1 The bandwidth

To get a good feeling for the significance of the bandwidth, a good example is the idealized wave packet

$$f(x) = \mathrm{e}^{-(\kappa x)^2} \cos k_c x \tag{4.1}$$

of approximate width $1/\kappa$ containing oscillations of length $2\pi/k_c$. The Fourier transform is

$$\hat{f}(k) = \frac{1}{2\pi}\int_{-\infty}^{\infty} f(x)\mathrm{e}^{-\mathrm{i}kx}\,\mathrm{d}x = \frac{1}{4\kappa\sqrt{\pi}}\left(\mathrm{e}^{-(\frac{k+k_c}{2\kappa})^2} + \mathrm{e}^{-(\frac{k-k_c}{2\kappa})^2}\right). \tag{4.2}$$

Thus the Fourier transform is centered around $k = \pm k_c$ with a width κ that is the inverse of the length of the wave packet. A natural definition of bandwidth could now be

$$\delta_k = \frac{\kappa}{k_c}. \tag{4.3}$$

If the width of the packet is much greater than the length of a basic wave oscillation, $\delta_k \ll 1$, it is natural to identify two characteristic scales for x; a fast scale associated with rapid oscillations $x_0 = k_0 x$ and a slow scale associated with the envelope $x_1 = \kappa x = \delta_k x_0$. Thus we may write

$$f(x) = f(x_0, x_1) = \mathrm{e}^{-x_1^2} \cos x_0. \tag{4.4}$$

4.2 Derivation of higher-order nonlinear Schrödinger equations

Based on the previous considerations, we now propose the "optimal" assumptions for the wave field in which the Draupner wave occurred,

$$\epsilon \approx 0.1, \qquad \delta_k, \delta_\omega = O(\epsilon^{1/2}), \qquad h^{-1} = O(1). \tag{4.5}$$

These "optimal" assumptions lead to rather complicated higher-order equations. Some limiting cases of this exercise have been done. For finite-depth narrowbanded waves Brinch–Nielsen and Jonsson (1986) assumed

$$\epsilon \approx 0.1, \qquad \delta_k, \delta_\omega = O(\epsilon), \qquad h^{-1} = O(1). \tag{4.6}$$

For wider-banded deep-water waves Trulsen and Dysthe (1996, 1997) assumed

$$\epsilon \approx 0.1, \qquad \delta_k, \delta_\omega = O(\epsilon^{1/2}), \qquad h^{-1} = O(\epsilon^{1/2}). \tag{4.7}$$

For narrowbanded deep-water waves, Dysthe (1979) first considered infinite depth, while Lo and Mei (1985, 1987) extended to deep-water assuming

$$\epsilon \approx 0.1, \qquad \delta_k, \delta_\omega = O(\epsilon), \qquad h^{-1} = O(\epsilon). \tag{4.8}$$

In the following we pursue the simplified assumptions (4.8) which are sufficient to capture important physics of the sea surface. In section 5 it is demonstrated how these results can be extended to broader bandwidth in a powerful way.

The following harmonic expansions for the velocity potential and surface displacement are employed

$$\phi = \epsilon\bar{\phi} + \frac{1}{2}\left(A_1' \mathrm{e}^{\mathrm{i}\theta} + \epsilon A_2' \mathrm{e}^{2\mathrm{i}\theta} + \epsilon^2 A_3' \mathrm{e}^{3\mathrm{i}\theta} + \cdots + \text{c.c.}\right), \tag{4.9}$$

$$\eta = \epsilon\bar{\eta} + \frac{1}{2}\left(B \mathrm{e}^{\mathrm{i}\theta} + \epsilon B_2 \mathrm{e}^{2\mathrm{i}\theta} + \epsilon^2 B_3 \mathrm{e}^{3\mathrm{i}\theta} + \cdots + \text{c.c.}\right). \tag{4.10}$$

Here c.c. denotes the complex conjugate. The phase is $\theta = x - t$ after having oriented the x-axis in the direction of the characteristic wave vector $\mathbf{k}_c$. The slow drift $\bar{\phi}$ and set-down $\bar{\eta}$ as well as the complex harmonic amplitudes $A_1', A_2', A_3', \ldots, B, B_2, B_3, \ldots$ are functions of the slow modulation variables $\epsilon\mathbf{r}$ and ϵt. In the assumed case of deep water $h^{-1} = O(\epsilon)$ the induced current $\bar{\phi}$ also depends on the slow vertical coordinate ϵz while the variables $A_1', A_2', A_3', \ldots$ also depend on the basic vertical coordinate z. For finite depth $h^{-1} = O(1)$ it would be necessary to renormalize $\bar{\phi}$ and assume that it too depends on the basic vertical coordinate.

The vertical dependence of the harmonic amplitudes A_n' for $n \geqslant 1$ is found from the Laplace equation and the bottom boundary condition,

$$\frac{\partial^2 A_n'}{\partial z^2} - n^2 A_n' + 2\epsilon \mathrm{i} n \frac{\partial A_n'}{\partial x} + \epsilon^2 \frac{\partial^2 A_n'}{\partial x^2} + \epsilon^2 \frac{\partial^2 A_n'}{\partial y^2} = 0 \quad \text{for} \quad -h < z, \tag{4.11}$$

$$\frac{\partial A_n'}{\partial z} = 0 \quad \text{at} \quad z = -\epsilon^{-1} h. \tag{4.12}$$

The vertical dependence can then be found by the perturbation expansion

$$A_n' = A_{n,0} + \epsilon A_{n,1} + \epsilon^2 A_{n,2} + \ldots \tag{4.13}$$

The harmonic amplitudes are found to decay exponentially on the basic vertical scale, so the bottom condition is essentially at infinite depth. The leading-order solution is

$$A_{n,0} = A_n \mathrm{e}^{nz} \tag{4.14}$$

which evaluates to A_n at $z = 0$. A_n is not a function of z. With the surface boundary condition

$$A_{n,j} = 0 \quad \text{at} \quad z = 0 \quad \text{for} \quad j = 1, 2, \ldots$$

we get solutions at higher orders

$$A_{n,1} = -\mathrm{i}z \frac{\partial A_n}{\partial x} \mathrm{e}^{nz}, \tag{4.15}$$

$$A_{n,2} = \left(-\frac{z^2}{2}\frac{\partial^2 A_n}{\partial x^2} - \frac{z}{2n}\frac{\partial^2 A_n}{\partial y^2}\right) \mathrm{e}^{nz}. \tag{4.16}$$

In the following we use the notation $A \equiv A_1$, and we express all results either in terms of the complex first-harmonic amplitude of the velocity potential A or the complex first-harmonic amplitude of the surface displacement B.

A review of literature on higher order nonlinear Schrödinger equations over the past three decades leads to a certain degree of confusion because authors often have a lack of awareness of whether they are expressing the equations in terms of the surface displacement (B) or the velocity potential (A). At the cubic NLS level the equations are identical and the distinction is not so important. At higher order the choice of A or B leads to qualitatively different equations with remarkably different properties. The spatial evolution equations in terms of A have particularly nice properties for analytical considerations, while the spatial evolution equations in terms of B are more convenient for most practical applications. The choice may also have important consequences for the stability and accuracy of numerical schemes.

In the following sections we summarize the higher-order equations for A and B, for spatial and temporal evolution. In passing between temporal and spatial evolution, we have made the assumption that the induced flow $\bar{\phi}$ is only due to nonlinearly bound response to the free first-harmonic waves.

4.3 Deep water time evolution in terms of velocity potential

This was the form first derived by Dysthe (1979) for infinite depth, and by Lo and Mei (1985, 1987) for finitely deep water. This is the form that corresponds to the empirical kinematics theory of Grue et al. (2003) reviewed in Chap. 4. The evolution equations are

$$\frac{\partial A}{\partial t}+\frac{1}{2}\frac{\partial A}{\partial x}+\frac{\mathrm{i}}{8}\frac{\partial^2 A}{\partial x^2}-\frac{\mathrm{i}}{4}\frac{\partial^2 A}{\partial y^2}+\frac{\mathrm{i}}{2}|A|^2A$$

$$-\frac{1}{16}\frac{\partial^3 A}{\partial x^3}+\frac{3}{8}\frac{\partial^3 A}{\partial x\partial y^2}+\frac{7}{4}|A|^2\frac{\partial A}{\partial x}-\frac{1}{4}A\frac{\partial |A|^2}{\partial x}+\mathrm{i}A\frac{\partial\bar{\phi}}{\partial x}=0 \quad \text{at} \quad z=0, \tag{4.17}$$

$$\frac{\partial\bar{\phi}}{\partial z}=\frac{1}{2}\frac{\partial|A|^2}{\partial x} \quad \text{at} \quad z=0, \tag{4.18}$$

$$\frac{\partial^2\bar{\phi}}{\partial x^2}+\frac{\partial^2\bar{\phi}}{\partial y^2}+\frac{\partial^2\bar{\phi}}{\partial z^2}=0 \quad \text{for} \quad -h<z, \tag{4.19}$$

$$\frac{\partial\bar{\phi}}{\partial z}=0 \quad \text{at} \quad z=-h. \tag{4.20}$$

The reconstruction formulas are

$$\bar{\eta}=\frac{1}{2}\frac{\partial\bar{\phi}}{\partial x}, \tag{4.21}$$

$$B=\mathrm{i}A+\frac{1}{2}\frac{\partial A}{\partial x}+\frac{\mathrm{i}}{8}\frac{\partial^2 A}{\partial x^2}-\frac{\mathrm{i}}{4}\frac{\partial^2 A}{\partial y^2}+\frac{\mathrm{i}}{8}|A|^2A, \tag{4.22}$$

$$A_2=0, \tag{4.23}$$

$$B_2=-\frac{1}{2}A^2+\mathrm{i}A\frac{\partial A}{\partial x}, \tag{4.24}$$

$$A_3=0, \tag{4.25}$$

$$B_3=-\frac{3\mathrm{i}}{8}A^3. \tag{4.26}$$

4.4 Deep water space evolution in terms of the velocity potential

This form was employed by Lo and Mei (1985, 1987). The evolution equations are

$$\frac{\partial A}{\partial x} + 2\frac{\partial A}{\partial t} + \mathrm{i}\frac{\partial^2 A}{\partial t^2} - \frac{\mathrm{i}}{2}\frac{\partial^2 A}{\partial y^2} + \mathrm{i}|A|^2 A - \frac{\partial^3 A}{\partial t \partial y^2} - 8|A|^2\frac{\partial A}{\partial t} - 4\mathrm{i}A\frac{\partial \bar{\phi}}{\partial t} = 0 \quad \text{at} \quad z=0, \tag{4.27}$$

$$\frac{\partial \bar{\phi}}{\partial z} = -\frac{\partial |A|^2}{\partial t} \quad \text{at} \quad z=0, \tag{4.28}$$

$$4\frac{\partial^2 \bar{\phi}}{\partial t^2} + \frac{\partial^2 \bar{\phi}}{\partial y^2} + \frac{\partial^2 \bar{\phi}}{\partial z^2} = 0 \quad \text{for} \quad -h < z, \tag{4.29}$$

$$\frac{\partial \bar{\phi}}{\partial z} = 0 \quad \text{at} \quad z = -h. \tag{4.30}$$

Equation (4.27) has the exceptional property that there is no term proportional to $A\partial|A|^2/\partial t$, giving it qualitatively different properties from the other three forms of the MNLS equation, as explained in section 6.1.

The reconstruction formulas are

$$\bar{\eta} = -\frac{\partial \bar{\phi}}{\partial t}, \tag{4.31}$$

$$B = \mathrm{i}A - \frac{\partial A}{\partial t} - \frac{3\mathrm{i}}{8}|A|^2 A, \tag{4.32}$$

$$A_2 = 0, \tag{4.33}$$

$$B_2 = -\frac{1}{2}A^2 - 2\mathrm{i}A\frac{\partial A}{\partial t}, \tag{4.34}$$

$$A_3 = 0, \tag{4.35}$$

$$B_3 = -\frac{3\mathrm{i}}{8}A^3. \tag{4.36}$$

4.5 Deep water time evolution in terms of the surface elevation

The evolution equations are

$$\frac{\partial B}{\partial t} + \frac{1}{2}\frac{\partial B}{\partial x} + \frac{\mathrm{i}}{8}\frac{\partial^2 B}{\partial x^2} - \frac{\mathrm{i}}{4}\frac{\partial^2 B}{\partial y^2} + \frac{\mathrm{i}}{2}|B|^2 B - \frac{1}{16}\frac{\partial^3 B}{\partial x^3} + \frac{3}{8}\frac{\partial^3 B}{\partial x \partial y^2} + \frac{5}{4}|B|^2\frac{\partial B}{\partial x} + \frac{1}{4}B\frac{\partial |B|^2}{\partial x} + \mathrm{i}B\frac{\partial \bar{\phi}}{\partial x} = 0 \quad \text{at} \quad z=0, \tag{4.37}$$

$$\frac{\partial \bar{\phi}}{\partial z} = \frac{1}{2}\frac{\partial |B|^2}{\partial x} \quad \text{at} \quad z=0, \tag{4.38}$$

$$\frac{\partial^2 \bar{\phi}}{\partial x^2} + \frac{\partial^2 \bar{\phi}}{\partial y^2} + \frac{\partial^2 \bar{\phi}}{\partial z^2} = 0 \quad \text{for} \quad -h < z, \tag{4.39}$$

$$\frac{\partial \bar{\phi}}{\partial z} = 0 \quad \text{at} \quad z = -h. \tag{4.40}$$

The reconstruction formulas are

$$\bar{\eta} = \frac{1}{2}\frac{\partial \bar{\phi}}{\partial x}, \tag{4.41}$$

$$A = -\mathrm{i}B + \frac{1}{2}\frac{\partial B}{\partial x} + \frac{3\mathrm{i}}{8}\frac{\partial^2 B}{\partial x^2} - \frac{\mathrm{i}}{4}\frac{\partial^2 B}{\partial y^2} + \frac{\mathrm{i}}{8}|B|^2 B, \tag{4.42}$$

$$A_2 = 0, \tag{4.43}$$

$$B_2 = \frac{1}{2}B^2 - \frac{\mathrm{i}}{2}B\frac{\partial B}{\partial x}, \tag{4.44}$$

$$A_3 = 0, \tag{4.45}$$

$$B_3 = \frac{3}{8}B^3. \tag{4.46}$$

4.6 Deep water space evolution in terms of the surface elevation

The evolution equations are

$$\frac{\partial B}{\partial x} + 2\frac{\partial B}{\partial t} + \mathrm{i}\frac{\partial^2 B}{\partial t^2} - \frac{\mathrm{i}}{2}\frac{\partial^2 B}{\partial y^2} + \mathrm{i}|B|^2 B \\ - \frac{\partial^3 B}{\partial t \partial y^2} - 6|B|^2\frac{\partial B}{\partial t} - 2B\frac{\partial |B|^2}{\partial t} - 4\mathrm{i}B\frac{\partial \bar{\phi}}{\partial t} = 0 \quad \text{at} \quad z = 0, \tag{4.47}$$

$$\frac{\partial \bar{\phi}}{\partial z} = -\frac{\partial |B|^2}{\partial t} \quad \text{at} \quad z = 0, \tag{4.48}$$

$$4\frac{\partial^2 \bar{\phi}}{\partial t^2} + \frac{\partial^2 \bar{\phi}}{\partial y^2} + \frac{\partial^2 \bar{\phi}}{\partial z^2} = 0 \quad \text{for} \quad -h < z, \tag{4.49}$$

$$\frac{\partial \bar{\phi}}{\partial z} = 0 \quad \text{at} \quad z = -h. \tag{4.50}$$

The reconstruction formulas are

$$\bar{\eta} = -\frac{\partial \bar{\phi}}{\partial t}, \tag{4.51}$$

$$A = -\mathrm{i}B - \frac{\partial B}{\partial t} + \mathrm{i}\frac{\partial^2 B}{\partial t^2} - \frac{3\mathrm{i}}{8}|B|^2 B, \tag{4.52}$$

$$A_2 = 0, \tag{4.53}$$

$$B_2 = \frac{1}{2}B^2 + \mathrm{i}B\frac{\partial B}{\partial t}, \tag{4.54}$$

$$A_3 = 0, \tag{4.55}$$

$$B_3 = \frac{3}{8}B^3. \tag{4.56}$$

4.7 Finite depth

When the depth is finite $h \approx 1$, all the numerical coefficients in the above equations become functions of $s = \tanh h$. The induced flow $\bar{\phi}$ must be renormalized to a lower order. The equations for the slow response $(\bar{\eta}, \bar{\phi})$ change their qualitative nature such as to support free long waves. Furthermore, several new types of terms enter the evolution equations for the short waves. Brinch–Nielsen and Jonsson (1986) derived equations for the temporal evolution of the velocity potential A, corresponding to those summarized in section 4.3. Sedletsky (2003) derived fourth-harmonic contributions also for the temporal evolution of the velocity potential A.

We limit to just two particularly interesting observations, the following two reconstruction formulas that can be extracted from Brinch–Nielsen and Jonsson (1986)

$$B_2 = \frac{3 - s^2}{4s^3} B^2 + \ldots \tag{4.57}$$

$$B_3 = \frac{3(3 - s^2)(3 + s^4)}{64s^6} B^3 + \ldots \tag{4.58}$$

In the limit of infinite depth the two coefficients become $1/2 = 0.5$ and $3/8 = 0.375$, respectively, while for the target depth $h = 2.0$ for the Draupner wave field the coefficients are 0.58 and 0.47, respectively. It is evident that nonlinear contributions to the reconstruction of the surface profile are more important for smaller depths.

5 Exact linear dispersion

The above equations are based on the simplified assumptions (4.8), and the perturbation analysis was carried out up to $O(\epsilon^4)$ for the evolution equations.

The "optimal" assumptions (4.5) for the Draupner wave field require wider bandwidth than that strictly allowed by the above equations. Carrying out the perturbation analysis to $O(\epsilon^{7/2})$ with the "optimal" bandwidth assumption leads to a large number of additional linearly dispersive terms, but no new nonlinear terms, see Trulsen and Dysthe (1996).

It is not difficult to account for the exact linear dispersive part of the evolution equations using pseudo-differential operators. This was shown for infinite depth by Trulsen et al. (2000). Here we make the trivial extension to any depth.

The full linearized solution of (3.6)–(3.9) can be obtained by Fourier transform. The surface displacement can thus be expressed as

$$\eta(\mathbf{r}, t) = \int \hat{\eta}(\mathbf{k}, t) e^{i\mathbf{k}\cdot\mathbf{r}} \, d\mathbf{k} = \frac{1}{2} \int b(\mathbf{k}) e^{i(\mathbf{k}\cdot\mathbf{r} - \omega(\mathbf{k})t)} \, d\mathbf{k} + \text{c.c.} \tag{5.1}$$

Here the normalized wave vector is $\mathbf{k} = \hat{\mathbf{x}} + \boldsymbol{\lambda}$ where $\hat{\mathbf{x}}$ is the unit vector in the x-direction and $\boldsymbol{\lambda} = (\lambda, \mu)$ is the modulation wave vector. The frequency $\omega(\mathbf{k})$ is given by the linear dispersion relation in normalized form

$$\omega^2 = k \tanh kh. \tag{5.2}$$

Writing this in the style of the first-harmonic term of the harmonic perturbation expansion (4.10) we get

$$\eta(\mathbf{r},t) = \frac{1}{2}B(\mathbf{r},t)\mathrm{e}^{\mathrm{i}(x-t)} + \text{c.c.} \tag{5.3}$$

where the complex amplitude $B(\mathbf{r},t)$ is defined by

$$B(\mathbf{r},t) = \int \hat{B}(\boldsymbol{\lambda},t)\mathrm{e}^{\mathrm{i}\boldsymbol{\lambda}\cdot\mathbf{r}}\,\mathrm{d}\boldsymbol{\lambda} = \int b(\hat{\mathbf{x}}+\boldsymbol{\lambda})\mathrm{e}^{\mathrm{i}[\boldsymbol{\lambda}\cdot\mathbf{r}-(\omega(\hat{\mathbf{x}}+\boldsymbol{\lambda})-1)t]}\,\mathrm{d}\boldsymbol{\lambda}. \tag{5.4}$$

We see that the Fourier transform $\hat{B}$ satisfies the equation

$$\frac{\partial \hat{B}}{\partial t} + \mathrm{i}\left[\omega(\hat{\mathbf{x}}+\boldsymbol{\lambda}) - 1\right]\hat{B} = 0. \tag{5.5}$$

From a computational point of view, equation (5.5) is trivial to compute in the Fourier space, the last term being a multiplication of $\hat{B}$ with an algebraic expression of the wave vector. However, taking the inverse Fourier transform of equation (5.5) the evolution equation for B can be formally written as

$$\frac{\partial B}{\partial t} + L(\partial_x, \partial_y)B = 0 \tag{5.6}$$

with

$$L(\partial_x,\partial_y) = \mathrm{i}\left\{\left(\left[(1-\mathrm{i}\partial_x)^2 - \partial_y^2\right]^{1/2}\tanh\left[(1-\mathrm{i}\partial_x)^2 - \partial_y^2\right]^{1/2} h\right)^{1/2} - 1\right\}. \tag{5.7}$$

On infinite depth this reduces to

$$L(\partial_x,\partial_y) = \mathrm{i}\left\{\left[(1-\mathrm{i}\partial_x)^2 - \partial_y^2\right]^{1/4} - 1\right\}. \tag{5.8}$$

Linear evolution equations at all orders can now be obtained by expanding (5.6) in powers of the derivatives. Hence we recover the linear part of the classical cubic NLS equation up to second order, the linear part of the MNLS equation of Dysthe (1979) up to third order, and the linear part of the broader bandwidth MNLS equation of Trulsen and Dysthe (1996) up to fifth order.

Alternatively, an evolution equation for space evolution along the x-direction can be written as

$$\frac{\partial B}{\partial x} + \mathcal{L}(\partial_t, \partial_y)B = 0 \tag{5.9}$$

where $\mathcal{L}$ results from expressing the linear dispersion relation for the wavenumber as a function of the frequency. This is difficult to do in closed form for finite depth, but for infinite depth it reduces to

$$\mathcal{L}(\partial_t,\partial_y) = -\mathrm{i}\left\{\left[(1+\mathrm{i}\partial_t)^4 + \partial_y^2\right]^{1/2} - 1\right\}. \tag{5.10}$$

Expanding (5.9) in powers of the derivatives we recover the linear part of the broader bandwidth MNLS equation for space evolution in Trulsen and Dysthe (1997) up to fifth order. Equation (5.10) has the interesting property that for long-crested waves ($\partial_y = 0$) the pseudo differential operator becomes an ordinary differential operator with only two terms.

The linear constant coefficient equations (5.6) and (5.9) are most easily solved numerically in Fourier transform space so there is no need to approximate the operators with truncated power series expansions.

After having derived exact linear evolution equations, and noticing that they are the linear parts of the corresponding nonlinear evolution equations derived earlier, the higher order nonlinear Schrödinger equations with exact linear dispersion are easily obtained. The replacement for equations (4.17), (4.27), (4.37) and (4.47) become, respectively,

$$\frac{\partial A}{\partial t} + L(\partial_x, \partial_y)A + \frac{\mathrm{i}}{2}|A|^2 A + \frac{7}{4}|A|^2 \frac{\partial A}{\partial x} - \frac{1}{4} A \frac{\partial |A|^2}{\partial x} + \mathrm{i} A \frac{\partial \bar{\phi}}{\partial x} = 0 \quad \text{at} \quad z = 0, \tag{5.11}$$

$$\frac{\partial A}{\partial x} + \mathcal{L}(\partial_t, \partial_y)A + \mathrm{i}|A|^2 A - 8|A|^2 \frac{\partial A}{\partial t} - 4\mathrm{i} A \frac{\partial \bar{\phi}}{\partial t} = 0 \quad \text{at} \quad z = 0, \tag{5.12}$$

$$\frac{\partial B}{\partial t} + L(\partial_x, \partial_y)B + \frac{\mathrm{i}}{2}|B|^2 B + \frac{5}{4}|B|^2 \frac{\partial B}{\partial x} + \frac{1}{4} B \frac{\partial |B|^2}{\partial x} + \mathrm{i} B \frac{\partial \bar{\phi}}{\partial x} \quad \text{at} \quad z = 0, \tag{5.13}$$

$$\frac{\partial B}{\partial x} + \mathcal{L}(\partial_t, \partial_y)B + \mathrm{i}|B|^2 B - 6|B|^2 \frac{\partial B}{\partial t} - 2B \frac{\partial |B|^2}{\partial t} - 4\mathrm{i} B \frac{\partial \bar{\phi}}{\partial t} \quad \text{at} \quad z = 0, \tag{5.14}$$

where we employ the versions of L and $\mathcal{L}$ for infinite depth, given by equations (5.8) and (5.10), respectively. Notice that equation (5.12) is qualitatively different from the other three equations, giving it qualitatively different properties from the other three forms. This is explained in section 6.1.

The Zakharov integral equation (Zakharov 1968), reviewed in Chap. 3, contains exact linear dispersion both at the linear and the cubic nonlinear orders. The above MNLS equations with exact linear dispersion are limiting cases of the Zakharov integral equation after employing a bandwidth constraint only on the cubic nonlinear part. If a bandwidth constraint is applied on both the linear and nonlinear parts, the MNLS equations in sections 4.3–4.6 can also be derived as special cases of the Zakharov equation. Stiassnie (1984) first showed how the temporal MNLS equation for velocity potential (4.17) can be obtained from the Zakharov integral equation in a systematic way. Kit and Shemer (2002) showed how the spatial MNLS equations for velocity potential (4.27) and surface elevation (4.47) can be obtained by the same approach.

6 Properties of the higher order nonlinear Schrödinger equations

We got four different forms of the MNLS equation. In order to perform a common analysis to extract some of their important properties, we define a generic form of the MNLS equation with generalized coefficients (their values can be found in sections 4.3–

4.6)

$$\frac{\partial A}{\partial t} + c_1 \frac{\partial A}{\partial x} + \mathrm{i}c_2 \frac{\partial^2 A}{\partial x^2} + \mathrm{i}c_3 \frac{\partial^2 A}{\partial y^2} + \mathrm{i}c_4 |A|^2 A$$
$$+ c_5 \frac{\partial^3 A}{\partial x^3} + c_6 \frac{\partial^3 A}{\partial x \partial y^2} + c_7 |A|^2 \frac{\partial A}{\partial x} + c_8 A^2 \frac{\partial A^*}{\partial x} + \mathrm{i}c_9 A \frac{\partial \bar{\phi}}{\partial x} = 0 \quad \text{at} \quad z = 0, \tag{6.1}$$

$$\frac{\partial \bar{\phi}}{\partial z} = \beta \frac{\partial |A|^2}{\partial x} \quad \text{at} \quad z = 0, \tag{6.2}$$

$$\alpha^2 \frac{\partial^2 \bar{\phi}}{\partial x^2} + \frac{\partial^2 \bar{\phi}}{\partial y^2} + \frac{\partial^2 \bar{\phi}}{\partial z^2} = 0 \quad \text{for} \quad -h < z, \tag{6.3}$$

$$\frac{\partial \bar{\phi}}{\partial z} = 0 \quad \text{at} \quad z = -h. \tag{6.4}$$

6.1 Conservation laws

Let the following two quantities be defined,

$$I = \int |A|^2 \,\mathrm{d}\mathbf{r} = (2\pi)^2 \int |\hat{A}|^2 \,\mathrm{d}\mathbf{k}, \tag{6.5}$$

$$J = \int \left(\frac{\mathrm{i}}{2} A \frac{\partial A^*}{\partial x} + \text{c.c.} \right) \mathrm{d}\mathbf{r} = (2\pi)^2 \int k_x |\hat{A}|^2 \,\mathrm{d}\mathbf{k}, \tag{6.6}$$

where we have used the Fourier transform in the form $A = \int \hat{A} \exp(\mathrm{i}\mathbf{k} \cdot \mathbf{r}) \,\mathrm{d}\mathbf{k}$, and where we have also invoked Parseval's theorem as explained in the Appendix (see Trulsen and Dysthe 1997).

It is readily shown that I is conserved

$$\frac{\mathrm{d}I}{\mathrm{d}t} = 0. \tag{6.7}$$

On the other hand, J is conserved when the coefficient c_8 vanishes

$$\frac{\mathrm{d}J}{\mathrm{d}t} = c_8 \int \left(-\mathrm{i} \left(A \frac{\partial A^*}{\partial x} \right)^2 + \text{c.c.} \right) \mathrm{d}\mathbf{r}. \tag{6.8}$$

Thus the spatial MNLS equation for the velocity potential, section 4.4, is exceptional since it satisfies a conservation law that not satisfied by the other three forms of the MNLS equation.

It may be useful to notice that the even though J is not conserved, this does not imply that the physical energy or the linear momentum of the waves are not conserved.

6.2 Modulational instability of Stokes waves

Equations (6.1)–(6.4) have a particularly simple exact uniform wave solution known as the Stokes wave. The solution is given by

$$A = A_0 e^{-ic_4|A_0|^2 t} \qquad \text{and} \qquad \bar{\phi} = 0. \tag{6.9}$$

Figure 6 shows the reconstructed surface elevation

$$\eta = \epsilon \cos\theta + \frac{\epsilon^2}{2}\cos 2\theta + \frac{3\epsilon^3}{8}\cos 3\theta$$

at three separate orders for the much exaggerated steepness $\epsilon = 0.3$.

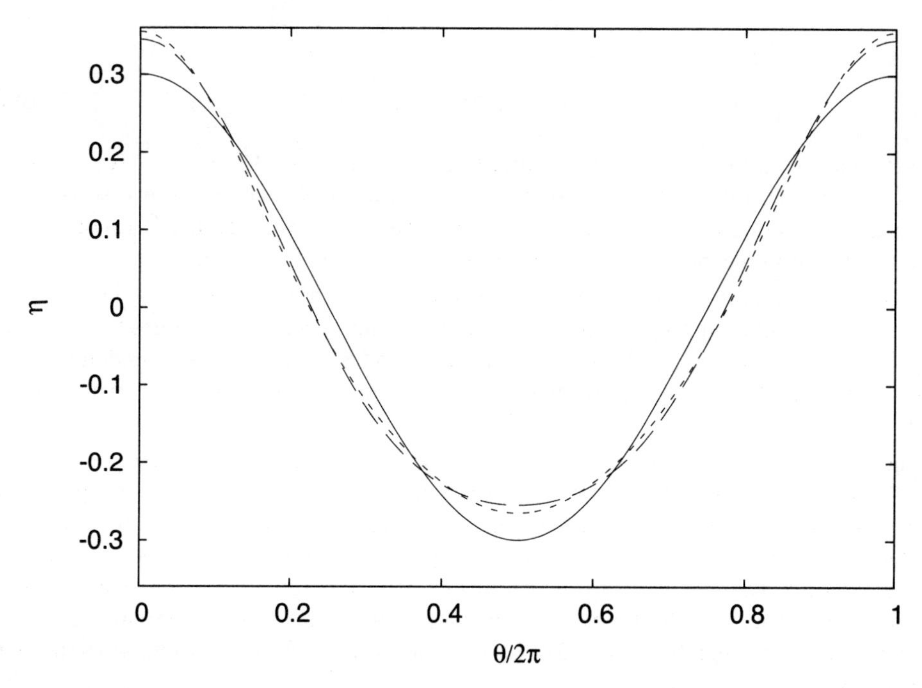

Figure 6. Stokes wave of steepness $\epsilon = 0.3$ reconstructed at three orders: —, first order; – –, second order, $\cdots$, third order.

The stability of the Stokes wave can be found by perturbing it in amplitude and phase. The stability analysis for the cubic NLS equation is reviewed in Chap. 3, here we give the analysis for the MNLS equation. We set

$$A = A_0(1 + a + i\theta)e^{-ic_4|A_0|^2 t}. \tag{6.10}$$

After linearization in a, θ and $\bar{\phi}$, and assuming plane-wave solutions

$$\begin{pmatrix} a \\ \theta \\ \bar{\phi} \end{pmatrix} = \begin{pmatrix} \hat{a} \\ \hat{\theta} \\ \hat{\phi} \end{pmatrix} e^{i(\lambda x + \mu y - \Omega t)} + \text{c.c.}, \tag{6.11}$$

we find that the perturbation behaves according to the dispersion relation

$$\Omega = P \pm \sqrt{Q\left(Q - 2c_4|A_0|^2 + \frac{2c_9\beta\lambda^2|A_0|^2 \coth Kh}{K}\right) + c_8^2|A_0|^4\lambda^2} \tag{6.12}$$

where

$$P = c_1\lambda - c_5\lambda^3 - c_6\lambda\mu^2 + c_7|A_0|^2\lambda, \tag{6.13}$$

$$Q = c_2\lambda^2 + c_3\mu^2 \tag{6.14}$$

and

$$K = \sqrt{\alpha^2\lambda^2 + \mu^2}. \tag{6.15}$$

A perturbation is unstable if the modulational "frequency" Ω has positive imaginary part. The growth rate of the instability is defined as $\mathrm{Im}\Omega$. The growth rate is seen to be symmetric about the λ-axis and the μ-axis, therefore we only need to consider the growth rates of perturbations in the first quadrant of the (λ, μ)-plane.

Instability diagrams for the NLS equation are found in Chap. 3. They have the important and unfortunate property that the instability region is unlimited toward large perturbation wavenumbers. This implies that the NLS equation has a tendency to leak energy out of its domain of validity when applied to short-crested seas (Martin and Yuen 1980).

Figures 7 and 8 show growth rates for the MNLS equation for target steepness $\epsilon = 0.12$ on infinite depth, while figures 9 and 10 show growth rates for the MNLS equation for target steepness $\epsilon = 0.12$ and target depth $h = 2.0$. For time evolution it makes no difference selecting equations for A or B. For space evolution it makes a lot of difference. Numerical simulation of the spatial evolution equation for A is more likely to suffer energy leakage through the unstable growth of rapidly oscillating modulations. In any case the instability regions for the MNLS equation are much more confined than for the NLS equation.

It is remarkable that the most attractive model for analytical work (due to the number of conservation laws satisfied) turns out to be the least attractive model for numerical work!

The most unstable perturbation is collinear with the principal evolution direction for sufficiently deep water. When the depth becomes smaller than a threshold, which depends on the steepness, the most unstable perturbation bifurcates into an oblique direction. The criterion for bifurcation was discussed by Trulsen and Dysthe (1996).

As far as the sea state of the Draupner wave is concerned, it follows that a Stokes wave with the same characteristic wavenumber and amplitude would have its most unstable perturbations obliquely to the main propagation direction, see figure 9. This result is most likely of little practical importance since the bandwidth of the Draupner wave field is so wide that the waves are modulationally stable (Alber 1978; Crawford et al. 1980).

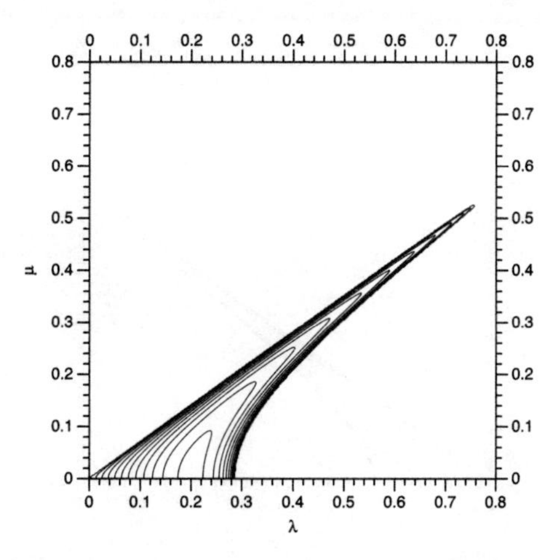

Figure 7. Growth rate for time evolution of MNLS equations for A (section 4.3) and B (section 4.5) for $\epsilon = 0.12$ and $h = \infty$. The maximum growth rate 0.0057 is achieved at (0.20,0). Contour interval 0.0005.

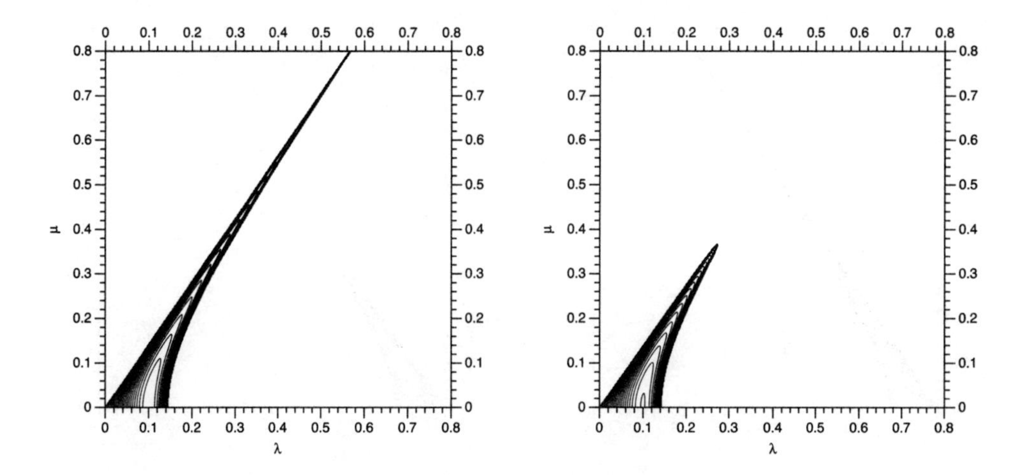

Figure 8. Growth rate for space evolution of MNLS equations for A (left, section 4.4) and B (right, section 4.6) for $\epsilon = 0.12$ and $h = \infty$. For the A equations the maximum growth rate 0.011 is achieved at (0.10,0). For the B equations the maximum growth rate 0.011 is achieved at (0.098, 0). Contour interval 0.0005.

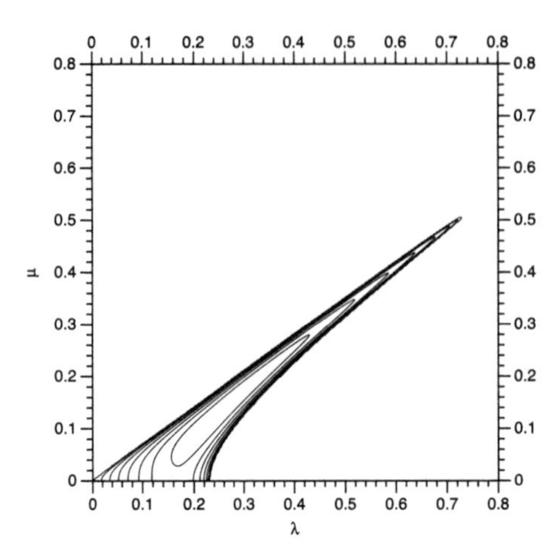

Figure 9. Growth rate for time evolution of MNLS equations for A (section 4.3) and B (section 4.5) for $\epsilon = 0.12$ and $h = 2.0$. The maximum growth rate 0.0039 is achieved at (0.27,0.14) corresponding to waves at an angle 6.5° with the principal propagation direction. Contour interval 0.0005.

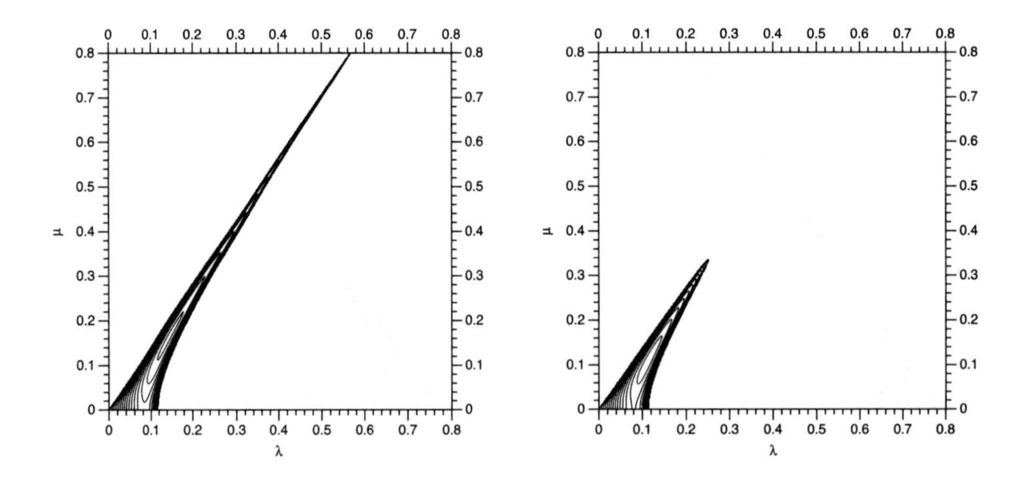

Figure 10. Growth rate for space evolution of MNLS equations for A (left, section 4.4) and B (right, section 4.6) for $\epsilon = 0.12$ and $h = 2.0$. For the A equations the maximum growth rate 0.0081 is achieved at (0.14,0.16). For the B equations the maximum growth rate 0.0072 is achieved at (0.11, 0.11). The definition of angle is not meaningful since the second axis denotes frequency. Contour interval 0.0005.

7 An application of the higher-order nonlinear Schrödinger equations: Deterministic wave forecasting

The equations in section 4.6 for the spatial evolution of the surface elevation can be used to forecast or hindcast waves deterministically. From measured time series of the surface elevation η, the evolution equations can be initialized with the complex amplitude B extracted by bandpass filtering the Fourier transform $\hat{\eta}$ in a neighborhood of ω_c.

Trulsen and Stansberg (2001) and Trulsen (2005) reported application to bichromatic waves and irregular laboratory waves. We here review results from an experiment with bichromatic waves in the 270 m long and 10.5 m wide towing tank at Marintek in Trondheim, Norway. Figure 11 shows a sketch of the arrangement of wave probes. Measurements at probe 1 are used for initialization, and detailed comparisons between simulated and measured waves are presented here for probes 2, 4, 5, 7 and 10. These probes were located at distances 9.3 m, 40 m, 80 m, 120 m, 160 m and 200 m from the wave maker, respectively. The bichromatic waves were generated with nominal periods 1.9 s and 2.1 s, both with nominal wave heights 0.16 m. The steepness of the portion of the time series at probe 1 used for initialization is estimated to be $\epsilon = 0.11$. We use a time domain of length 204.8 s. Figure 12 shows comparisons between simulated and measured time series in a small time window that moves with the linear group velocity down along the towing tank.

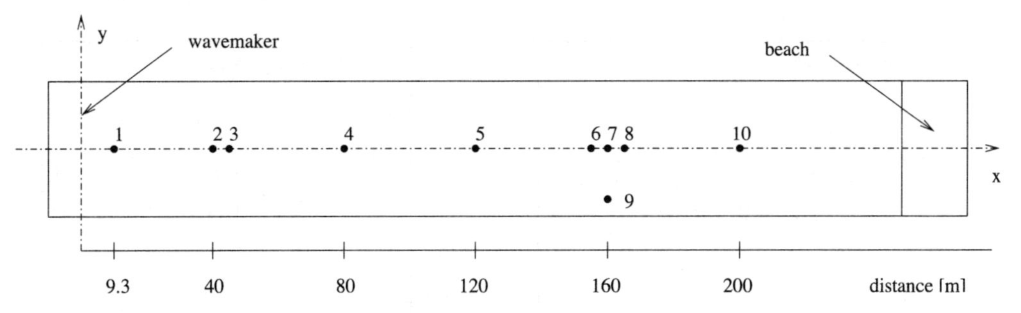

Figure 11. Sketch of towing tank with wave probes.

The same approach was also attempted to hindcast the wave field at Draupner (Trulsen 2001). This simulation most likely has little to do with reality since we are neglecting the fact that the waves were not perfectly long-crested, however there is some empirical evidence that the loads on the platform at least were unidirectional, see Karunakaran et al. (1997).

The predicted time histories at 50 meter intervals upstream and downstream are shown in figure 13. At 500 m upstream of Draupner E there appears to be a large wave group about one minute before the freak wave hits the platform. This wave group appears to split up into a short and large leading group that runs away from a longer and smaller trailing group that becomes rather diffuse as it approaches the platform. The freak wave that hits the platform is in the middle of the short leading group. After the impact with the platform, this wave group broadens, becomes less steep, and slows down

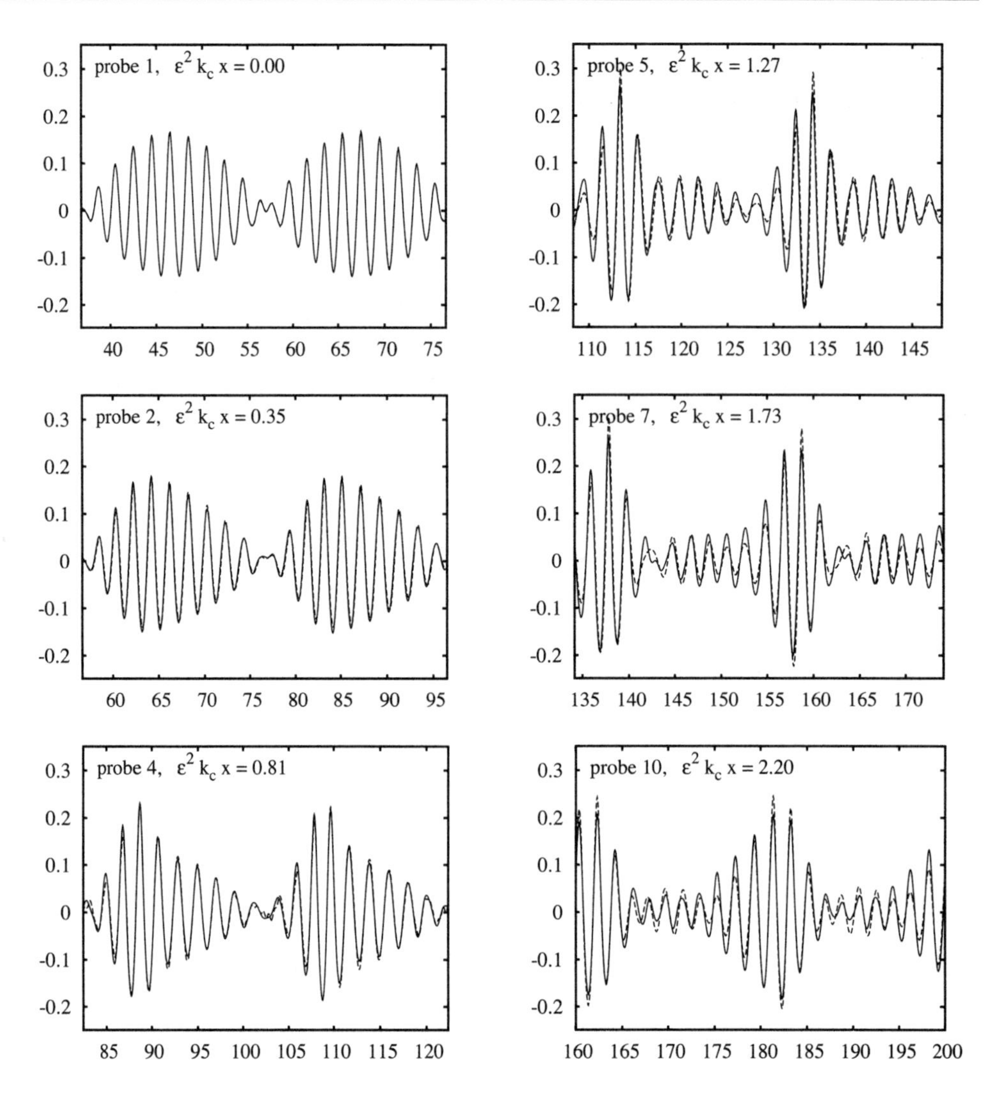

Figure 12. Deterministic forecasting of bichromatic waves: Simulation (solid curves) and experimental measurements (dotted curves) at six stations along a towing tank sketched in figure 11. The measured time series at the first station was used for initialization. Horizontal axis is time in seconds. Vertical axis is normalized surface displacement $k_c\eta$.

slightly. A large trough (a "hole" in the ocean) probably occurred slightly upstream of the platform.

As far as the occurrence of the Draupner "New Year Wave" is concerned, few certain conclusions can be drawn since we do not have detailed information about directional distribution, i.e. details about short-crestedness. We may conclude that if the waves had been long-crested, then the extreme wave likely did not occur suddenly and unexpectedly. An imaginary observer on the platform, looking north, would probably have seen large waves (a wall of water) approaching the platform for at least a minute or so, including an apparent hole in the ocean. This description is indeed similar to stories told by seafarers who survived ship founderings due to large waves, however, such stories have often not been believed in the past.

8 Stochastic description of surface waves

Our concern is to account for the vertical displacement $\eta(\mathbf{r},t)$ of the water surface as a function of horizontal position $\mathbf{r}=(x,y)$ and time t. A quick look at any water surface under the action of wind suggests that the previous deterministic description of uniform waves is highly unsatisfactory. Given otherwise identical conditions (wind, etc.) it is our everyday experience that the waves are random and unpredictable. We may therefore think of the surface displacement as a *stochastic process* with several possible outcomes or *realizations.* A collection of realizations we denote an *ensemble.* The view elaborated here is that for any single realization, the spatiotemporal evolution is deterministically given as a solution of nonlinear evolution equations such as those previously derived. Another point of view not elaborated here is that the continued action of random wind etc. requires the spatiotemporal evolution to be stochastic to a certain level as well. There is an underlying assumption behind the view employed here that the effect of wind is much slower than the rapid deterministic modulations described by the above deterministic evolution equations.

In reality we have of course only one realization of the waves on the water surface. However, in our minds we can imagine several realizations, and on the computer we can simulate several realizations. In reality it may also be possible to achieve several essentially independent realizations even if we do not have several identical copies of the sea. In the case that we suffice with measurements during a limited time interval only, we may achieve several independent realizations by measuring waves over several different limited time intervals under otherwise identical conditions (weather, etc.), provided the limited time intervals are sufficiently separated in time. Alternatively, if we suffice with measurements at a group of wave probes only, we may achieve several independent realizations of such time series from several different groups of wave probes provided they are sufficiently far apart. The process of achieving a realization we denote as an *experiment.*

We define a *stochastic variable* Z as a rule that assigns a number z_o to every outcome o of an experiment.

We define a *stochastic process* $Z(t)$ as a rule that assigns a function $z_o(t)$ to every outcome o of an experiment. Of fundamental interest to us, we shall describe the vertical displacement of the water surface as a stochastic process $Z(\mathbf{r},t)$ defined as a rule that assigns an actual sea surface $\eta(\mathbf{r},t)=z_o(\mathbf{r},t)$ to every outcome o of an experiment.

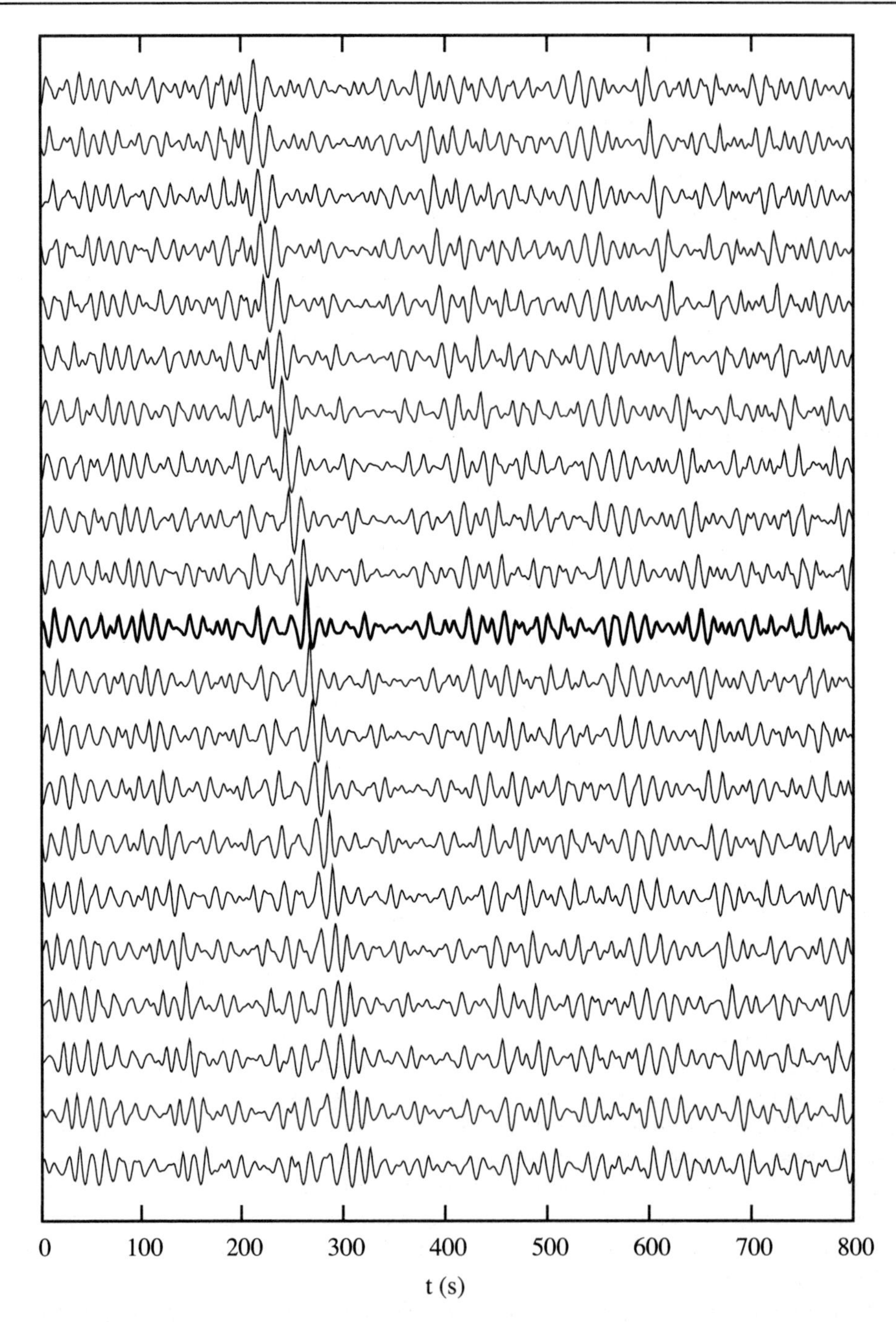

Figure 13. Hindcasted spatiotemporal evolution of the Draupner wave field. The simulation is presented from upstream to downstream as time series displayed at equidistant intervals of 50 meters. The measured time series used for initialization of the hindcast is in bold. The zero lines for hindcasted time series are shifted vertically in comparison with the measured time series.

There are now four different interpretations of a stochastic process $Z(t)$:
1. We consider the process $Z(t)$ for all times t and all outcomes o.
2. Given an outcome o, we consider a time series $z_o(t)$ for all times t.
3. Given a time t_1, we consider a stochastic variable $Z(t_1)$ for all outcomes o.
4. Given an outcome o and a time t_1, we consider a number $z_o(t_1)$.

The reader who desires a thorough introduction to the stochastic description of the sea, the books by Goda (2000) and Ochi (1998) are recommended. For an introduction to stochastic analysis, see the book by Papoulis (1991). In the following we give a relatively elementary introduction to essential parts of this theory.

9 Theory for stochastic variables

9.1 Theory for a single stochastic variable

With the notation $\{Z \le z\}$ we refer to the collection of all outcomes z_o of the stochastic variable Z such that $z_o \le z$. The probability for this collection of outcomes defines the *cumulative distribution function*

$$F(z) \equiv P\{Z \le z\}$$

where $P\{\cdot\}$ reads the "probability of $\{\cdot\}$". The cumulative probability function has the properties that
1. $F(-\infty) = 0$,
2. $F(\infty) = 1$,
3. $F(z)$ is an increasing function, $F(z_1) \le F(z_2)$ for $z_1 < z_2$.

The probability that an outcome is between a lower and an upper bound is $P\{a \le Z \le b\} = F(b) - F(a)$. Similarly, the probability that an outcome is in an interval of infinitesimal width is $P\{z \le Z \le z + \mathrm{d}z\} = F(z + \mathrm{d}z) - F(z) \approx \frac{\mathrm{d}F}{\mathrm{d}z}dz = f(z)\mathrm{d}z$. We define the *probability density function* as

$$f(z) \equiv \frac{\mathrm{d}F}{\mathrm{d}z}.$$

The probability density function has the properties that
1. $f(z) \ge 0$,
2. $\int_{-\infty}^{\infty} f(z)\,\mathrm{d}z = 1$,
3. $F(z) = \int_{-\infty}^{z} f(\xi)\,\mathrm{d}\xi$.

We define the *mode* of a stochastic variable Z to be the value of z such that the probability density function $f(z)$ achieves its maximum. This is the most probable outcome for the stochastic variable Z. If the probability density has a single maximum, then the variable Z is said to be *unimodal.*

We define the *median* of a stochastic variable Z to be the value of z such that the cumulative distribution function $F(z) = 0.5$. It is equally probable that the stochastic variable Z gives an outcome smaller than or greater than the median.

The *expected value* μ of a stochastic variable Z is defined as the weighted average

$$\mu = E[Z] = \int_{-\infty}^{\infty} z f(z)\,\mathrm{d}z.$$

The expected value of a function $g(Z)$ of the stochastic variable Z is the weighted average

$$E[g(Z)] = \int_{-\infty}^{\infty} g(z) f(z)\,\mathrm{d}z.$$

It is seen that the expected value operator is linear. Suppose we have two functions $g(Z)$ and $h(Z)$ and two constants a and b, then we have

$$E[ag(Z) + bh(Z)] = aE[g(Z)] + bE[h(Z)].$$

The *variance* σ^2 of a stochastic variable Z is defined by

$$\sigma^2 = \mathrm{Var}[Z] = E[(Z-\mu)^2] = \int_{-\infty}^{\infty} (z-\mu)^2 f(z)\,\mathrm{d}x.$$

By the linearity of the expected value operator, this can be written

$$\sigma^2 = E[(Z-\mu)^2] = E[Z^2 - 2\mu Z + \mu^2] = E[Z^2] - 2\mu E[Z] + \mu^2 = E[Z^2] - \mu^2.$$

The *standard deviation* σ of a stochastic variable is defined as the square root of the variance.

The *nth moment* of a stochastic variable Z is defined as

$$m_n = E[Z^n] = \int_{-\infty}^{\infty} z^n f(z)\,\mathrm{d}x,$$

while the *nth centered moment* is defined as

$$E[(Z-\mu)^n] = \int_{-\infty}^{\infty} (z-\mu)^n f(z)\,\mathrm{d}x.$$

The variance can thus be defined as the second centered moment of the stochastic variable. We see that $\mu = m_1$ and $\sigma^2 = m_2 - m_1^2$.

The *skewness* γ of a stochastic variable is defined as the third centered moment normalized by the cube of the standard deviation

$$\gamma = \frac{E[(Z-\mu)^3]}{\sigma^3} = \frac{m_3 - 3\sigma^2\mu - \mu^3}{\sigma^3}.$$

The *kurtosis* κ of a stochastic variable is defined as the fourth centered moment normalized by the square of the variance

$$\kappa = \frac{E[(Z-\mu)^4]}{\sigma^4} = \frac{m_4 - 4\gamma\sigma^3\mu - 6\sigma^2\mu^2 - \mu^4}{\sigma^4}.$$

Given a stochastic variable Z we can transform it into a new stochastic variable $\tilde{Z}$ as a function of Z. Particularly useful is the transformation given by

$$\tilde{Z} = \frac{Z - \mu}{\sigma}. \tag{9.1}$$

The cumulative distribution function of $\tilde{Z}$ is

$$F_{\tilde{Z}}(\tilde{z}) = P\{\tilde{Z} \leq \tilde{z}\} = P\{\frac{Z-\mu}{\sigma} \leq \tilde{z}\} = P\{Z \leq \mu + \sigma\tilde{z}\} = F_Z(\mu + \sigma\tilde{z}).$$

The probability density function of $\tilde{Z}$ is

$$f_{\tilde{Z}}(\tilde{z}) = \frac{\mathrm{d}F_{\tilde{Z}}(\tilde{z})}{\mathrm{d}\tilde{z}} = \frac{\mathrm{d}F_Z(\mu + \sigma\tilde{z})}{\mathrm{d}\tilde{z}} = \sigma f_Z(\mu + \sigma\tilde{z}).$$

For the special transformation (9.1) with μ and σ^2 being the mean and variance of Z, we notice that the mean of $\tilde{Z}$ is

$$E[\tilde{Z}] = E[\frac{Z-\mu}{\sigma}] = \frac{1}{\sigma}E[Z] - \frac{\mu}{\sigma} = 0,$$

the variance of $\tilde{Z}$ is

$$E[\tilde{Z}^2] = E[\frac{(Z-\mu)^2}{\sigma^2}] = \frac{1}{\sigma^2}E[(Z-\mu)^2] = 1,$$

while the skewness and kurtosis of $\tilde{Z}$ are the same as the skewness and kurtosis of Z, respectively.

The *characteristic function* $\phi(k)$ of a stochastic variable Z is defined as the expected value of $\mathrm{e}^{\mathrm{i}kz}$,

$$\phi(k) = E[\mathrm{e}^{\mathrm{i}kz}] = \int_{-\infty}^{\infty} f(z)\mathrm{e}^{\mathrm{i}kz}\,\mathrm{d}z,$$

with inverse transform

$$f(z) = \frac{1}{2\pi}\int_{-\infty}^{\infty} \phi(k)\mathrm{e}^{-\mathrm{i}kz}\,\mathrm{d}k.$$

The variable k used here should not be confused with the wavenumber of surface waves. We can formally expand the complex exponential function in a power series

$$\mathrm{e}^{\mathrm{i}kz} = 1 + \mathrm{i}kz - \frac{1}{2}k^2z^2 - \frac{\mathrm{i}}{6}k^3z^3 + \dots$$

and thus the characteristic function can be written as a superposition of all the moments

$$\phi(k) = 1 + \mathrm{i}km_1 - \frac{1}{2}k^2m_2 - \frac{\mathrm{i}}{6}k^3m_3 + \dots$$

In fact, the nth moment can conveniently be achieved by differentiating the characteristic function n times and evaluating the result at $k = 0$.

Example: Gaussian or normal distribution. A stochastic variable Z is said to be normally distributed with mean μ and variance σ^2 with the probability density function given by

$$f(z) = \frac{1}{\sqrt{2\pi}\sigma} \mathrm{e}^{-\frac{(z-\mu)^2}{2\sigma^2}}.$$

The cumulative distribution function is

$$F(z) = \int_{-\infty}^{z} \frac{1}{\sqrt{2\pi}\sigma} \mathrm{e}^{-\frac{(\xi-\mu)^2}{2\sigma^2}} \, \mathrm{d}\xi = \frac{1}{2} + \frac{1}{2} \operatorname{erf}(\frac{z-\mu}{\sqrt{2}\sigma})$$

where $\operatorname{erf}(z) = \frac{2}{\sqrt{\pi}} \int_0^z \mathrm{e}^{-t^2} \, dt$ is the error function. The stochastic variable Z is unimodal (has a single maximum) with mode $z_{\text{mode}} = \mu$. The median is $z_{\text{median}} = \mu$ since $\operatorname{erf}(0) = 0$. The mean is

$$E[Z] = \int_{-\infty}^{\infty} \frac{z}{\sqrt{2\pi}\sigma} \mathrm{e}^{-\frac{(z-\mu)^2}{2\sigma^2}} \, \mathrm{d}z = \mu,$$

the variance is

$$E[(Z-\mu)^2] = \int_{-\infty}^{\infty} \frac{(z-\mu)^2}{\sqrt{2\pi}\sigma} \mathrm{e}^{-\frac{(z-\mu)^2}{2\sigma^2}} \, \mathrm{d}z = \sigma^2.$$

the skewness is

$$\gamma = \frac{E[(Z-\mu)^3]}{\sigma^3} = \frac{1}{\sigma^3} \int_{-\infty}^{\infty} \frac{(z-\mu)^3}{\sqrt{2\pi}\sigma} \mathrm{e}^{-\frac{(z-\mu)^2}{2\sigma^2}} \, \mathrm{d}z = 0.$$

and the kurtosis is

$$\kappa = \frac{E[(Z-\mu)^4]}{\sigma^4} = \frac{1}{\sigma^4} \int_{-\infty}^{\infty} \frac{(z-\mu)^4}{\sqrt{2\pi}\sigma} \mathrm{e}^{-\frac{(z-\mu)^2}{2\sigma^2}} \, \mathrm{d}z = 3.$$

The characteristic function is

$$\phi(k) = \mathrm{e}^{-\frac{\sigma^2 k^2}{2} + \mathrm{i}\mu k}.$$

In the limit of zero variance, $\sigma \to 0$, the probability density function converges to a Dirac delta function $f(z) = \delta(z-\mu)$ and the characteristic function has unit magnitude $|\phi(k)| = 1$.

Example: Uniform distribution. A stochastic variable Θ is said to be uniformly distributed on the interval $[0, 2\pi]$ with the probability density function given by

$$f(\theta) = \begin{cases} \frac{1}{2\pi} & 0 \le \theta \le 2\pi \\ 0 & \text{otherwise} \end{cases}$$

The cumulative distribution function is

$$F(\theta) = \begin{cases} 0 & \theta < 0 \\ \frac{\theta}{2\pi} & 0 \le \theta \le 2\pi \\ 1 & \theta > 2\pi \end{cases}$$

The mode (maximum) is not well defined since $f(\theta)$ does not have an isolated extremal point. The median is $\theta_{\text{median}} = \pi$ because $F(\pi) = 0.5$. The mean is $\mu = E[\Theta] = \pi$. The variance is $\sigma^2 = E[(\Theta - \mu))^2] = \pi^2/3$. The skewness is $\gamma = 0$. The kurtosis is $\kappa = 9/5$. The characteristic function is

$$\phi(k) = \frac{1}{2\pi i k}(\mathrm{e}^{2\pi i k} - 1).$$

Example: Rayleigh distribution. A stochastic variable R is said to be Rayleigh distributed with the probability density function given by

$$f(r) = \begin{cases} \alpha r \mathrm{e}^{-\frac{\alpha r^2}{2}} & r \geq 0 \\ 0 & r < 0 \end{cases}$$

where α is a parameter. Find the mode, median, mean, variance, standard deviation, skewness, kurtosis and characteristic function!

Example: Exponential distribution. A stochastic variable Z is said to be exponentially distributed with the probability density function given by

$$f(z) = \begin{cases} \alpha \mathrm{e}^{-\alpha z} & z \geq 0 \\ 0 & z < 0 \end{cases}$$

where α is a parameter. Find the mode, median, mean, variance, standard deviation, skewness, kurtosis and characteristic function!

9.2 Theory for several stochastic variables

Let X and Y be two stochastic variables. With the notation $\{X \leq x \text{ and } Y \leq y\}$ we refer to the collection of all outcomes of the two stochastic variables X and Y such that $X \leq x$ and $Y \leq y$ simultaneously. The probability for this collection of outcomes defines the *joint cumulative distribution function* $F(x, y) \equiv P\{X \leq x \text{ and } Y \leq y\}$. The joint cumulative probability function has the properties that

1. $F(-\infty, -\infty) = 0$,
2. $F(\infty, \infty) = 1$,
3. $F(x_1, y_1) \leq F(x_2, y_2)$ for $x_1 \leq x_2$ and $y_1 \leq y_2$.

The *marginal cumulative distribution functions* are defined as

$$F_X(x) = F(x, \infty) \qquad \text{and} \qquad F_Y(y) = F(\infty, y).$$

The *joint probability density function* is defined as

$$f(x, y) \equiv \frac{\partial^2 F(x, y)}{\partial x \partial y}$$

and has the properties that

1. $f(x, y) \geq 0$,

2. $\int_{-\infty}^{\infty} \int_{-\infty}^{\infty} f(x,y)\,\mathrm{d}x\,\mathrm{d}y = 1$,
3. $F(x,y) = \int_{-\infty}^{x} f(\xi)\,\mathrm{d}\xi$.

The *marginal probability density functions* are defined as

$$f_X(x) = \int_{-\infty}^{\infty} f(x,y)\,\mathrm{d}y = \frac{\partial F_X(x)}{\partial x} \qquad \text{and} \qquad f_Y(y) = \int_{-\infty}^{\infty} f(x,y)\,\mathrm{d}x = \frac{\partial F_Y(y)}{\partial y}.$$

Two stochastic variables X and Y are said to be *statistically independent* if the joint probability density function can be factorized $f(x,y) = f_X(x) f_Y(y)$.

The mean values of X and Y are now

$$\mu_X = E[X] = \int_{-\infty}^{\infty} \int_{-\infty}^{\infty} x f(x,y)\,\mathrm{d}x\,\mathrm{d}y = \int_{-\infty}^{\infty} x f_X(x)\,\mathrm{d}x,$$

$$\mu_Y = E[Y] = \int_{-\infty}^{\infty} \int_{-\infty}^{\infty} y f(x,y)\,\mathrm{d}x\,\mathrm{d}y = \int_{-\infty}^{\infty} y f_Y(y)\,\mathrm{d}y.$$

The *covariance* of two stochastic variables X and Y is defined as

$$\mathrm{Cov}[X,Y] = E[(X-\mu_X)(Y-\mu_Y)] = E[XY] - \mu_X\mu_Y.$$

The *correlation coefficient* of the two stochastic variables is defined as the ratio

$$\frac{\mathrm{Cov}[X,Y]}{\sigma_X\sigma_Y}.$$

We show that the absolute value of the correlation coefficient is not greater than one, or equivalently $|\,\mathrm{Cov}[X,Y]| \le \sigma_X\sigma_Y$.

Proof. Look at

$$E[(a(X-\mu_X) + (Y-\mu_Y))^2] = a^2\sigma_X^2 + 2a\,\mathrm{Cov}[X,Y] + \sigma_Y^2 \ge 0.$$

Solving for a when this expression is zero we get

$$a = \frac{-\,\mathrm{Cov}[X,Y] \pm \sqrt{\mathrm{Cov}[X,Y]^2 - \sigma_X^2\sigma_Y^2}}{\sigma_X^2}.$$

Observing that there cannot be two distinct real solutions for a, the radicand must be non-positive $\mathrm{Cov}[X,Y]^2 \le \sigma_X^2\sigma_Y^2$. □

Two stochastic variables X and Y are said to be *uncorrelated* if $\mathrm{Cov}[X,Y] = 0$, which is equivalent to the correlation coefficient being equal to zero, and which is equivalent to $E[XY] = E[X]E[Y]$.

We now observe that if two stochastic variables are statistically independent, then they are uncorrelated and their covariance is zero. The converse is not always true.

Generalization to an arbitrary number of stochastic variables should now be obvious.

Example: Multi normal distribution. The n stochastic variables $X_1, X_2, \ldots, X_n$ are said to be jointly normally distributed with the probability density function given by

$$f(x_1, \ldots, x_n) = \frac{\sqrt{|A|}}{(2\pi)^{n/2}} \exp\left(-\frac{1}{2}\sum_{j,l}(x_j - \mu_j)a_{jl}(x_l - \mu_l)\right)$$

where $C^{-1} = A = \{a_{jl}\}$ is a symmetric positive definite matrix and $|A|$ denotes the determinant of A. The mean values are $E[X_j] = \mu_j$, the covariances are $\mathrm{Cov}[X_j, X_l] = E[(X_j - \mu_j)(X_l - \mu_l)] = c_{jl}$ where $A^{-1} = C = \{c_{jl}\}$ is the covariance matrix. The characteristic function is given by the n-dimensional Fourier transform

$$\phi(k_1, \ldots, k_n) = E[\exp(\mathrm{i}(k_1 X_1 + \ldots + k_n X_n))] = \exp\left(-\frac{1}{2}\sum_{j,l} k_j c_{jl} k_l + \mathrm{i}\sum_j k_j \mu_j\right).$$

It is useful to consider the limit when the variance $\sigma_j^2 = c_{jj}$ of one of the variables X_j goes to zero. In this case the the covariance matrix becomes singular and the probability density function becomes a generalized function. However, the characteristic function is still a well defined ordinary function, and it may therefore be more convenient to take the characteristic function as the definition of the multi normal distribution.

Suppose that the variable X_1 is uncorrelated with all the other variables X_j. Then the covariances $c_{j,1} = c_{1,j} = 0$ for all $j \neq 1$, and thus the n-dimensional characteristic function can be factored $\phi(k_1, k_2, \ldots, k_n) = \phi_1(k_1)\tilde{\phi}(k_2, \ldots, k_n)$. From the multi-dimensional inverse Fourier transform, it follows that the probability density function can be factored likewise. Hence for jointly normally distributed stochastic variables, statistical independence is equivalent to uncorrelatedness.

9.3 The Central Limit Theorem

Suppose that a stochastic variable Y is a superposition of n statistically independent variables X_j with mean values μ_j and variances σ_j^2, respectively,

$$Y = X_1 + X_2 + \ldots + X_n.$$

Due to the assumption of statistical independence we have for the joint probability density function

$$f(x_1, x_2, \ldots, x_n) = f_1(x_1) f_2(x_2) \ldots f_n(x_n).$$

The expected value of Y is

$$E[Y] = E\left[\sum_j X_j\right] = \sum_j E[X_j] = \sum_j \mu_j = \mu.$$

The variance of Y is

$$\mathrm{Var}[Y] = E[(Y - E[Y])^2] = E\left[\left(\sum_j (X_j - \mu_j)\right)^2\right]$$
$$= \sum_j E[(X_j - \mu_j)^2] + \sum_{j \neq l} E[(X_j - \mu_j)(X_l - \mu_l)] = \sum_j \sigma_j^2 = \sigma^2$$

where the sum over all $j \neq l$ vanishes because of statistical independence. We have defined μ as the sum of the means and σ^2 as the sum of the variances.

Now define the transformed variable

$$Z = \frac{Y - \mu}{\sigma} = \sum_{j=1}^{n} \frac{X_j - \mu_j}{\sigma}$$

such that $E[Z] = 0$ and $E[Z^2] = 1$. The characteristic function $\phi_{X_j}(k)$ is not known, however we can write down the first three terms of the power series expansion in k

$$\phi_{X_j}(k) = E[\mathrm{e}^{\mathrm{i}kX_j}] = 1 + \mathrm{i}kE[X_j] - \frac{k^2}{2}E[X_j^2] + \ldots = 1 + \mathrm{i}k\mu_j - \frac{k^2\sigma_j^2}{2} + \ldots.$$

Similarly, the characteristic function $\phi_Z(k)$ has a power series expansion in k

$$\phi_Z(k) = E[\mathrm{e}^{\mathrm{i}kZ}] = E\left[\exp\left(\mathrm{i}k\sum_{j=1}^{n}\frac{X_j - \mu_j}{\sigma}\right)\right] = \prod_{j=1}^{n} E\left[\exp\left(\mathrm{i}k\frac{X_j - \mu_j}{\sigma}\right)\right]$$
$$= \prod_{j=1}^{n}\left(1 - \frac{k^2\sigma_j^2}{2\sigma^2} + \ldots\right) = (1 - \frac{k^2}{2n})^n + R$$

where the transition from sum to product depends on statistical independence, and R is a remainder.

Now we let $n \to \infty$ and recall the limit

$$\left(1 - \frac{x}{n}\right)^n \to \mathrm{e}^{-x} \qquad \text{as } n \to \infty.$$

Provided the remainder term R vanishes as $n \to \infty$ it follows that

$$\phi_Z(k) \to \mathrm{e}^{-\frac{k^2}{2}}$$

and thus the asymptotic probability density of the transformed variable Z is

$$f(z) \to \frac{1}{\sqrt{2\pi}}\mathrm{e}^{-\frac{z^2}{2}},$$

which is the Gaussian distribution with mean 0 and variance 1.

If all the variables X_j are equally distributed it becomes particularly simple to demonstrate that $R = O(n^{-1/2})$. If the X_j are not equally distributed, sufficient conditions for the vanishing of R could depend on certain conditions on the variances σ_j^2 and higher moments of all X_j (see e.g. Papoulis 1991).

The central limit theorem states that when these conditions are met, the sum of a great number of statistically independent stochastic variables tends to a Gaussian stochastic variable with mean equal to the sum of the means and variance equal to the sum of the variances.

As far as the sea surface is concerned, if the surface elevation is a linear superposition of a great number of independent wave oscillations, then the surface elevation is expected to have a Gaussian distribution.

We can now understand an important reason why the Gaussian assumption should not be satisfied for sea surface waves: The spatiotemporal behavior is governed by a nonlinear evolution equation, therefore the individual wave oscillations will not be independent. For the MNLS equations described in earlier sections, there are two different reasons why the sea surface displacement should deviate from the Gaussian distribution: Firstly, the reconstruction of the sea surface involves the contributions from zeroth, second and higher harmonics, which are nonlinear wave oscillations that depend on the first harmonic linearly dispersive waves. Secondly, the spatiotemporal evolution equation for the first harmonic is nonlinear, and will introduce dependencies between the wave oscillations comprising the first harmonic.

Common practice in the stochastic modeling of weakly nonlinear sea surface waves is to assume that the first harmonic contribution to the wave field is Gaussian. This implies an assumption that the nonlinear Schrödinger equation itself does not introduce deviations from Gaussian statistics, but the nonlinear reconstruction formulas for the surface displacement do produce such deviations. This assumption has recently been checked experimentally and by Monte–Carlo simulations using the nonlinear evolution equations (Onorato et al. 2004; Socquet–Juglard et al. 2005). It is found that for unidirectional long-crested waves the nonlinear Schrödinger equation can produce significant deviation from Gaussian statistics, with an increased number of extreme waves. For more realistic short-crested waves, typical for the sea surface, the assumption that the first harmonic is Gaussian is surprisingly good, despite the nonlinearities of the nonlinear Schrödinger equation.

10 Theory for stochastic processes

We already mentioned four different ways to interpret a stochastic process $Z(t)$. Suppose t is time. At a fixed time t_1 the interpretation is a stochastic variable $Z(t_1)$. At two fixed times t_1 and t_2 the interpretation is two stochastic variables $Z(t_1)$ and $Z(t_2)$. At an arbitrary number of times t_j the interpretation is a collection of stochastic variables $Z(t_j)$. We seek a description of the joint distribution of these stochastic variables.

The *first order distribution* describes the behavior at one fixed time t_1

$$F(z_1; t_1) = P\{Z(t_1) \leq z_1\} \qquad \text{and} \qquad f(z_1; t_1) = \frac{\partial F(z_1; t_1)}{\partial z_1}.$$

The *second order distribution* describes the joint behavior at two fixed times t_1 and t_2

$$F(z_1, z_2; t_1, t_2) = P\{Z(t_1) \leq z_1 \text{ and } Z(t_2) \leq z_2\}$$

and

$$f(z_1, z_2; t_1, t_2) = \frac{\partial^2 F(z_1, z_2; t_1, t_2)}{\partial z_1 \partial z_2}.$$

Several compatibility relations follow, e.g. $F(z_1; t_1) = F(z_1, \infty; t_1, t_2)$, etc.

In principle we can proceed to derive the *n-th order distribution* for the joint behavior at n fixed times $F(z_1, z_2, \ldots, z_n; t_1, t_2, \ldots, t_n)$, however, the first and second order distributions will suffice in the following.

Care should be exercised not to be confused by our double use of the word "order": The order of a distribution of a stochastic process refers to the number of distinct temporal (or spatial) locations employed for joint statistics. The order of nonlinearity refers to the power of the steepness that measures the importance of wave-wave interactions.

The expected value of the stochastic process is

$$\mu(t) = E[Z(t)] = \int_{-\infty}^{\infty} z f(z; t) \, \mathrm{d}z.$$

We define the *autocorrelation function* as

$$R(t_1, t_2) = E[Z(t_1)Z(t_2)] = \int_{-\infty}^{\infty} \int_{-\infty}^{\infty} z_1 z_2 f(z_1, z_2; t_1, t_2) \, \mathrm{d}z_1 \, \mathrm{d}z_2.$$

The *mean power* of the process is defined as the second moment $R(t, t) = E[|Z(t)|^2]$. The *autocovariance function* is defined as

$$C(t_1, t_2) = E[(Z(t_1) - \mu(t_1))(Z(t_2) - \mu(t_2))] = R(t_1, t_2) - \mu(t_1)\mu(t_2).$$

A process is said to be *steady state* or *stationary* if the statistical properties are independent of translation of the origin, i.e. $Z(t)$ and $Z(t+\tau)$ have the same distribution. For the first order distribution we need $f(z_1; t_1) = f(z_1)$. For the second order distribution we need $f(z_1, z_2; t_1, t_2) = f(z_1, z_2; \tau)$ where $\tau = t_2 - t_1$. Similarly, the distributions at any higher order should only depend on the time intervals and not the absolute times.

A process is said to be *weakly stationary* if the expected value is constant with respect to time $E[Z(t)] = \mu$ and the autocorrelation function is independent of translation of the origin $R(t_1, t_2) = E[Z(t_1)Z(t_2)] = R(\tau)$ where $\tau = t_2 - t_1$.

A process $Z(t)$ is said to be *ergodic* (with respect to time-averaging) for the computation of some function $g(Z(t))$ if ensemble-averaging gives the same result as time-averaging, e.g.

$$E[g(Z(t))] \equiv \int_{-\infty}^{\infty} g(z) f(z; t) \, \mathrm{d}z = \lim_{T \to \infty} \frac{1}{2T} \int_{-T}^{T} g(z(t)) \, \mathrm{d}t.$$

It is obvious that ergodicity is only meaningful provided the process has some kind of stationarity. Ergodicity with respect to second order statistics, such as the mean and the autocorrelation, is meaningful for a weakly stationary process.

Often we suppose without further justification that ocean waves are both weakly stationary and ergodic. However, the sea state is known to change in time. It may still be a good approximation to assume that within a limited time, say a few hours, the sea state is nearly weakly stationary and ergodic.

For a complex stochastic process $Z(t)$, the autocorrelation function is defined as

$$R(t_1, t_2) = E[Z(t_1)Z^*(t_2)]$$

where the asterisk denotes the complex conjugate.

In the following we describe some properties of the autocorrelation function for weakly stationary processes:

- For a complex process $R(-\tau) = R^*(\tau)$, and for a real process $R(-\tau) = R(\tau)$

$$R(-\tau) = E[Z(t)Z^*(t-\tau)] = E[Z(t+\tau)Z^*(t)] = R^*(\tau).$$

- $R(0)$ is real and non-negative

$$R(0) = E[Z(t)Z^*(t)] = E[|Z(t)|^2] \geq 0.$$

- For a complex process $R(0) \geq |\text{Re}R(\tau)|$, and for a real process $R(0) \geq |R(\tau)|$.

Proof. Let a be a real number and look at

$$\begin{aligned} &E[|aZ(t) + Z(t+\tau)|^2] \\ &\quad = a^2 E[|Z(t)|^2] + aE[Z(t)Z^*(t+\tau)] + aE[Z^*(t)Z(t+\tau)] + E[|Z(t+\tau)|^2] \\ &\qquad = a^2 R(0) + 2a\text{Re}R(\tau) + R(0) \geq 0. \end{aligned}$$

Solving for a when this expression is zero we get

$$a = \frac{-\text{Re}R(\tau) \pm \sqrt{(\text{Re}R(\tau))^2 - R^2(0)}}{R(0)}.$$

Since there cannot be two distinct real solutions for a we have $R(0) \geq |\text{Re}R(\tau)|$. □

- $R(0)$ is the second moment of $Z(t)$. If $E[Z(t)] = 0$, then $R(0)$ is the variance of $Z(t)$.
- If $X(t)$ and $Y(t)$ are statistically independent processes with zero mean, and $Z(t) = X(t) + Y(t)$, then

$$R_{ZZ}(\tau) = E[(X(t) + Y(t))(X^*(t+\tau) + Y^*(t+\tau))] = R_{XX}(\tau) + R_{YY}(\tau).$$

The *cross-correlation* between two complex processes $X(t)$ and $Y(t)$ is defined as

$$R_{XY}(t_1, t_2) = E[X(t_1)Y^*(t_2)].$$

If the two processes are independent, then the cross-correlation is zero.

Example: Simple harmonic wave with random phase. Consider a simple harmonic wave with fixed amplitude a and arbitrary phase

$$\eta(x,t) = a\cos(kx - \omega t + \theta)$$

where θ is uniformly distributed on the interval $[0, 2\pi)$.

The expected value of the surface displacement is

$$\mu(x,t) = E[\eta(x,t)] = \int_0^{2\pi} a\cos(kx - \omega t + \theta)\frac{1}{2\pi}\,\mathrm{d}\theta = 0.$$

The autocorrelation function is

$$\begin{aligned} R(x,t,x+\rho,t+\tau) &= E[\eta(x,t)\eta(x+\rho,t+\tau)] \\ &= \int_0^{2\pi} a^2\cos(kx-\omega t+\theta)\cos(k(x+\rho)-\omega(t+\tau)+\theta)\frac{1}{2\pi}\,\mathrm{d}\theta \\ &= \frac{a^2}{2}\cos(k\rho - \omega\tau). \end{aligned}$$

The mean power of the process is $E[\eta^2(x,t)] = R(x,t,x,t) = \frac{a^2}{2}$. In fact we could have defined the amplitude of the process as $a = \sqrt{2E[\eta^2]}$.

Let us proceed to derive the first-order distribution of the process. We have

$$F(z;x,t) = P\{\eta(x,t) \le z\} = \begin{cases} 0 & z < -a \\ 1 - \frac{1}{\pi}\arccos\frac{z}{a} & |z| \le a \\ 1 & z > a \end{cases}$$

$$f(z;x,t) = \frac{\partial F(z;x,t)}{\partial z} = \begin{cases} \frac{1}{\pi a\sqrt{1-(\frac{z}{a})^2}} & |z| \le a \\ 0 & |z| > a \end{cases}$$

Notice that the first order distribution is independent of x and t. In fact, this can be seen even without deriving the exact distribution, upon making the substitution $\psi = kx - \omega t + \theta$ and noting that ψ is a stochastic variable uniformly distributed on an interval of length 2π. With this observation it becomes clear that the stochastic distributions at any order are independent of translation of the spatiotemporal origin, and hence the process is steady state or stationary.

The process is ergodic for computation of the mean and the autocorrelation by time averaging when $\omega \neq 0$, and by spatial averaging when $k \neq 0$, e.g.

$$\begin{aligned} &\lim_{T\to\infty}\frac{1}{2T}\int_{-T}^{T} a\cos(kx-\omega t+\theta)\,\mathrm{d}t \\ &\qquad = \lim_{T\to\infty}\frac{a}{2\omega T}(-\sin(kx-\omega T+\theta)+\sin(kx+\omega T+\theta)) = 0, \end{aligned}$$

$$\lim_{T\to\infty} \frac{1}{2T} \int_{-T}^{T} a^2 \cos(kx - \omega t + \epsilon) \cos(kx - \omega(t+\tau) + \epsilon)\, dt$$
$$= \lim_{T\to\infty} \frac{a^2}{2T} \int_{-T}^{T} \cos^2(kx - \omega t + \epsilon) \cos(\omega\tau) + \frac{1}{2} \sin 2(kx - \omega t + \epsilon) \sin(\omega\tau)\, dt$$
$$= \frac{a^2}{2} \cos \omega\tau.$$

However, if $k = 0$ or $\omega = 0$ the process is not ergodic for computation of these quantities by spatial or time averaging, respectively.

Example: Third order Stokes wave with random phase. The third-order normalized Stokes wave with arbitrary phase can be written as

$$\eta(\psi) = \varepsilon \cos \psi + \gamma_2 \varepsilon^2 \cos 2\psi + \gamma_3 \varepsilon^3 \cos 3\psi \tag{10.1}$$

where η is now the normalized surface displacement, the phase is $\psi = x - t + \theta$, and the coefficients γ_2 and γ_3 are found by reference to section 4. The stochastic variable θ is uniformly distributed on the interval $[0, 2\pi)$, and thus ψ is also uniformly distributed on some interval of length 2π. The steepness ε of the first-harmonic term should not be confused with the steepness $\epsilon = k_c \bar{a} = \sqrt{2E[\eta^2]}$ of the wave field.

The expected value of the normalized surface displacement is

$$E[\eta(\psi)] = \int_0^{2\pi} \eta(\psi) \frac{1}{2\pi}\, d\psi = 0.$$

The autocorrelation function is

$$R(\rho, \tau) = E[\eta(\psi)\eta(\psi + \rho - \tau)] = \int_0^{2\pi} \eta(\psi)\eta(\psi + \rho - \tau) \frac{1}{2\pi}\, d\psi$$
$$= \frac{\varepsilon^2}{2} \cos(\rho - \tau) + \frac{\gamma_2^2 \varepsilon^4}{2} \cos 2(\rho - \tau) + \frac{\gamma_3^2 \varepsilon^6}{2} \cos 3(\rho - \tau).$$

The mean power of the process is

$$\mathrm{Var}[\eta] = E[\eta^2] = R(0,0) = \frac{\varepsilon^2}{2} + \frac{\gamma_2^2 \varepsilon^4}{2} + \frac{\gamma_3^2 \varepsilon^6}{2}.$$

It follows that the relationship between the steepness of the first harmonic ε and the overall steepness of the wave field ϵ is

$$\epsilon^2 = \varepsilon^2 + \gamma_2^2 \varepsilon^4 + \gamma_3^2 \varepsilon^6 \qquad \text{and} \qquad \varepsilon^2 = \epsilon^2 - \gamma_2^2 \epsilon^4 + (2\gamma_2^4 - \gamma_3^2)\epsilon^6. \tag{10.2}$$

In any case, for the small Draupner overall steepness $\epsilon = 0.12$ and depths not smaller than the Draupner depth we find $\epsilon = \varepsilon$ within two digits accuracy.

We now want to derive the distribution of surface elevation, i.e. the first-order distribution of the stochastic process, accurate to third order in nonlinearity. This can be achieved by explicitly inverting the expression for surface elevation $\psi = \psi(\eta)$. We limit to deep water ($\gamma_2 = 1/2$, $\gamma_3 = 3/8$). The trick is to rewrite the left-hand side as

$$\eta = \varepsilon z + \frac{1}{2}\varepsilon^2 + \frac{3}{8}\varepsilon^3 s \tag{10.3}$$

where s is the sign of η and $|z| \leq 1$. Employing the perturbation expansion

$$\psi = \psi_0 + \varepsilon\psi_1 + \varepsilon^2\psi_2 + \ldots \tag{10.4}$$

we find

$$\psi_0 = \arccos z, \tag{10.5}$$

$$\psi_1 = -\sqrt{1 - z^2} \tag{10.6}$$

and

$$\psi_2 = \frac{3(z - s)}{8\sqrt{1 - z^2}}. \tag{10.7}$$

The cumulative probability distribution is therefore

$$F(\eta) = 1 - \frac{1}{\pi}\left(\arccos z - \varepsilon\sqrt{1 - z^2} + \frac{3\varepsilon^2(z - s)}{8\sqrt{1 - z^2}}\right) \tag{10.8}$$

and the probability density function is

$$f(\eta) = \begin{cases} \dfrac{1}{\varepsilon\pi\sqrt{1 - z^2}}\left(1 - \varepsilon z - \dfrac{3\varepsilon^2(1 - sz)}{8(1 - z^2)}\right) & \text{for } |z| < 1 \\ 0 & \text{for } |z| > 1 \end{cases} \tag{10.9}$$

where

$$z = \frac{\eta}{\varepsilon} - \frac{1}{2}\varepsilon - \frac{3}{8}\varepsilon^2 s. \tag{10.10}$$

The probability densities at various nonlinear orders are shown in figure 14 for an unrealistically high steepness $\varepsilon = 0.3$.

Example: Simple harmonic wave with random amplitude and phase. Consider a simple harmonic wave with arbitrary amplitude and phase, more specifically let

$$\eta(x, t) = a\cos(kx - \omega t) + b\sin(kx - \omega t)$$

where a and b are statistically independent Gaussian stochastic variables with common mean 0 and common variance σ^2.

The expected value of the surface displacement is

$$\mu(x, t) = E[\eta(x, t)] = E[a]\cos(kx - \omega t) + E[b]\sin(kx - \omega t) = 0.$$

The autocorrelation function is

$$R(x, t, x + \rho, t + \tau) = E[\eta(x, t)\eta(x + \rho, t + \tau)] = \sigma^2\cos(k\rho - \omega\tau).$$

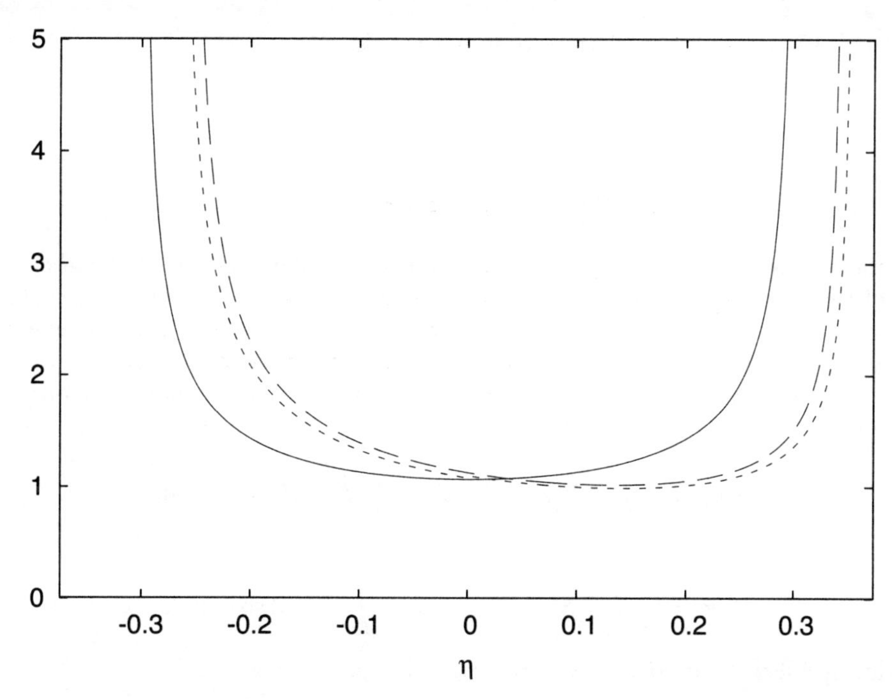

Figure 14. Probability density functions for Stokes waves of first-harmonic steepness $\varepsilon = 0.3$: —, linear; - -, second nonlinear order; $\cdots$, third nonlinear order.

The mean power is $E[\eta^2(x,t)] = R(x,t,x,t) = \sigma^2$.
The characteristic amplitude of the process is $\bar{a} = \sqrt{2E[\eta^2]} = \sqrt{2}\sigma$.
Exercise: Proceed to derive the first-order distribution of the process

$$F(z;x,t) = P\{\eta(x,t) \leq z\} \qquad \text{and} \qquad f(z;x,t) = \frac{\partial F(z;x,t)}{\partial z}.$$

11 The spectrum

11.1 Definition of frequency spectrum

The *frequency spectrum* $S(\omega)$ of a weakly stationary process is defined as the Fourier transform of the autocorrelation function $R(\tau)$. Whereas the Fourier transform is in principle undetermined by a multiplicative constant, the spectrum becomes uniquely defined by the constraint that the integral of the spectrum over the domain of the frequency axis that is used, should be equal to the variance of the process. Since our target process (the surface elevation) is real, the Fourier transform is complex conjugate symmetric about the origin, and it is enough to use only non-negative frequencies to represent the spectrum. The desired Fourier transform pair is then

$$S(\omega) = \frac{1}{\pi} \int_{-\infty}^{\infty} R(\tau) e^{i\omega\tau} \, d\tau, \tag{11.1}$$

$$R(\tau) = \frac{1}{2} \int_{-\infty}^{\infty} S(\omega) e^{-i\omega\tau} \, d\omega. \tag{11.2}$$

Some authors define the spectrum as the squared absolute value of the Fourier transform of the process. In this case the Fourier transform pair (11.1)–(11.2) is called the Wiener–Khintchine theorem. On the other hand, we show in section 11.3 that the squared absolute value of the Fourier transform is an estimator for the spectrum.

For a real process we recall that the autocorrelation function is real and even $R(-\tau) = R(\tau)$ and thus we may write

$$S(\omega) = \frac{1}{\pi} \int_{-\infty}^{\infty} R(\tau) \cos \omega\tau \, d\tau = \frac{2}{\pi} \int_{0}^{\infty} R(\tau) \cos \omega\tau \, d\tau$$

and

$$R(\tau) = \frac{1}{2} \int_{-\infty}^{\infty} S(\omega) \cos \omega\tau \, d\omega = \int_{0}^{\infty} S(\omega) \cos \omega\tau \, d\omega$$

and thus it follows that $S(\omega)$ is real and even. In particular we have

$$\int_{0}^{\infty} S(\omega) \, d\omega = R(0)$$

which shows that the normalization criterion is satisfied.

For application to a complex process we cannot expect any symmetry for the spectrum and would need to include both positive and negative frequencies, the appropriate transform pair being

$$S(\omega) = \frac{1}{2\pi} \int_{-\infty}^{\infty} R(\tau) e^{i\omega\tau} \, d\tau, \tag{11.3}$$

$$R(\tau) = \int_{-\infty}^{\infty} S(\omega) e^{-i\omega\tau} \, d\omega. \tag{11.4}$$

For a complex process we recall that $R(-\tau) = R^*(\tau)$ and thus

$$S(\omega) = \frac{1}{2\pi} \int_{-\infty}^{\infty} R(\tau) e^{i\omega\tau} \, d\tau = \frac{1}{2\pi} \int_{0}^{\infty} \left(R^*(\tau) e^{-i\omega\tau} + R(\tau) e^{i\omega\tau} \right) \, d\tau$$

which shows that $S(\omega)$ is real.

It can further be shown that for a real or complex weakly stationary process, the spectrum is non-negative $S(\omega) \geq 0$: Take a complex process $Z(t)$, a complex function $a(t)$ and a real number T, we trivially have

$$E\left[\left| \int_{-T}^{T} a(t) Z(t) \, dt \right|^2 \right] \geq 0.$$

Letting $a(t) = \frac{1}{\sqrt{4\pi T}} \exp(-\mathrm{i}\omega t)$, the left hand side can be written

$$\int_{-T}^{T} \int_{-T}^{T} a(t_1) a^*(t_2) E[Z(t_1) Z^*(t_2)] \, \mathrm{d}t_1 \mathrm{d}t_2 = \frac{1}{4\pi T} \int_{-T}^{T} \int_{-T}^{T} \mathrm{e}^{\mathrm{i}\omega(t_2 - t_1)} R(t_2 - t_1) \, \mathrm{d}t_1 \mathrm{d}t_2$$

where use has been made of weak stationarity. Introducing new variables $\tau = t_2 - t_1$ and $\xi = t_1$ this becomes

$$\frac{1}{4\pi T} \int_{-2T}^{2T} \int_{-T}^{T} \mathrm{e}^{\mathrm{i}\omega\tau} R(\tau) \, \mathrm{d}\xi \mathrm{d}\tau = \frac{1}{2\pi} \int_{-2T}^{2T} \mathrm{e}^{\mathrm{i}\omega\tau} R(\tau) \, \mathrm{d}\tau \stackrel{T \to \infty}{\longrightarrow} S(\omega) \geq 0.$$

Example: Periodic oscillation with random amplitude and phase. Take the real periodic process with period T

$$\eta(t) = \sum_{j=1}^{\infty} a_j \cos \omega_j t + b_j \sin \omega_j t \tag{11.5}$$

where $\omega_j = 2\pi j/T$ where a_j and b_j are statistically independent Gaussian stochastic variables with mean 0 and variance σ_j^2.

The mean is zero $E[\eta(t)] = 0$, the autocorrelation function is

$$R(\tau) = \sum_j \sigma_j^2 \cos \omega_j \tau,$$

and using the discrete Fourier transform (see the Appendix) the spectrum is

$$S(\omega_j) = \frac{2}{T} \int_0^T R(\tau) \mathrm{e}^{\mathrm{i}\omega_j \tau} \, \mathrm{d}\tau = \sigma_j^2. \tag{11.6}$$

The normalization criterion is seen to be satisfied by observing that

$$\sum_j S(\omega_j) = \sum_j \sigma_j^2 = R(0) = \mathrm{Var}[\eta(t)].$$

11.2 Definition of wave spectrum

The *wave spectrum* $S(\mathbf{k}, \omega)$ of a weakly stationary wave process $\eta(\mathbf{x}, t)$ is defined as the Fourier transform of the autocorrelation function $R(\boldsymbol{\rho}, \tau)$. Again the spectrum becomes uniquely defined by the constraint that the integral of the spectrum over the three-dimensional domain of the spectral axes that is used, should be equal to the variance of the process. For a real process like the surface elevation, the Fourier transform has one complex conjugate symmetry, and we therefore limit to non-negative frequencies only, $\omega \geq 0$, while the wave vector $\mathbf{k}$ is unconstrained. The desired Fourier transform pair is then

$$S(\mathbf{k}, \omega) = \frac{1}{4\pi^3} \int_{-\infty}^{\infty} \mathrm{d}\boldsymbol{\rho} \int_{-\infty}^{\infty} \mathrm{d}\tau \, R(\boldsymbol{\rho}, \tau) \mathrm{e}^{-\mathrm{i}(\mathbf{k} \cdot \boldsymbol{\rho} - \omega\tau)} \tag{11.7}$$

$$R(\boldsymbol{\rho},\tau) = \frac{1}{2}\int_{-\infty}^{\infty} \mathrm{d}\mathbf{k} \int_{-\infty}^{\infty} \mathrm{d}\omega\, S(\mathbf{k},\omega) \mathrm{e}^{\mathrm{i}(\mathbf{k}\cdot\boldsymbol{\rho}-\omega\tau)}. \tag{11.8}$$

Recall that for a real process $R(-\boldsymbol{\rho},-\tau) = R(\boldsymbol{\rho},\tau)$ and thus it follows that the spectrum is real and has the symmetry $S(-\mathbf{k},-\omega) = S(\mathbf{k},\omega)$. Then it also follows that the normalization criterion is satisfied

$$\frac{1}{2}\int_{-\infty}^{\infty} \mathrm{d}\mathbf{k} \int_{-\infty}^{\infty} \mathrm{d}\omega\, S(\mathbf{k},\omega) = R(\mathbf{0},0).$$

The *wave vector spectrum* $S(\mathbf{k})$ can now be defined as the projection of the wave spectrum into the wave vector plane

$$S(\mathbf{k}) = \int_{0}^{\infty} S(\mathbf{k},\omega)\,\mathrm{d}\omega = \frac{1}{2}\int_{-\infty}^{\infty} S(\mathbf{k},\omega)\,\mathrm{d}\omega = \frac{1}{4\pi^2}\int_{-\infty}^{\infty} R(\boldsymbol{\rho},0)\mathrm{e}^{-\mathrm{i}\mathbf{k}\cdot\boldsymbol{\rho}}\,\mathrm{d}\boldsymbol{\rho}.$$

The frequency spectrum $S(\omega)$ is recovered by projecting into the frequency axis

$$S(\omega) = \int_{-\infty}^{\infty} S(\mathbf{k},\omega)\,\mathrm{d}\mathbf{k} = \frac{1}{\pi}\int_{-\infty}^{\infty} R(\mathbf{0},\omega)\mathrm{e}^{\mathrm{i}\omega\tau}\,\mathrm{d}\tau.$$

The *wavenumber spectrum* $S(k)$ is achieved through the transformation

$$\left\{ \begin{array}{l} k_x = k\cos\theta \\ k_y = k\sin\theta \end{array} \right\} \qquad \text{with Jacobian} \qquad \frac{\partial(k_x,k_y)}{\partial(k,\theta)} = k.$$

The wavenumber $k = \sqrt{k_x^2 + k_y^2} \geq 0$ is by definition non-negative. The wavenumber spectrum is thus

$$S(k) = \int_{0}^{2\pi} \mathrm{d}\theta \int_{0}^{\infty} \mathrm{d}\omega\, S(\mathbf{k},\omega)k = \frac{1}{4\pi^2}\int_{0}^{\infty} \mathrm{d}\theta \int_{-\infty}^{\infty} \mathrm{d}\boldsymbol{\rho}\, R(\boldsymbol{\rho},0)k\mathrm{e}^{-\mathrm{i}\mathbf{k}\cdot\boldsymbol{\rho}}.$$

The *directional spectrum* $S(\theta)$ similarly becomes

$$S(\theta) = \int_{0}^{\infty} \mathrm{d}k \int_{0}^{\infty} \mathrm{d}\omega\, S(\mathbf{k},\omega)k.$$

Example: Linear waves with random phase. Look at the real process

$$\eta(x,t) = \sum_j a_j \cos(k_j x - \omega_j t + \theta_j)$$

where a_j are fixed scalars and θ_j are statistically independent stochastic variables uniformly distributed between 0 and 2π. The mean is zero $E[\eta(x,t)] = 0$, and the autocorrelation function is

$$R(\rho,\tau) = \sum_j \frac{1}{2}a_j^2 \cos(k_j\rho - \omega_j\tau).$$

The variance or mean power of the process is

$$R(0,0) = \sum_j \frac{1}{2} a_j^2 = \frac{1}{2} \bar{a}^2$$

where $\bar{a} = \sqrt{2R(0,0)}$ is the characteristic amplitude. The spectrum is

$$S(k,\omega) = \sum_j \frac{a_j^2}{2} \left(\delta(k+k_j)\delta(\omega+\omega_j) + \delta(k-k_j)\delta(\omega-\omega_j) \right)$$

where $\delta(\cdot)$ is the Dirac delta function.

Example: Linear waves with random amplitude and phase. Let us consider the real process

$$\eta(x,t) = \sum_j a_j \cos(k_j x - \omega_j t) + b_j \sin(k_j x + \omega_j t)$$

where a_j and b_j are statistically independent Gaussian stochastic variables with mean 0 and variance σ_j^2.

The mean is zero $E[\eta(x,t)] = 0$, and the autocorrelation function is

$$R(\rho,\tau) = \sum_j \sigma_j^2 \cos(k_j \rho - \omega_j \tau).$$

Notice that

$$\mathrm{Var}[\eta(x,t)] = R(0,0) = \sum_j \sigma_j^2 = \frac{1}{2} \bar{a}^2.$$

The spectrum is readily found

$$S(k,\omega) = \sum_j \sigma_j^2 \left(\delta(k+k_j)\delta(\omega+\omega_j) + \delta(k-k_j)\delta(\omega-\omega_j) \right).$$

11.3 An estimator for the spectrum

Using the real periodic process (11.5), let us now compute the Fourier transform of $\eta(t)$ using the Fourier transform (A.8)

$$\hat{\eta}_j = \frac{1}{T} \int_0^T \eta(t) \mathrm{e}^{\mathrm{i}\omega_j t} \, \mathrm{d}t = \frac{1}{2}(a_j + \mathrm{i} b_j).$$

Let us construct the quantity $\tilde{S}_j$

$$\tilde{S}_j = 2|\hat{\eta}_j|^2 = \frac{1}{2}(a_j^2 + b_j^2).$$

Can $\tilde{S}_j$ be used as an estimator for $S(\omega_j)$ found in equation (11.6)? To answer this question we compute the expected value

$$E[\tilde{S}_j] = \frac{1}{2} E[a_j^2 + b_j^2] = \sigma_j^2$$

which shows that it is an unbiased estimator. Then let us compute the variance

$$\mathrm{Var}[\tilde{S}_j] = E[(\tilde{S}_j - \sigma_j^2)^2] = E[\tilde{S}_j^2] - \sigma_j^4$$

where

$$E[\tilde{S}_j^2] = \frac{1}{4}E[a_j^4 + 2a_j^2 b_j^2 + b_j^4] = \frac{1}{2}E[a_j^4] + \frac{\sigma_j^4}{2}.$$

Invoking the Gaussian assumption we get $E[a_j^4] = 3\sigma_j^4$, and finally we find that

$$\mathrm{Var}[\tilde{S}_j] = \sigma_j^4,$$

or in other words, $\tilde{S}_j$ has standard deviation equal to its expected value! This implies that if $\tilde{S}_j$ is an estimator for the spectrum, but we should expect a messy looking result, just like what we actually found in figures 3 and 4.

11.4 The equilibrium spectrum

Phillips (1958) first argued that the high frequency tail of the spectrum could be expected to obey the power law ω^{-5}. Later various observations and theories have suggested that wind waves may be better characterized by the power law ω^{-4}. In figure 4 these two power laws are superposed the estimated unsmoothed spectrum.

Dysthe et al. (2003) showed that when irregular waves are evolved with the MNLS equation (they used the equations in section 4.5), the spectrum relaxes toward a quasi-stationary state on a time scale $(\epsilon^2\omega_0)^{-1}$, which is the time scale of Benjamin–Feir or modulational instability. Simulations of shortcrested irregular waves show that a power law ω^{-4} is established on the high-frequency side of the angularly integrated spectrum within the range of the first harmonic on this time scale. Moreover, the MNLS equation appears to be the most simplified equation that reproduces the ω^{-4} power law. Figures 15 and 16 show the development of an initially Gaussian spectrum F_B for the first-harmonic amplitude B

$$F_B(\mathbf{K}) = \frac{\epsilon^2}{2\pi\sigma^2}\exp\left(-\frac{K_x^2 + K_y^2}{2\sigma^2}\right),$$

where $\mathbf{K} = (K_x, K_y)$ is the modulation wave vector and σ is the spectral width, subject to the MNLS equation in section 4.5.

Onorato et al. (2002) showed that using the full Euler equations, the ω^{-4} power law is established over a much wider frequency range than just the first harmonic, and the establishment of the power law for the higher spectral range happens after a longer time than that of modulational instability.

It is worthwhile to stress that the quasi-stationary spectrum is established within the range of the first-harmonic after a time that is shorter than the domain of validity of the MLNS equation. This has the important implication that an idealized wave simulation that is carried beyond the validity range of the MNLS equation, and has not yet reached a quasi-stationary spectrum, probably has little to do with the real sea surface.

Alber (1978) and Crawford et al. (1980) showed that modulational instability vanishes when the background spectrum becomes wider than a threshold. Their results imply that

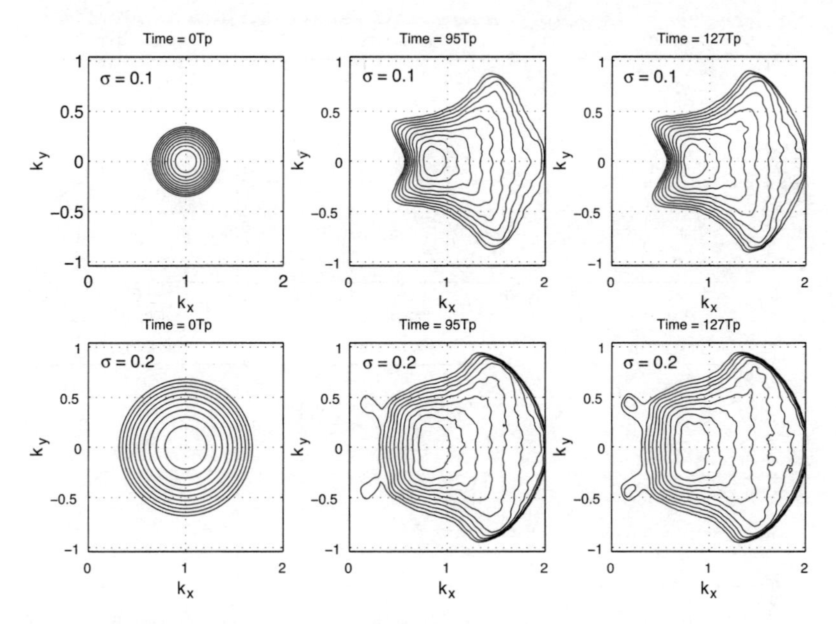

Figure 15. Spectral evolution with the MNLS equation. The contour interval 2.5 dB down to −25 dB. The steepness ϵ is equal to 0.1, and the spectral width is $\sigma = 0.1$ in the upper row and $\sigma = 0.2$ in the lower row. From Dysthe et al. (2003). Reproduced with permission from J. Fluid Mech.

no modulational instability should remain when the quasi-stationary state in figures 15 and 16 has been achieved. This has another important implication that idealized modulational instability should not be expected to act uninterrupted over longer times and distances than the validity range of the MNLS equation in a realistic sea.

12 Probability distributions of surface waves

12.1 Linear waves

Let the surface displacement be a linear superposition of simple-harmonic waves

$$\eta(\mathbf{x}, t) = \sum_j a_j \cos(\mathbf{k}_j \cdot \mathbf{r} - \omega_j t) + b_j \sin(\mathbf{k}_j \cdot \mathbf{r} - \omega_j t)$$

$$= \sum_j c_j \cos(\mathbf{k}_j \cdot \mathbf{r} - \omega_j t + \theta_j)$$

where

$$a_j = c_j \cos\theta_j, \qquad b_j = c_j \sin\theta_j$$

and $c_j \geq 0$.

Our standard assumptions have been that a_j and b_j are statistically independent Gaussian variables with mean 0 and standard deviation σ_j^2, while the frequencies depend

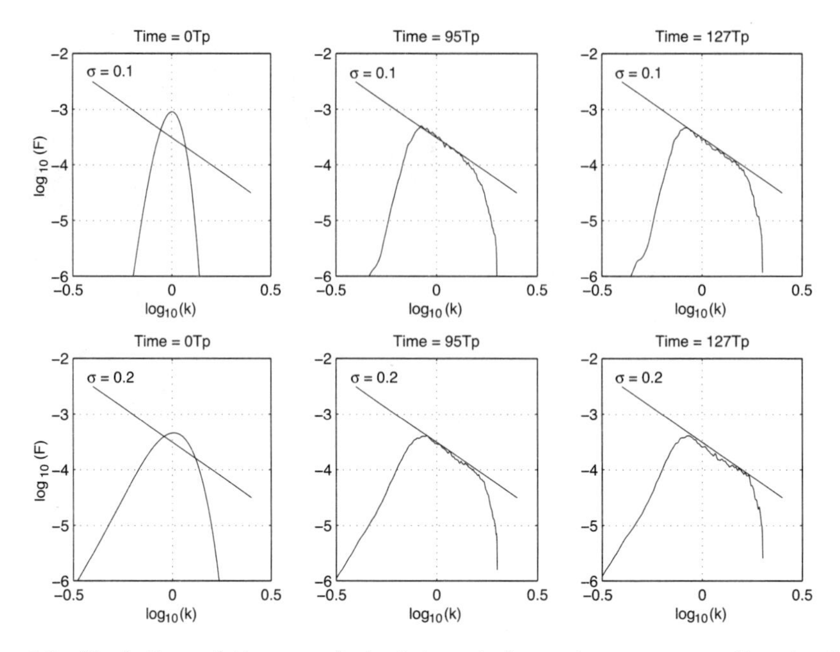

Figure 16. Evolution of the angularly integrated spectra corresponding to those in figure 15. The lines have slopes -2.5 corresponding to the power law ω^{-4}. From Dysthe et al. (2003). Reproduced with permission from J. Fluid Mech.

on the wave vectors through the linear dispersion relation $\omega_j = \omega(\mathbf{k}_j)$. The variance of the process is

$$\sigma^2 = \sum_j \sigma_j^2.$$

We now derive the probability distributions for the amplitudes c_j and phases θ_j.

The cumulative distribution function for the amplitude c_j is

$$F_{c_j}(z) = P\{\sqrt{a_j^2 + b_j^2} \leq z\} = \int_0^z \frac{r}{\sigma_j^2} \mathrm{e}^{-\frac{r^2}{2\sigma_j^2}}\, \mathrm{d}r$$

and the probability density is

$$f_{c_j}(z) = \begin{cases} \dfrac{z}{\sigma_j^2} \mathrm{e}^{-\frac{z^2}{2\sigma_j^2}} & z \geq 0 \\ 0 & z < 0 \end{cases}$$

which is the Rayleigh distribution.

The cumulative distribution for the phase θ_j

$$F_{\theta_j}(\theta) = P\{\theta_j \leq \theta\} = \int_0^\theta \frac{1}{2\pi}\, \mathrm{d}\psi$$

and the probability density is

$$f_{\theta_j}(\theta) = \begin{cases} \dfrac{1}{2\pi} & 0 \le \theta \le 2\pi \\ 0 & \text{otherwise} \end{cases}$$

which is the uniform distribution.

The energy density of the spectral component $(\mathbf{k}_j, \omega_j)$ is defined by

$$\epsilon_j = \frac{1}{2}(a_j^2 + b_j^2)$$

and has the probability density

$$f_{\epsilon_j}(z) = \begin{cases} \dfrac{1}{\sigma_j^2} e^{-\frac{z}{\sigma_j^2}} & z \ge 0 \\ 0 & z < 0 \end{cases}$$

which is the exponential distribution.

12.2 Linear narrowbanded waves

In section 4.2 we introduced harmonic perturbation expansions for weakly nonlinear narrowbanded waves. This can now be related to the superposition of simple harmonic waves in the previous section 12.1

$$\begin{aligned}\eta(\mathbf{r}, t) &= \sum_j a_j \cos(\mathbf{k}_j \cdot \mathbf{r} - \omega_j t) + b_j \sin(\mathbf{k}_j \cdot \mathbf{r} - \omega_j t) = \sum_j c_j \cos(\mathbf{k}_j \cdot \mathbf{r} - \omega_j t + \theta_j) \\ &= \frac{1}{2}\left(B e^{\mathrm{i}(k_c x - \omega_c t)} + \text{c.c.}\right) = |B| \cos(k_c x - \omega_c t + \arg B).\end{aligned}$$

According to the narrowband assumption, the complex amplitude B has slow dependence on $\mathbf{r}$ and t. Necessarily, the narrowband assumption implies that σ_j^2 rapidly decays to zero for values of $\mathbf{k}_j$ not close to $\mathbf{k}_c = (k_c, 0)$. The magnitude $|B|$ is called the *linear envelope* of the process.

In the limit of extremely small bandwidth, we may consider that only one index j contributes to the sum, and thus the limiting distribution for $|B|$ is the Rayleigh distribution

$$f_{|B|}(z) = \frac{z}{\sigma^2} e^{-\frac{z^2}{2\sigma^2}} \quad \text{for} \quad z \ge 0. \tag{12.1}$$

In this limiting case, the probability distribution of crest heights is identical to the probability distribution of the upper envelope.

In the limit that the bandwidth goes to zero, the wave height is twice the crest height, $H = 2|B|$, and is also Rayleigh distributed

$$f_H(z) = \frac{z}{4\sigma^2} e^{-\frac{z^2}{8\sigma^2}} \quad \text{for} \quad z \ge 0. \tag{12.2}$$

Let us consider the distribution of the $1/N$ highest waves. The probability density for wave height (12.2) is shown in figure 17. The threshold height H_* that divides the $1/N$ highest waves from the smaller waves is the solution of

$$\int_{H_*}^{\infty} \frac{z}{4\sigma^2} \mathrm{e}^{-\frac{z^2}{8\sigma^2}} \, \mathrm{d}z = \frac{1}{N}$$

which is $H_* = \sqrt{8 \ln N}\sigma$. The probability distribution for the $1/N$ highest waves is

$$f_{H \geq H_*}(z) = N \frac{z}{4\sigma^2} \mathrm{e}^{-\frac{z^2}{8\sigma^2}} \quad \text{for} \quad z \geq H_*$$

and the mean height of the $1/N$ highest waves is

$$H_{1/N} = N \int_{H_*}^{\infty} \frac{z^2}{4\sigma^2} \mathrm{e}^{-\frac{z^2}{8\sigma^2}} \, \mathrm{d}z = \left[\sqrt{8 \ln N} + \sqrt{2\pi} N \operatorname{erfc} \sqrt{\ln N} \right] \sigma$$

where $\operatorname{erfc} = \frac{2}{\sqrt{\pi}} \int_z^{\infty} \mathrm{e}^{-t^2} \, dt$ is the complementary error function. If we set $N = 3$ then we get $H_{1/3} = 4.0043\sigma$ which should be compared with the definition $H_s = 4\sigma$.

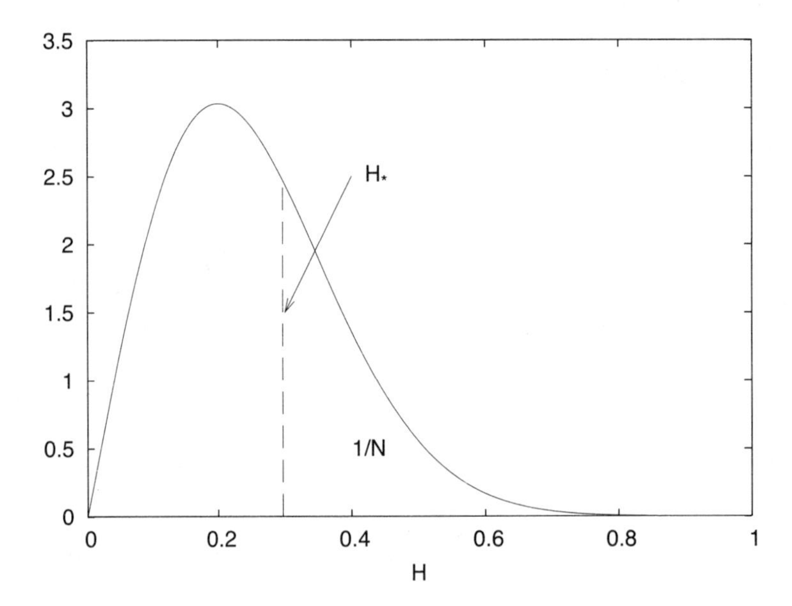

Figure 17. Highest $1/N$ waves, threshold height H_*, definition sketch using Rayleigh distribution ($\sigma = 0.1$, $N = 3$).

12.3 Second order nonlinear narrowbanded waves with Gaussian first harmonic

At the second nonlinear order we must account for two types of nonlinear contributions. One is the effect of the cubic nonlinear term in the nonlinear Schrödinger equation, and another is the second harmonic contribution to the reconstruction of the wave profile. If the first type of contribution can be neglected, then it is reasonable to assume that the first harmonic contribution to the harmonic perturbation expansion has a Gaussian distribution. Let us consider the expansion

$$\eta(\mathbf{r},t) = \frac{1}{2}\left(B\mathrm{e}^{\mathrm{i}(k_c x-\omega_c t)} + \gamma B^2 \mathrm{e}^{2\mathrm{i}(k_c x-\omega_c t)} + c.c.\right)$$

where γ is a constant that can be found with reference to section 4.

It now makes a lot of sense to define nonlinear upper and lower envelopes

$$e_U = |B| + \gamma |B|^2,$$

$$e_L = -|B| + \gamma |B|^2.$$

For small bandwidth the distribution of crest height is the same as the distribution of the upper envelope e_U, while the distribution of trough depth is the same as the distribution of lower envelope e_L. For the limit of vanishing bandwidth the distribution of wave height is the same as the distribution of the distance between upper and lower envelope, $e_U - e_L = 2|B|$, thus the Rayleigh distribution (12.2) is valid to second nonlinear order.

Assuming that $|B|$ is Rayleigh distributed, we get the distribution for the second order nonlinear upper envelope

$$f_{e_U}(z) = \frac{1}{2\gamma\sigma^2}\left(1 - \frac{1}{\sqrt{1+4\gamma z}}\right)\exp\{\frac{\sqrt{1+4\gamma z}-1-2\gamma z}{(2\gamma\sigma)^2}\} \quad \text{for} \quad z>0. \tag{12.3}$$

This distribution was first derived by Tayfun (1980).

The distribution for the second order nonlinear lower envelope becomes

$$f_{e_L}(z) = \begin{cases} \frac{1}{2\gamma\sigma^2}\left(1+\frac{1}{\sqrt{1+4\gamma z}}\right)\exp(-\frac{\sqrt{1+4\gamma z}+1+2\gamma z}{(2\gamma\sigma)^2}) & \text{for} \quad z \geqslant 0, \\ \frac{1}{2\gamma\sigma^2}\left\{\left(1+\frac{1}{\sqrt{1+4\gamma z}}\right)\exp(-\frac{\sqrt{1+4\gamma z}+1+2\gamma z}{(2\gamma\sigma)^2}) \right. & \\ \left. -\left(1-\frac{1}{\sqrt{1+4\gamma z}}\right)\exp(\frac{\sqrt{1+4\gamma z}-1-2\gamma z}{(2\gamma\sigma)^2})\right\} & \text{for} \quad -\frac{1}{4\gamma}<z<0. \end{cases} \tag{12.4}$$

This distributions was given in a slightly incorrect form by Tung and Huang (1985).

For application to the Draupner wave field, we employ the normalized standard deviation $\sigma = 0.086$, and recall that γ is 0.5 for infinite depth and 0.58 for the target depth. The probability densities for linear and nonlinear crest height and trough depths are seen in figures 18 and 19, respectively. The probability density for nonlinear lower envelope above the mean water level is totally negligible, and is not included in the figure.

Figures 18 and 19 suggest that there are far fewer freak troughs than freak crests.

12.4 Broader bandwidth and non-Gaussian first harmonic

The Tayfun distributions in the previous section should be expected to fail as the effect of broader bandwidth and nonlinear spatiotemporal evolution becomes important. However, this turns out not to be true! Linear crest distributions for broader banded

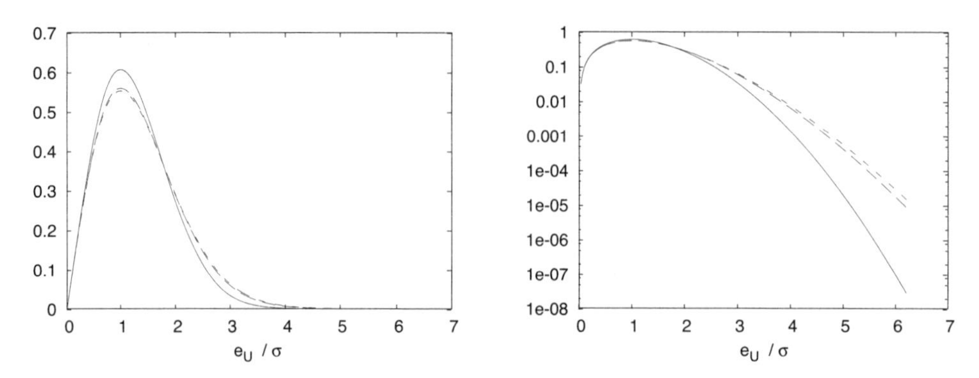

Figure 18. Probability density functions for crest height corresponding to the standard deviation of the Draupner wave field. Linear second axis left, logarithmic second axis right. —, linear Rayleigh distribution; – –, second-order nonlinear Tayfun distribution for infinite depth; ···, second-order nonlinear Tayfun distribution for target depth.

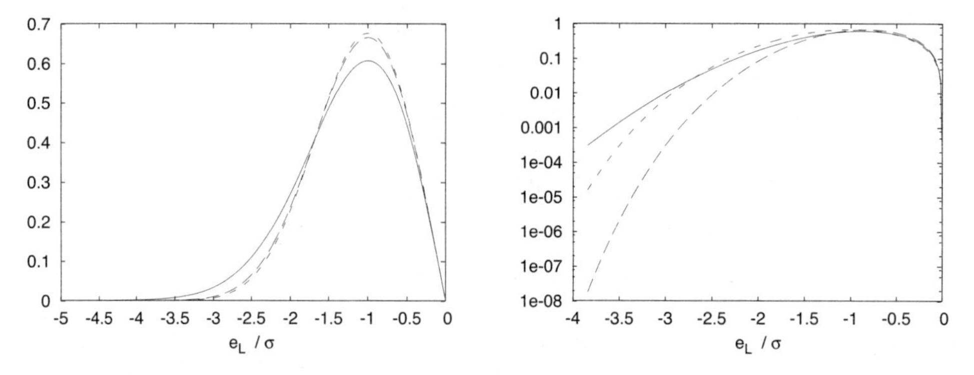

Figure 19. Probability density functions for trough depth corresponding to the standard deviation of the Draupner wave field. Linear second axis left, logarithmic second axis right. —, linear Rayleigh distribution; – –, second-order nonlinear Tayfun distribution for infinite depth; ···, second-order nonlinear Tayfun distribution for target depth.

waves were dealt with in a classical theory by Cartwright and Longuet–Higgins (1956). More novel is the problem to assess the influence of nonlinearity in the spatiotemporal evolution equations. It should be anticipated that nonlinearity in the evolution equation will break the Gaussian distribution of the first-harmonic free waves, and therefore the Rayleigh distribution of the linear crest heights as well.

For long-crested waves, nonlinearity in the spatiotemporal evolution does indeed produce deviation from the Gaussian/Rayleigh distribution for the first-harmonic free waves. Moreover, there is evidence that increased intensity of extreme waves results when steep narrowbanded waves undergo rapid spectral change, establishing a broader-banded equilibrium spectrum. Numerical evidence has been provided by Onorato et al. (2002), Janssen (2003) and Socquet–Juglard et al. (2005), in good agreement with experimental observations by Onorato et al. (2004).

For directionally distributed, short-crested waves, numerical evidence found by Onorato et al. (2002) and Socquet–Juglard et al. (2005) suggests that there is much less increase in intensity of extreme waves during rapid spectral change, and there is much less deviation from the Gaussian/Rayleigh distribution for the first-harmonic free waves, despite nonlinear spatiotemporal evolution. Socquet–Juglard et al. (2005) find evidence that the distribution of crest height of the reconstructed sea surface is quite well described by the second-order nonlinear Tayfun distribution of the previous section 12.3. This is indeed remarkable: A directionally distributed nonlinearly evolving sea with non-vanishing bandwidth is very well described by the weakly non-Gaussian statistical theory for vanishing bandwidth!

13 Return periods and return values

The *return period* is defined as the expected time that an extreme event, exceeding some threshold, shall occur once. The *return value* is the threshold that is exceeded once.

For example, for waves with constant period T_0 the 100-year wave height H_{100} is defined by the relationship

$$P\{H \geq H_{100}\} = \frac{T_0}{100\text{years}}.$$

For practical design applications to field sites, assessment of return periods and return values requires the knowledge of the joint distribution of periods and heights over all the sea states that characterize the site, and requires systematic observations over extended time. That is beyond the scope of the present discussion, however, we shall consider the return period within a single sea state that is assumed to last forever. We are now finally in a position to say something non-trivial about the Draupner "New Year Wave":

13.1 How unusual is the Draupner "New Year Wave"?

Let us make the thought experiment that the sea state of the Draupner wave time series is extended to arbitrary duration. Let us compute the return period for the "New Year Wave" according to the distributions derived above. The characteristic period was previously found to be $T_c = 12$ s. The standard deviation was found to be $\sigma = 2.98$ m.

The exceedance probability for the Draupner wave height $H_{\mathrm{D}} = 25.6$ m is according to the Rayleigh distribution (12.2)

$$P\{H \geq H_{\mathrm{D}}\} = \exp(-\frac{H_{\mathrm{D}}^2}{8\sigma^2}) = 9.86 \cdot 10^{-5}.$$

This is a 34 hour event. This estimate holds at first and second nonlinear order, and does not depend on the depth for these two nonlinear orders. The exceedance probability for the Draupner crest height $\eta_{\mathrm{D}} = 18.5$ m is according to the Rayleigh distribution (12.1)

$$P\{\eta \geq \eta_{\mathrm{D}}\} = \exp(-\frac{\eta_{\mathrm{D}}^2}{2\sigma^2}) = 4.28 \cdot 10^{-9}.$$

This is a 89 year event. This estimate holds at linear order irrespective of the assumed depth.

Using the second-order nonlinear Tayfun distribution (12.3) the exceedance probability for the Draupner crest height is

$$P\{\eta \geq \eta_{\mathrm{D}}\} = \exp(\frac{\sqrt{1+4\gamma\eta_{\mathrm{D}}} - 1 - 2\gamma\eta_{\mathrm{D}}}{(2\gamma\sigma)^2}) = \begin{cases} 2.40 \cdot 10^{-6} & \text{for infinite depth,} \\ 4.36 \cdot 10^{-6} & \text{for target depth.} \end{cases}$$

These are 58 and 32 day events, for infinite and target depth, respectively. It is indeed true that the "New Year Wave" crest height is much more spectacular than the wave height!

It will be useful to recapitulate the conditions for validity of these estimates. The Tayfun and Rayleigh distributions that have been applied are in principle valid for wave fields with vanishing bandwidth and with Gaussian first harmonic. A conclusion of Socquet–Juglard et al. (2005) is that the Tayfun distribution can also be applied with great accuracy for wave fields on infinite depth with non-vanishing bandwidth on the condition that the waves are sufficiently short-crested. For application to the Draupner wave field, one should in principle also account for finite depth, and we have in principle no direct measurement of the actual crest lengths.

Many people have suggested that extreme waves, such as the Draupner wave, can be produced by exotic third-order nonlinear behavior such as "breather solutions" of the nonlinear Schrödinger equation (Dysthe and Trulsen 1999; Osborne et al. 2000, etc.). These solutions typically depend on Benjamin–Feir instability of nearly uniform wave trains. To the extent that such exotic wave solutions are relevant in the ocean, they are in fact already accounted for in the analysis of Socquet–Juglard et al. (2005) suggesting the applicability of the Tayfun distributions.

14 Conclusion

This chapter is aimed at readers with some background in nonlinear wave theory, but who lacks essential background in stochastic analysis. A case study of an observed freak wave is first used to motivate why the modified nonlinear Schrödinger equation is a good candidate for Monte–Carlo simulations. Then the necessary stochastic theory is introduced at an elementary level to understand the results from the Monte–Carlo simulations. These simulations yield nontrivial physical insight about the sea state and occurrence probability of the observed freak wave.

A Continuous and discrete Fourier transforms

A.1 Continuous Fourier transform of a function on an infinite interval

The Fourier transform pair of a function $f(t)$ is

$$\hat{f}(\omega) = \int_{-\infty}^{\infty} f(t)\mathrm{e}^{\mathrm{i}\omega t}\,\mathrm{d}t, \tag{A.1}$$

$$f(t) = \frac{1}{2\pi}\int_{-\infty}^{\infty} \hat{f}(\omega)\mathrm{e}^{-\mathrm{i}\omega t}\,\mathrm{d}\omega. \tag{A.2}$$

It is useful to recall the integral form of the Dirac delta function

$$\delta(t) = \frac{1}{2\pi}\int_{-\infty}^{\infty} \mathrm{e}^{\mathrm{i}\omega t}\,\mathrm{d}\omega. \tag{A.3}$$

Parseval's theorem states that

$$\int_{-\infty}^{\infty} |f(t)|^2\,\mathrm{d}t = \frac{1}{2\pi}\int_{-\infty}^{\infty} |\hat{f}(\omega)|^2\,\mathrm{d}\omega \tag{A.4}$$

which is derived substituting either (A.1) or (A.2) and then using (A.3).

A.2 Fourier series of a function on a finite length interval

Given a function $f(t)$ on a finite interval of length T. We want to write

$$f(t) = \sum_{j=-\infty}^{\infty} \hat{f}_j \mathrm{e}^{-\mathrm{i}\omega_j t} \tag{A.5}$$

where $\omega_j = 2\pi j/T$. Using the L_2 inner product

$$\langle f(t), g(t)\rangle = \int_0^T f(t)g^*(t)\,\mathrm{d}t, \tag{A.6}$$

and looking at the system of complex exponentials, we find that

$$\langle \mathrm{e}^{\mathrm{i}\omega_j t}, \mathrm{e}^{\mathrm{i}\omega_l t}\rangle = T\delta_{j,l} \tag{A.7}$$

where $\delta_{j,l}$ is the Kronecker delta function. Thus the complex exponentials are orthogonal.

Taking the inner product of (A.5) and a complex exponential we get

$$\hat{f}_j = \frac{1}{T}\int_0^T f(t)\mathrm{e}^{\mathrm{i}\omega_j t}\,\mathrm{d}t \tag{A.8}$$

Parseval's theorem states that

$$\int_0^T |f(t)|^2\,\mathrm{d}t = T\sum_{j=-\infty}^{\infty} |\hat{f}_j|^2 \tag{A.9}$$

which is derived substituting (A.8) into the left-hand side of (A.9) and then using (A.7).

A.3 Discrete Fourier Transform (DFT) of a finite series

Given a sequence f_n of N numbers, with $n = 0, 1, \ldots, N-1$. We want to write

$$f_n = \sum_{j=0}^{N-1} \tilde{f}_j e^{-i\omega_j t_n} \tag{A.10}$$

where $\omega_j = 2\pi j/T$ and $t_n = nT/N$. The t_n are known as collocation points. Using the l_2 inner product

$$\langle f_n, g_n \rangle = \sum_{n=0}^{N-1} f_n g_n^*, \tag{A.11}$$

and looking at the system of complex exponentials, we find that

$$\langle e^{i\omega_j t_n}, e^{i\omega_l t_n} \rangle = N\delta_{j,l} \tag{A.12}$$

where $\delta_{j,l}$ is the Kronecker delta function. Thus the complex exponentials are orthogonal.

Taking the inner product of (A.10) and a complex exponential we get

$$\tilde{f}_j = \frac{1}{N} \sum_{n=0}^{N-1} f_n e^{i\omega_j t_n} \tag{A.13}$$

Parseval's theorem states that

$$\sum_{n=0}^{N-1} |f_n|^2 = N \sum_{j=0}^{N-1} |\tilde{f}_j|^2 \tag{A.14}$$

which is derived substituting either (A.13) or (A.10) and then using (A.12).

In the DFT inverse transform (A.10) we may rotate the sum-indices cyclically, after making the periodic extension $\tilde{f}_{j+N} = \tilde{f}_j$. For reconstruction at the collocation points we may start at an arbitrary first index α

$$\sum_{j=\alpha}^{\alpha+N-1} \tilde{f}_j e^{-i\omega_j t_n} = \sum_{j=0}^{N-1} \tilde{f}_j e^{-i\omega_j t_n}. \tag{A.15}$$

For interpolation between the collocation points, using a value of t different from t_n, it is desirable to minimize oscillatory behavior. This is achieved by setting $\alpha \approx N/2$.

Bibliography

I. E. Alber (1978) The effects of randomness on the stability of two-dimensional surface wavetrains. Proc. R. Soc. Lond. A, Vol. 363, 525–546.

U. Brinch-Nielsen and I. G. Jonsson (1986) Fourth order evolution equations and stability analysis for Stokes waves on arbitrary water depth. Wave Motion, Vol. 8, 455–472.

D. E. Cartwright and M. S. Longuet-Higgins (1956) The statistical distribution of the maxima of a random function. Proc. R. Soc. Lond. A, Vol. 237, 212–232.

D. R. Crawford, P. G. Saffman and H. C. Yuen (1980) Evolution of a random inhomogeneous field of nonlinear deep-water gravity waves. Wave Motion, Vol. 2, 1–16.

K. B. Dysthe (1979) Note on a modification to the nonlinear Schrödinger equation for application to deep water waves. Proc. R. Soc. Lond. A, Vol. 369, 105–114.

K. B. Dysthe and K. Trulsen (1999) Note on breather type solutions of the NLS as models for freak-waves. Physica Scripta, Vol. T82, 48–52.

K. B. Dysthe, K. Trulsen, H. E. Krogstad and H. Socquet-Juglard (2003) Evolution of a narrow band spectrum of random surface gravity waves. J. Fluid Mech., Vol. 478, 1–10.

Y. Goda (2000) Random seas and design of maritime structures. World Scientific.

J. Grue, D. Clamond, M. Huseby and A. Jensen (2003) Kinematics of extreme waves in deep water. Appl. Ocean Res., Vol. 25, 355–366.

S. Haver (2004) A possible freak wave event measured at the Draupner jacket Januar 1 1995. http://www.ifremer.fr/web-com/stw2004/rw/fullpapers/walk_on_haver.pdf. In Rogue Waves 2004, pages 1–8.

P. A. E. M. Janssen (2003) Nonlinear four-wave interactions and freak waves. J. Phys. Ocean., Vol. 33, 863–884.

D. Karunakaran, M. Bærheim, and B. J. Leira (1997) Measured and simulated dynamic response of a jacket platform. In C. Guedes-Soares, M. Arai, A. Naess, and N. Shetty, editors, Proceedings of the 16th International Conference on Offshore Mechanics and Arctic Engineering, Vol. II, 157–164. ASME.

E. Kit and L. Shemer (2002) Spatial versions of the Zakharov and Dysthe evolution equations for deep-water gravity waves. J. Fluid Mech., Vol. 450, 201–205.

V. P. Krasitskii (1994 On reduced equations in the Hamiltonian theory of weakly nonlinear surface-waves. J. Fluid Mech., Vol. 272, 1–20.

E. Lo and C. C. Mei (1985) A numerical study of water-wave modulation based on a higher-order nonlinear Schrödinger equation. J. Fluid Mech., Vol. 150, 395–416.

E. Y. Lo and C. C. Mei (1987) Slow evolution of nonlinear deep water waves in two horizontal directions: A numerical study. Wave Motion, Vol. 9, 245–259.

D. U. Martin and H. C. Yuen (1980) Quasi-recurring energy leakage in the two-space-dimensional nonlinear Schrödinger equation. Phys. Fluids, Vol. 23, 881–883.

M. K. Ochi (1998) Ocean waves. Cambridge.

M. Onorato, A. R. Osborne and M. Serio (2002) Extreme wave events in directional, random oceanic sea states. Phys. Fluids, Vol. 14, L25–L28.

M. Onorato, A. R. Osborne, M. Serio, L. Cavaleri, C. Brandini and C. T. Stansberg (2004) Observation of strongly non-gaussian statistics for random sea surface gravity waves in wave flume experiments. Phys. Rev. E, Vol. 70, 1–4.

M. Onorato, A. R. Osborne, M. Serio, D. Resio, A. Pushkarev, V. E. Zakharov and C. Brandini (2002) Freely decaying weak turbulence for sea surface gravity waves. Phys. Rev. Lett., Vol. 89, 144501–1–144501–4.

A. R. Osborne, M. Onorato and M. Serio (2000) The nonlinear dynamics of rogue waves and holes in deep-water gravity wave trains. Physics Letters A, Vol. 275, 386–393.

A. Papoulis (1991) Probability, random variables, and stochastic processes. McGraw-Hill, 3rd edition.

O. M. Phillips (1958) The equilibrium range in the spectrum of wind-generated waves. J. Fluid Mech., Vol. 4, 426–434.

Y. V. Sedletsky (2003) The fourth-order nonlinear Schrödinger equation for the envelope of Stokes waves on the surface of a finite-depth fluid. J. Exp. Theor. Phys., Vol. 97, 180–193.

H. Socquet-Juglard, K. Dysthe, K. Trulsen, H. E. Krogstad and J. Liu (2005) Probability distributions of surface gravity waves during spectral changes. J. Fluid Mech., Vol. 542, 195–216.

M. Stiassnie (1984) Note on the modified nonlinear Schrödinger equation for deep water waves. Wave Motion, Vol. 6, 431–433.

M. A. Tayfun (1980) Narrow-band nonlinear sea waves. J. Geophys. Res., Vol. 85, 1548–1552.

K. Trulsen (2001) Simulating the spatial evolution of a measured time series of a freak wave. In M. Olagnon and G. Athanassoulis, editors, Rogue Waves 2000, 265–273. Ifremer.

K. Trulsen (2005) Spatial evolution of water surface waves. In Proc. Ocean Wave Measurement and Analysis, Fifth International Symposium WAVES 2005, number 127, 1–10.

K. Trulsen and K. B. Dysthe (1996) A modified nonlinear Schrödinger equation for broader bandwidth gravity waves on deep water. Wave Motion, Vol. 24, 281–289.

K. Trulsen and K. B. Dysthe (1997) Freak waves — a three-dimensional wave simulation. In Proceedings of the 21st Symposium on Naval Hydrodynamics, 550–560. National Academy Press.

K. Trulsen and K. B. Dysthe (1997) Frequency downshift in three-dimensional wave trains in a deep basin. J. Fluid Mech., Vol. 352, 359–373.

K. Trulsen, I. Kliakhandler, K. B. Dysthe and M. G. Velarde (2000) On weakly nonlinear modulation of waves on deep water. Phys. Fluids, Vol. 12, 2432–2437.

K. Trulsen and C. T. Stansberg (2001) Spatial evolution of water surface waves: Numerical simulation and experiment of bichromatic waves. In Proc. 11th International Offshore and Polar Engineering Conference, Vol. 3, 71–77.

C. C. Tung and N. E. Huang (1985) Peak and trough distributions of nonlinear waves. Ocean Engineering, Vol. 12, 201–209.

V. E. Zakharov (1968) Stability of periodic waves of finite amplitude on the surface of a deep fluid. J. Appl. Mech. Tech. Phys., Vol. 9, 190–194.

Freak Waves Phenomenon: Physical Mechanisms and Modelling

Christian Kharif [*] and Efim Pelinovsky [†]

[*] Institut de Recherche sur les phénomènes Hors Equilibre
Marseille, France
[†] Institute of Applied Physics
Niznhy Novgorod, Russia

Abstract The main physical mechanisms responsible for the formation of huge waves known as freak waves are described and analyzed. Data of observations in the marine environment as well as laboratory experiments are briefly discussed. They demonstrate that freak waves may appear in deep and shallow waters. The mathematical definition of these huge waves is based on statistical parameters, namely the significant height. As linear models of freak waves the following mechanisms are considered: Wave-current interaction and dispersion enhancement of transient wave packets. When the nonlinearity of the water waves is introduced, these mechanisms remain valid but should be modified and new mechanisms such as modulational instability and soliton-collisions become good candidates to explain the freak wave occurrence. Specific numerical simulations were performed within the framework of classical nonlinear evolution equations: The Nonlinear Schrödinger equation, the Davey-Stewartson system, the Korteweg-de Vries equation, the Kadomtsev-Petviashvili equation, the Zakharov equation and the fully nonlinear potential equations. Their results show the main features of the physical mechanisms of freak wave phenomenon.

1 Introduction

Freak, rogue, or giant waves correspond to large-amplitude waves surprisingly appearing on the sea surface (wave from nowhere). Such waves can be accompanied by deep troughs (holes), which occur before and/or after the large crest. As it is pointed out by Lawton (2001) the freak waves have been part of marine folklore for centuries. Seafarers speak of walls of water, or of holes in the sea, or of several successive high waves (three sisters), which appear without warning in otherwise benign conditions. But since the seventies of the last century, oceanographers have started to believe them. Observations gathered by the oil and shipping industries suggest there really is something like a true monster of the deep that devors ships and sailors without mercy or warning. Over the two last decades dozens of super-carriers-cargo ships over 200m long have been lost at sea. There are several definitions for such surprisingly huge water waves. Haver and Andersen (2000) put the question, what are freak waves: Are they extremely rare realizations of a typical slightly non-Gaussian population? - or - Are they typical realizations from a rare strongly non-Gaussian population? Up to now there is no definitive consensus about a unique definition of freak wave. Sometimes, the definition of the freak waves includes that such waves are too high, too asymmetric and too steep. More popular now is the amplitude criterion of a freak wave: its height should exceed twice the significant height. Due to the rare character of the rogue waves their prediction based on data analysis with use of statistical methods is not too productive. Owing to the non-Gaussian and non-stationary character of the water wave fields on sea surface, it is a very tricky task to compute the probability density function of freak waves. So our approach to the problem is aimed at describing the deterministic mechanisms responsible for the occurrence of these huge waves. During the last 30 years the various physical models of the rogue wave phenomenon have been intensively developed and many laboratory experiments conducted. The main objective of these studies is to better understand the physics of the huge wave occurrence and its relation to environmental conditions (wind and atmospheric pressure, bathymetry and current field) and to provide the design of freak wave needed for engineering purposes.

2 Freak wave observations

Recently, Lawton (2001) gave a large collection of freak wave observations from ships. In particular, 22 super-carriers were lost due to collision with freak waves for 1969-1994 in the Pacific and Atlantic causing 525 fatalities (see figure 1). At least, the 12 events of the ships collisions with freak waves were recorded after 1952 in the Indian Ocean, near the Agulhas Current, coast off South Africa (Smith (1976), Lavranov (1998a), Lavranov (1998b)). During the closure of the Suez Canal a number of ships, particularly oil tankers, have reported extensive damage caused by freak waves off the south-east coast of South Africa (Mallory (1974), Sturm (1974), Sanderson (1974)). The Agulhas Current flows in a south-west direction along the African coast. Energetic swell, generated at higher latitudes propagating against this current is amplified by local concentration of wave energy. The following text by Graham (2000) describes the event which occurred in shallow water 4th November 2000 with the NOAA vessel:

At 11:30 a.m. last Saturday morning (November 4, 2000) the 56-foot research vessel R/V Ballena capsized in a rogue wave south of Point Arguello, California. The Channel Islands National Marine Sanctuary's research vessel was engaged in a routine side-scan sonar survey for the U.S. Geological Survey of the seafloor along the 30-foot-depth contour approximately 1/4 nautical mile from the shore. The crew of the R/V Ballena, all of whom survived, consisted of the captain, NOAA Corps officer LCdr. Mark Pickett, USGS research scientist Dr. Guy Cochrane, and USGS research assistant, Mike Boyle. According to National Oceanic & Atmospheric Administration spokesman Matthew Stout, the weather was good, with clear skies and glassy swells. The forecasted swell was 7 feet and the actual swell appeared to be 5-7 feet. At approximately 11:30 a.m., Pickett and Boyle said they observed a 15-foot swell begin to break 100 feet from the vessel. The wave crested and broke above the vessel, caught the Ballena broadside, and quickly overturned her. All crew members were able to escape the overturned vessel and deploy the vessel's life raft. The crew attempted to paddle to the shore, but realized the possibility of nav-

igating the raft safely to shore was unlikely due to strong near-shore currents. The crew abandoned the life raft approximately 150 feet from shore and attempted to swim to safety. After reaching shore, Pickett swam back out first to assist Boyle to safety and again to assist Cochrane safely to shore. The crew climbed the rocky cliffs along the shore. The R/V Ballena is total loss.

The photo in figure 2 showing a freak wave corresponds to the following description given by Chase *(http://bell.mma.edu/ achase/NS-221-Big-Wave.html)*
A substantial gale was moving across Long Island, sending a very long swell down our way, meeting the Gulf Stream. We saw several rogue waves during the late morning on the horizon, but thought they were whales jumping. It was actually a nice day with light breezes and no significant sea. Only the very long swell, of about 15 feet high and probably 600 to 1000 feet long. This one hit us at the change of the watch at about noon. The photographer was an engineer, and this was the last photo on his roll of film. We were on the wing of the bridge, with a height of eye of 56 feet, and this wave broke over our heads. This shot was taken as we were diving down off the face of the second of a set of three waves, so the ship just kept falling into the trough, which just kept opening up under us. It bent the foremast (shown) back about 20 degrees, tore the foreword firefighting station (also shown) off the deck (rails, monitor, platform and roll) and threw it against the face of the house. It also bent all the catwalks back severely. Later that night, about 19:30, another hit the after house, hitting the stack and sending solid water down into the engine room through the forced draft blower intakes.

The photo in figure 2 and previous descriptions show the main features of the freak wave phenomenon: The rapid occurrence of large amplitude solitary pulses or a group of large amplitude waves on the almost calm sea in shallow as well as in deep water. Furthermore they highlight the steep front aspect of the freak waves.

The instrumental data of the freak wave registration are obtained for different oil platforms. Figure 1 in Ch. 2 of the volume shows the famous time record of the New Year Wave in the North Sea which occurred at Draupner (Statoil operated jacket platform, Norway) 1st January 1995, see Haver and Andersen (2000). The water depth, h, is 70m, the characteristic period of freak wave is 12s; so the wavelength is about 220m according to the linear dispersion relation. The important parameter of dispersion kh is $kh \sim 2$, that means that the recorded freak wave can be considered as a wave propagating in finite depth. Nonlinearity of this wave is characterized by the steepness ak (a and k are the amplitude and wave number respectively) and its value is 0.37. Therefore the freak wave is strongly nonlinear. Sand *et al.* (1990) have collected data of freak wave observations in the North Sea (depth 20-40m) between 1969 and 1985. Maximum ratio of freak wave height, H_f, to the significant wave height, H_s, reached 3 (Hanstholm, Danish Sector, depth 20m, H_s=2m, H_f=6m). Such an event is defined as a freak wave phenomenon in shallow water. Recently, Mori, Liu and Yasuda (2002) published an analysis of freak wave observations (at least 14 times with the wave height exceeding 10m) in the Japan Sea (Yura Harbor, 43m depth) during 1986-1990. Maximum ratio, $\frac{H_f}{H_s}$ reached 2.67.

Data given above demonstrate that freak waves can occur on arbitrary depth (deep, intermediate, shallow) with or without strong current. The main features of this phenomenon are: rare occurrence, short-lived character, solitary-like shape (or a group of several waves), high nonlinearity and quasi-plane wave fronts.

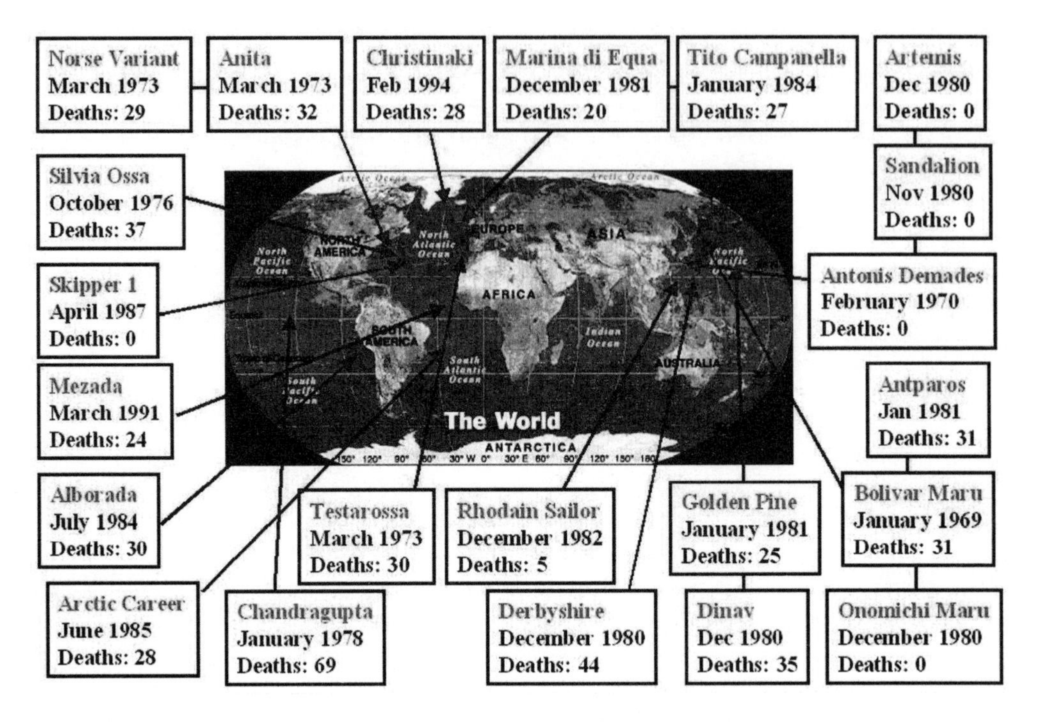

Figure 1. Statistics of super tanker collisions with freak waves between 1968-1994. From Kharif and Pelinovsky (2003).

Freak waves have been also mechanically generated in water wave tanks (Brown and Jensen (2001), Clauss (2002), Contento, Codiglia and D'Este (2001), Johannessen and Swan (2001), Stansberg (2001) and Johannessen and Swan (2003)). Very recently Giovanangeli, Kharif and Pelinovsky (2004a) and Giovanangeli, Kharif and Pelinovsky (2004b) conducted a series of experiment in the large wind-wave tank of IRPHE to study the effect of the wind on the formation of rogue waves.

3 A brief description of the main physical mechanisms of freak wave generation

There are a number of physical mechanisms producing the occurrence of freak waves. Extreme wave events can be due to refraction (presence of variable currents or bottom topography), dispersion (frequency modulation), wave instability (Benjamin-Feir instability), soliton interactions, etc. that may focus the wave energy into a small area. All these different mechanisms will be developed in detail in the next sections. The goal of this section is to give a brief survey on the generation of freak waves. The most recent reviews on freak waves are those of Dysthe (2001a) and Kharif and Pelinovsky (2003).

Figure 2. Photo of a freak wave event.

3.1 Wave-current interaction

The most popular example of freak waves is that corresponding to abnormal waves appearing suddenly off the south-east coast of South Africa when the dominant wind-generated waves meet a counter-current (Agulhas Current). Smith (1976) suggested that giant waves, as observed on the Agulhas Current, occur where the wave groups are reflected by the current. The local behavior of the wave amplitude is modelled by the nonlinear Schrödinger equation. Using a more global linear approach, Lavranov (1998b) investigated the problem by taking into account the initial value of wave parameters. They are defined by the wave transformation on the large scale of the Agulhas Current: To compute the wave transformation, the real values of the current are introduced in the wave action equation. White and Fornberg (1998) considered the formation of a freak wave when a regular ocean swell traverses a region of deep water with random fluctuations. It is assumed that freak waves are produced by caustics resulting from ray focusing when the swell interacts with the random current. This latter approach is linear. Dysthe (2001a) suggested that even a small directional distribution of the incoming wave field will smear out the caustics and thus the occurrence of freak waves. Following the paper by White and Fornberg (1998), Dysthe (2001b) derived a very nice result showing that the curvature of rays is a simple function of the current vorticity and group velocity. Very recently, Wu and Yao (2004) reported experimental results of limiting freak waves on currents.

3.2 Geometrical focusing

Refraction of surface waves can be due to underwater topography as well. The result is spatial variations of the kinematic (frequency and wavenumber) and dynamic (amplitude or energy) properties of the wave packets. This problem can be solved by using the well-known equations of the ray theory (geometrical optics) and freak waves occurrence corresponds to caustics. The behavior of the rays in basins with real topography is rather complicated and many caustics are formed (see figure 7 of the paper by Kharif and Pelinovsky (2003)).

Note that freak waves can arise from wave-current interactions in water of varying depth. It means that refraction effects due to sea bottom and current both are working.

3.3 Spatio-temporal focusing

If initially short wave packets with small group velocities are located in front of long wave packets having large group velocities, then during the stage of evolution, long waves will overtake short waves. A large-amplitude wave can occur at some fixed time because of the superposition of all the waves merging at the same location. Afterward, the long waves will be in front of the short waves, and the amplitude of the wave train will decrease. This scenario of frequency modulated wave groups is described by Pelinovsky, Talipova and Kharif (2000) within the framework of the shallow water theory. Later Slunyaev *et al.* (2002) using the Davey-Stewartson system, investigated the 3D dispersive wave focusing in a nonlinear medium on finite depth.

3.4 Modulational instability

It is well known that periodic nonlinear uniform wave trains suffer an instability known as the Benjamin-Feir instability, which occurs in the form of growing modulations. These modulations which evolve into short groups of steep waves correspond to a nonlinear focusing of the wave energy. The evolving wave trains experience modulations that reach a maximum corresponding the formation of steep waves (assumed to be freak waves), followed by demodulations. This phenomenon characterized by modulation-demodulation cycles is the famous Fermi-Pasta-Ulam recurrence (FPU recurrence). When the steepness of the initial wave train is beyond a given threshold it is observed that the energy becomes focused into a short group of steep waves which contains a wave which becomes too steep and therefore breaks. The formation of freak wave through modulational instability has been investigated by several authors, among them are Henderson, Peregrine and Dold (1999), Dysthe and Trulsen (1999), Osborne, Onorato and Serio (2000), Calini and Schober (2002), Slunyaev *et al.* (2002), Dyachenko and Zakharov (2005).

3.5 Soliton interactions

Soliton interaction as a possible model for extreme waves in shallow water has been suggested by Peterson *et al.* (2003) and Soomere and Engelbrecht (2005). They considered the interaction of two long-crested shallow water waves within the framework of the two-soliton solution of the Kadomstev-Petviashvili equation. It was found that extreme surface elevation exceeds several times the amplitude of the incoming waves over a small area. In deep water Clamond and Grue (2002) and Clamond, Francius, Grue and Kharif (2006) showed that strong interactions between envelope-solitons may produce freak wave event. They performed long time simulations based on fully nonlinear equations.

3.6 Wind effect

If the wind is not directly responsible for the formation of freak waves, it can sustain them through the Jeffreys' sheltering theory. Jeffreys (1925) assumed that the energy transfer from the wind to the waves was exclusively caused by form drag associated with flow separation occurring on the leeward side of wave crests with re-attachment further down on the leeward slopes. This mechanism as been observed by Giovanangeli, Kharif and Pelinovsky (2004a) and Giovanangeli, Kharif and Pelinovsky (2004b) in the wind-wave tank of IRPHE.

4 Freak wave definition

In the simplest model of ocean waves, the sea elevation is considered as a summation of sinusoidal waves of different frequency with random phases and amplitudes. In the linear approximation, the random wave field is considered as a stationary random Gaussian process with the following probability density distribution

$$f(\eta) = \frac{1}{\sqrt{2\pi\sigma}} \exp(-\frac{\eta^2}{2\sigma^2}), \tag{4.1}$$

where η is the sea surface elevation with zero mean level, $< \eta >= 0$, and σ^2 is the variance computed from the frequency spectrum, $S(\omega)$

$$\sigma^2 =< \eta^2 >= \int_0^\infty S(\omega) d\omega. \tag{4.2}$$

Typically, the wind wave spectrum is assumed to be narrow, thus the cumulative probability function of the wave heights will be given by the Rayleigh distribution

$$P(H) = \exp(-\frac{H^2}{8\sigma^2}). \tag{4.3}$$

The probability that the wave heights will exceed a certain value, H, is given by (4.3).

One specific wave height frequently used in oceanography and ocean engineering is the significant wave height, H_s. This concept was introduced by Sverdrup and Munk (1947) who defined H_s as the average of the highest one-third of wave heights. This wave height is close to the mean wave height estimated by human eye. Using the Rayleigh distribution Massel (1996) showed that H_s is given by

$$H_s = (3\sqrt{2\pi}\mathrm{erfc}(\sqrt{\ln 3} + 2\sqrt{2\ln 3})\sigma \approx 4\sigma, \tag{4.4}$$

where erfc(.) is the complementary error function. So H_s is four times the standard deviation σ. Equation (4.3) can be rewritten as follows

$$P(H) = \exp(-\frac{2H^2}{{H_s}^2}). \tag{4.5}$$

Mathematically, a wave is considered to be a freak wave if its height, H_f, satisfies the condition

$$H_f > 2H_s. \tag{4.6}$$

5 Governing equations

The fluid is assumed to be inviscid and the motion irrotational, such as the velocity $\boldsymbol{u}$ may be expressed as the gradient of a potential ϕ: $\boldsymbol{u} = \nabla\phi$. If the fluid is assumed to be incompressible, such that $\nabla\ \mathbf{.u} = 0$, the equation that holds throughout the fluid is the Laplace's equation

$$\nabla^2\phi = 0 \quad \text{for} \quad -h < z < \eta(x,y,t). \tag{5.1}$$

The x and y coordinates are taken to be horizontal plane, the z axis vertically upwards. The bottom is located at $z = -h(x,y)$. The bottom condition is

$$\frac{\partial\phi}{\partial x}\frac{\partial h}{\partial x} + \frac{\partial\phi}{\partial y}\frac{\partial h}{\partial y} + \frac{\partial\phi}{\partial z} = 0 \quad \text{on} \quad z = -h(x,y). \tag{5.2}$$

The kinematic requirement that a particle on the free surface remains on it is expressed by

$$\frac{\partial\eta}{\partial t} + \frac{\partial\phi}{\partial x}\frac{\partial\eta}{\partial x} + \frac{\partial\phi}{\partial y}\frac{\partial\eta}{\partial y} - \frac{\partial\phi}{\partial z} = 0 \quad \text{on} \quad z = \eta(x,y,t). \tag{5.3}$$

The dynamic boundary condition can be written

$$\frac{\partial\phi}{\partial t} + \frac{1}{2}(\nabla\phi)^2 + g\eta + \frac{p_a}{\rho} = 0 \quad \text{on} \quad z = \eta(x,y,t), \tag{5.4}$$

where g is the gravitational acceleration, p_a the pressure at the free surface and ρ the density of the fluid. In water of infinite depth, the kinematic boundary equation (5.2) is replaced by $\nabla\phi \to 0$ as $z \to -\infty$. Surface tension effects are ignored.

During the last four decades, the need for a satisfactory description of the sea surface leads the scientists to develop initially approximate dynamic models of nonlinear wave interactions. Usually these models are derived from the fully nonlinear equations (5.1)-(5.4) by means of the multiple scaling technique. For finite depth (including shallow water or long water waves) the most popular approximate models are the KdV equation, the Boussinesq equation, the KP equation, the Benney-Roskes system or the Davey-Stewartson system and the Zakharov equation. Except the KdV equation all these models are three-dimensional. Let a, k and h be the amplitude of the wave, its wavenumber and the depth, respectively. The celebrated KdV equation is applicable when the parameters $\frac{a}{h}$ and $(kh)^2$ are small with $\frac{a}{h} = O(k^2h^2)$. This means that there is a balance between the competing weakly nonlinear and weakly dispersive effects. The KP equation which is a generalization of the KdV equation, describes the evolution of weakly nonlinear, weakly dispersive and weakly three-dimensional water waves. Let $\mathbf{k} = (k,l)$ be a representative horizontal wavenumber vector, weak three-dimensionality means here that $(\frac{l}{k})^2 = O(\frac{a}{h}) = O(k^2h^2)$. The KdV, Boussinesq and KP equations describe the evolution of weakly nonlinear long waves. Consider now a train of waves of wave number k and steepness $\epsilon = ak$ ($\epsilon << 1$), subject to three-dimensional large scale variations identified by the small parameter μ. The choice of $\epsilon = \mu$ brings into balance nonlinearity and dispersion and yields the Benney-Roskes system or the Davey-Stewartson system. The wave packet propagates in the x-direction with a slowly evolving structure in both x- and y-directions. In these models the slow spatial modulation of the fundamental wavenumber induces a weak slowly varying mean motion that is of prime importance in stability analysis. If one assumes that the envelope propagation is 1D and the mean flow is zero the Davey-Stewartson system reduces to the classical nonlinear Schrödinger equation (NLS equation). The Benney-Roskes and Davey-Stewartson systems are third-order in wave steepness evolution equations describing the behavior of wave packets correctly at the early stages (time and length scales $O(\epsilon^{-2})$). Brinch-Nielsen and Jonsson (1986) extended to arbitrary depth the fourth-order in wave steepness evolution equation of Dysthe (1979) that is applicable over time and length scales $O(\epsilon^{-3})$. The previous models that describe modulational evolution of wave trains assume a narrow-spectral width. Using the Hamiltonian formalism, Zakharov (1968) derived an integral equation for nonlinear deep water waves. Krasitskii (1994) revisited the Zakharov equation by taking into account symmetry properties that the kernels must satisfy. Stassnie and Shemer (1984) extended the derivation to water waves on finite depth. One can emphasize that the above extended nonlinear Schrödinger equations can be derived from the Zakharov equation under the assumption of narrow-banded spectrum. All these approximate models have been used to study freak waves.

6 Linear approaches to the problem

When dealing with linear or weakly nonlinear models, it is sometimes not required to solve equations (5.1)-(5.4) or approximate models derived from this set of equations. Conservation laws may be sufficient to describe the kinematics and dynamics of the wave packets evolving into freak waves.

6.1 Wave trains in inhomogeneous moving media

Wave kinematics. The transformation of water waves by currents and bottom topography is a significant physical process in many coastal areas and where wind waves meet major ocean currents. Herein wave trains are assumed to be a system of almost sinusoidal propagating waves with a dominant frequency ω, wavenumber vector $\mathbf{k}$, and amplitude a. All these quantities are assumed to be slowly varying functions of $\mathbf{x} = (x, y)$ and t, where $\mathbf{x}$ corresponds to the horizontal coordinates. This means that appreciable changes are apparent only over many periods and wavelengths. The frequency and wavenumber vector are derived from a phase function $\chi(\mathbf{x}, t)$ by

$$\omega = -\chi_t, \tag{6.1}$$

$$\mathbf{k} = \nabla_1 \chi, \tag{6.2}$$

with $\nabla_1 = (\partial/\partial x, \partial/\partial y)$. At each point, ω and $\mathbf{k}$ satisfy the linear dispersion relation

$$\omega = W(\mathbf{k}, \mathbf{x}, t), \tag{6.3}$$

where the space and time variability of the medium is characterized by the presence of $\mathbf{x}$ and t. Attention is paid to linear waves for which ω and the components of $\mathbf{k}$ are real. From (6.1) and (6.2) it is found that

$$\frac{\partial \mathbf{k}}{\partial t} + \nabla_1 \omega = 0, \tag{6.4}$$

$$\nabla_1 \times \mathbf{k} = \mathbf{0}. \tag{6.5}$$

The components of the group velocity are defined by

$$C_{g_j} = \frac{\partial W}{\partial k_j} \quad j = 1, 2. \tag{6.6}$$

Using (6.4), (6.5) and (6.6), the spatio-temporal evolution of the wavenumber vector may be written in characteristic form as

$$\frac{\mathrm{d}k_i}{\mathrm{d}t} = -\frac{\partial W}{\partial x_i} \quad \text{on} \quad \frac{\mathrm{d}x_i}{\mathrm{d}t} = \frac{\partial W}{\partial k_i}, \tag{6.7}$$

where

$$\frac{\mathrm{d}}{\mathrm{d}t} = \frac{\partial}{\partial t} + C_{g_j} \frac{\partial}{\partial x_j}. \tag{6.8}$$

It can be shown as well that

$$\frac{\mathrm{d}\omega}{\mathrm{d}t} = \frac{\partial W}{\partial t} \quad \text{on} \quad \frac{\mathrm{d}x_i}{\mathrm{d}t} = \frac{\partial W}{\partial k_i}. \tag{6.9}$$

If the medium is homogeneous ($\partial W/\partial x_i = 0$), then $\mathbf{k}$ is a constant vector along the characteristic curves defined as

$$\frac{\mathrm{d}x_i}{\mathrm{d}t} = C_{g_i}. \tag{6.10}$$

If the medium is steady ($\partial W/\partial t = 0$), ω is constant along the characteristic curves (rays).

Wave dynamics. Equations (6.7) and (6.9) are kinematic results, depending only on the existence of a phase function and a local dispersion relation (6.3).
If the wave train described by $\eta(\mathbf{x}, t) = a(\mathbf{x}, t) \exp[\mathrm{i}\chi(\mathbf{x}, t)] + c.c.$ propagates on unsteady and nonhomogeneous current $\mathbf{U}(\mathbf{x}, t)$ its frequency ω is

$$\omega = \mathbf{k} \cdot \mathbf{U}(\mathbf{x}, t) + \sigma(\mathbf{k}) = W(\mathbf{k}, \mathbf{x}, t), \tag{6.11}$$

where $\sigma(\mathbf{k})$ is the intrinsic frequency, i.e. the frequency relative to the water or the frequency as seen by an observer moving with the current. The wavenumber vector $\mathbf{k}$ depends on $\mathbf{x}$ and t. In the linear case, Bretherton and Garrett (1969) derived the following conservation equation

$$\frac{\partial}{\partial t}(\frac{E}{\sigma}) + \nabla_x \cdot [\mathbf{C}_g(\frac{E}{\sigma})] = 0, \tag{6.12}$$

where $E(\mathbf{x},t)$ is proportional to the mean wave energy density per unit surface, $\mathbf{C}_g$ is the group velocity and the subscript x denotes differentiation with respect to horizontal space coordinates. The wave energy density and the group velocity are given by

$$E(\mathbf{x},t) = \rho g \mid a(\mathbf{x},t) \mid^2, \tag{6.13}$$

$$\mathbf{C}_g = \nabla_k W = \mathbf{U} + \nabla_k \sigma, \tag{6.14}$$

where $\nabla_k \sigma$ is the intrinsic group velocity, i.e. the velocity relative to the water.

Equation (6.12) has the general form of a conservation law in which the local rate of change of a density is determined by a flux of that density. This means that the integral over all space of the wave action density E/σ is conserved in time.

Kharif (1990) investigated the modulations in wavenumber and amplitude of short waves propagating over the surface of a longer gravity waves on deep water. He showed that this problem can be studied either by integrating the ray equations (6.9) coupled with the action conservation principle (6.12) or by considering the linear stability of the long wave to superharmonic perturbations.

Longuet-Higgins and Stewart (1961, 1964) investigated the propagation of surface gravity waves on a slowly varying nonuniform current. They used a general equation governing the energy evolution of surface waves in which the third term corresponds to the energy transfer between waves and current,

$$\frac{\partial E}{\partial t} + \nabla_x \cdot (\mathbf{C}_g E) + \frac{1}{2} S_{ij} (\frac{\partial U_i}{\partial x_j} + \frac{\partial U_j}{\partial x_i}) = 0, \tag{6.15}$$

where S_{ij} is the radiation stress tensor associated with averaged momentum fluxes in a sinusoidal wave train.

Bretherton and Garrett (1969) showed that equations (6.12) and (6.15) are equivalent. In a nonuniform media the wave energy density is not conserved while the wave action density is conserved.

Equation (6.12) is the linearized version of the following equation derived by Whitham (1965) within the framework of a variational approach

$$\frac{\partial}{\partial t}(\frac{\partial \mathcal{L}}{\partial \omega}) - \nabla_x \cdot (\nabla_k \mathcal{L}) = 0, \tag{6.16}$$

where $\mathcal{L}$ is the averaged Lagrangian over the phase of the fluctuating motion.

By defining $A = \partial \mathcal{L}/\partial \omega$ the wave action density and $\mathbf{B} = -\nabla_k \mathcal{L}$, the wave action flux, equation (6.16) can be re-written

$$\frac{\partial A}{\partial t} + \nabla_x \cdot \mathbf{B} = 0. \tag{6.17}$$

For linear waves $A = E/\sigma$ and $\mathbf{B} = \mathbf{C}_g A$.

A generalization to a superposition of a large number of wave components was introduced by Willebrand (1975): $\eta(\mathbf{x},t) = \sum_n a_n(\mathbf{x},t) \exp \mathrm{i}\chi_n(\mathbf{x},t)$. The amplitude $a_n(\mathbf{x},t)$ and phase $\chi_n(\mathbf{x},t)$ that are slowly varying function of $\mathbf{x}$ and t, satisfy $a_{-n} = a_n^*$ and $\chi_{-n} = -\chi_n$. Locally the wave trains are like plane waves with frequency

$\omega_n = -\partial\chi_n/\partial t$ and wave number $\mathbf{k}_n = \nabla_x\chi_n$. Note that the summation index n is not directly associated with a fixed wave number. The elevation can be written as $\eta(\mathbf{x},t) = \sum_n b_n(\mathbf{x},t)\exp \mathrm{i}[\mathbf{k}_n(\mathbf{x},t)\cdot\mathbf{x} - \omega_n(\mathbf{x},t)t]$ where the slowly varying mean phase has been included in b_n.

In the discrete formulation of the elevation there are many wave components, each of which may slowly change its wavenumber while in the continuous representation we have, at any instant of time, wave components for all $\mathbf{k}$.

Willebrand noted that the conservation of wave action holds for every wave components separately

$$\frac{\partial A_n}{\partial t} + \nabla_x \cdot [(\nabla_k W)A_n] = 0, \tag{6.18}$$

where $A_n = F_n/\sigma_n$ and $F_n = 2 \mid a_n \mid^2$. He showed that this discrete equation can be written in a continuous form as well for $A(\mathbf{k},\mathbf{x},t)$,

$$\frac{\partial}{\partial t}A(\mathbf{k},\mathbf{x},t) + (\nabla_k W)\cdot(\nabla_x A(\mathbf{k},\mathbf{x},t)) = 0, \tag{6.19}$$

with $A(\mathbf{k},\mathbf{x},t)\mathrm{d}\mathbf{k} = \sum_n^{\mathbf{dk}} A_n$. The superscript $\mathbf{dk} = \mathrm{d}k_1\mathrm{d}k_2$ on the summation sign indicates that the sum is taken only over values of n for which $\mathbf{k}_n$ lies in the range $\mathbf{dk}$ around a fixed vector wave number $\mathbf{k}$. Introducing now derivatives for fixed $\mathbf{k}(\mathbf{x},t)$ (6.19) becomes

$$\frac{\partial}{\partial t}A(\mathbf{k},\mathbf{x},t) + (\nabla_k W)\cdot\nabla_x A(\mathbf{k},\mathbf{x},t) - (\nabla_x W)\cdot\nabla_k A(\mathbf{k},\mathbf{x},t) = 0, \tag{6.20}$$

and in terms of the wave spectrum E,

$$[\frac{\partial}{\partial t} + (\nabla_k W)\cdot\nabla_x - (\nabla_x W)\cdot\nabla_k](\frac{F}{\sigma})(\mathbf{k},\mathbf{x},t) = 0, \tag{6.21}$$

where $F(\mathbf{k},\mathbf{x},t)\mathrm{d}\mathbf{k} = \sum_n^{\mathbf{dk}} F_n$ is proportional to the spectral wave energy density.

We assume that the sea surface is the superposition of a great number of wave packets. Each packet has a dominant wavenumber vector $\mathbf{k}$ which propagates with the corresponding group velocity. The space location of the dominant wavenumber vector is $\mathbf{x}$. Hence the wave packet is characterized by $\mathbf{k}$ and $\mathbf{x}$ which satisfy the following eikonal equations or Hamilton equations (see equations (6.7) and (6.9)):

$$\mathrm{d}\mathbf{x}/\mathrm{d}t = \nabla_k W(\mathbf{k},\mathbf{x}),$$

$$\mathrm{d}\mathbf{k}/\mathrm{d}t = -\nabla_x W(\mathbf{k},\mathbf{x}).$$

Hence equation (6.20) can be re-written as follows

$$\frac{\partial}{\partial t}A(\mathbf{k},\mathbf{x},t) + \frac{\mathrm{d}\mathbf{x}}{\mathrm{d}t}\cdot\nabla_x A(\mathbf{k},\mathbf{x},t) + \frac{\mathrm{d}\mathbf{k}}{\mathrm{d}t}\cdot\nabla_k A(\mathbf{k},\mathbf{x},t) = 0. \tag{6.22}$$

6.2 Wave-current interaction

As emphasized previously abnormal waves are often observed in different regions of the world ocean, where strong currents exist. First, let us consider the simplest situation when waves and current are collinear. The current is $\mathbf{U}(\mathbf{x},t) = (U(x),0)$ and the wavenumber vector $\mathbf{k}(\mathbf{x},t) = (k(x),0)$. This kind of current must be accompanied by a vertical velocity component to satisfy continuity equation. The medium is steady, hence $\omega = const.$ so that

$$\omega = kU + \sigma = k_0U_0 + \sigma_0 = const. \tag{6.23}$$

where the index signifies the value at a reference point (generally still water).

For deep water and $U_0 = 0$ it is shown that

$$\frac{c}{c_0} = \frac{1}{2}\left[1 + \left(1 + 4\frac{U}{c_0}\right)^{1/2}\right], \tag{6.24}$$

with $c^2 = g/k$ and $c_0^2 = g/k_0$. See also Grue and Palm (1985). Thus an opposing current ($U < 0$) reduces the phase velocity and shortens the waves. The critical velocity which corresponds to vanishing of the square root in (6.24) is $U = -c_0/4 = -c/2$. The spatial integration of (6.12) gives

$$C_g\frac{E}{\sigma} = const. \tag{6.25}$$

where $C_g = U + \partial\sigma/\partial k$ so that

$$\frac{a^2}{a_0^2} = \frac{\sigma}{\sigma_0}\frac{U_0 + \frac{\partial\sigma_0}{\partial k_0}}{U + \frac{\partial\sigma}{\partial k}}, \tag{6.26}$$

or

$$\frac{a}{a_0} = \left[\frac{4}{(1+\sqrt{1+4\frac{U}{c_0}})(1+4\frac{U}{c_0}+\sqrt{1+4\frac{U}{c_0}})}\right]^{\frac{1}{2}}. \tag{6.27}$$

From (6.26) we see that the amplitude becomes unbounded when $U = -\partial\sigma/\partial k$, i.e. when the local current is equal and opposite to the group velocity relative to the water ($U = -c/2 = -c_0/4$). Figure 3 displays the change of a/a_0 as a function of U/c_0. Longuet-Higgins and Stewart (1961) considered the two-dimensional propagation of waves on deep water traversing a steady nonuniform current parallel to the y-axis, which velocity field is $\mathbf{U}(\mathbf{x}) = [0, V(x), 0]$ with $\partial V/\partial y = \partial V/\partial z = 0$. The wavenumber and amplitude of the waves are supposed also to be independent of y. The angle which the waves make locally with the x-axis is denoted by θ. From kinematical considerations it results that the wave number in the y-direction, $\mid \mathbf{k} \mid \sin\theta$, is independent of x,

$$\mid \mathbf{k} \mid \sin\theta = \mid \mathbf{k}_0 \mid \sin\theta_0, \tag{6.28}$$

where the index denotes values when the transverse velocity vanishes. From (6.11) with $\sigma^2 = g \mid \mathbf{k} \mid$, and (6.28) it can be shown that

$$\sin\theta = \frac{\sin\theta_0}{[1 - (V/c_0)\sin\theta_0]^2}, \tag{6.29}$$

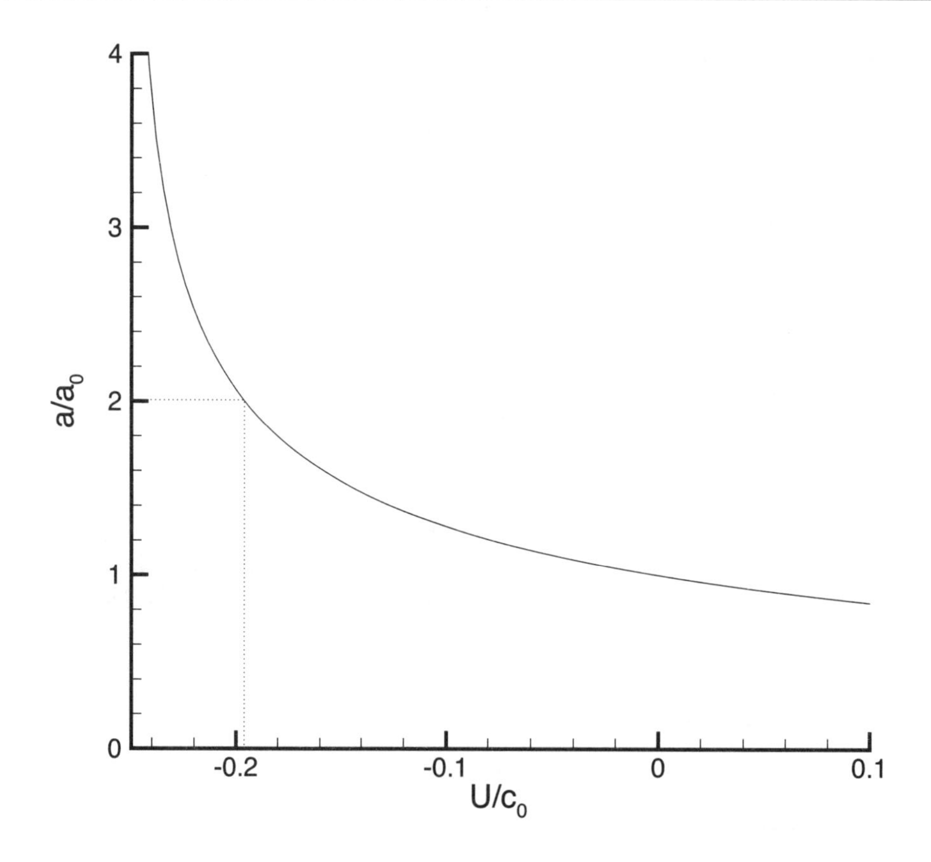

Figure 3. Change of amplitude due to collinear current.

where $c_0 = \omega_0 / \mid \mathbf{k}_0 \mid$ (with $\omega_0 = \omega$ since the current is steady).

Since $\sin\theta$ cannot exceed unity, there is clearly an upper limit for which the solution exists

$$\frac{V}{c_0} \leq \frac{1 - (\sin\theta_0)^{\frac{1}{2}}}{\sin\theta_0}. \tag{6.30}$$

The upper limit corresponds to $\theta = \pi/2$, and the waves are totally reflected by the current. From (6.12) and (6.28) we obtain for a medium independent of time and x

$$\frac{E}{E_0} = \frac{\sin 2\theta_0}{\sin 2\theta}. \tag{6.31}$$

Hence the relative amplification of the waves is

$$\frac{a}{a_0} = (\frac{\sin 2\theta_0}{\sin 2\theta})^{\frac{1}{2}}. \tag{6.32}$$

This amplification factor for waves crossing a shearing current V at an oblique angle θ is given in figure 4, for various angles of entry θ_0.

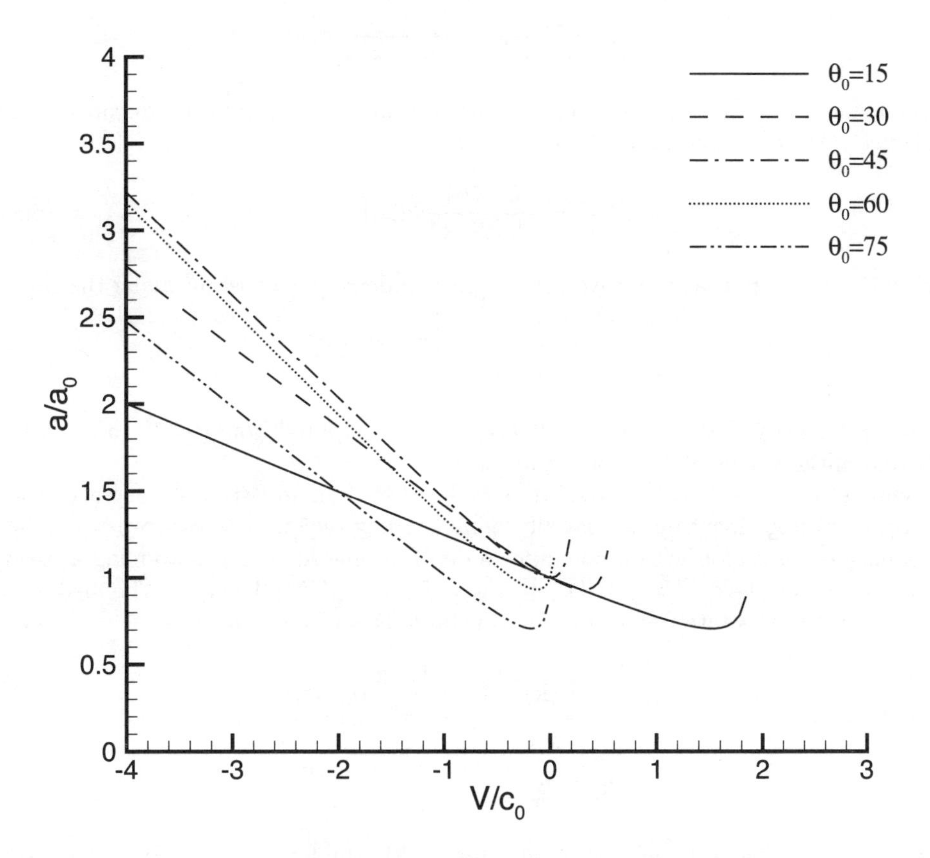

Figure 4. The amplification factor for waves crossing a shearing current V at an oblique angle θ, for various angles of entry θ_0.

Refraction of waves interacting with currents and/or underwater topography slowly varying, has been investigated by Tayfun, Dalrymple and Yang (1976) who considered random waves given by : $\eta(\mathbf{x},t) = \mathrm{Re}\left(\sum_n a_n(\mathbf{x},t) \exp \mathrm{i}[\mathbf{k}_n(\mathbf{x},t)\cdot\mathbf{x} - \omega_n(\mathbf{x},t)t + \mu_n]\right)$ where μ_n is a random phase uniformly distributed over $[0, 2\pi]$. The variance of the elevation written in a continuous form is

$$< \eta^2(\mathbf{x},t) >= \int_{\mathbf{k}} F(\mathbf{k},\mathbf{x},t)\mathrm{d}\mathbf{k}. \tag{6.33}$$

In practice one often works with the frequency directional spectrum

$$< \eta^2(\mathbf{x}, t) >= \int_\omega \int_\theta \Phi(\omega, \theta, \mathbf{x}, t)\omega \mathrm{d}\omega \mathrm{d}\theta, \tag{6.34}$$

where θ is the direction of $\mathbf{k}$ relative to x and W is given by (6.11) with

$$\sigma(\mathbf{k}) = \sqrt{g \mid \mathbf{k} \mid \tanh \mid \mathbf{k} \mid h}. \tag{6.35}$$

h is the nonuniform still water depth. The distribution Φ represents the so-called directional spectral density and is connected to F by

$$F(\mathbf{k}) = \frac{\omega}{\mid \mathbf{k} \mid} \frac{\partial W}{\partial \mid \mathbf{k} \mid} \Phi(\omega, \theta). \tag{6.36}$$

From (6.19) it results that the wave action spectral density is constant along the rays

$$\frac{F}{\sigma} = \frac{\omega}{\mid \mathbf{k} \mid} \frac{\partial W}{\partial \mid \mathbf{k} \mid} \frac{\Phi(\omega, \theta)}{\sigma} = const. \tag{6.37}$$

The spectral density $F(\mathbf{k}, \mathbf{x}, t)$ or $\Phi(\omega, \theta, \mathbf{x}, t)$ can be computed from (6.37), (6.7), (6.11), (6.35) and initial values of F (or Φ), $\mathbf{x}$ and $\mathbf{k}$.

Tayfun *et al.* (1976) considered the simple case of steady state conservative random waves propagating from a spatially homogeneous region such as deep water into an inhomogeneous region with a nonuniform depth profile $h(x)$, and traversing a steady nonuniform current field, $[\mathbf{U}(x), W(x, z)]$ where $\mathbf{U}(\mathbf{x}) = [U(x), V(x), 0]$. The medium is steady, hence the absolute frequency ω is constant. From (6.37) we obtain

$$F(\mathbf{k}) = (1 - \frac{\mathbf{U} \cdot \mathbf{k}}{\omega}) F_0(\mathbf{k}_0), \tag{6.38}$$

$$\Phi(\omega, \theta) = \frac{\mid \mathbf{k} \mid}{\mid \mathbf{k}_0 \mid} \frac{\frac{\partial W}{\partial |\mathbf{k}_0|}}{\frac{\partial W}{\partial |\mathbf{k}|}} (1 - \frac{\mathbf{U} \cdot \mathbf{k}}{\omega}) \Phi_0(\omega, \theta_0), \tag{6.39}$$

where $\frac{\partial W}{\partial |\mathbf{k}_0|}$ and $\frac{\partial W}{\partial |\mathbf{k}|}$ are the group velocities in the uniform and nonuniform media respectively. More precisely $\frac{\partial W}{\partial |\mathbf{k}|}$ is the group velocity in the propagation direction $\mathbf{k}/ \mid \mathbf{k} \mid$ (see equation (6.43)). Equations (6.38) and (6.39) describe the spatial transformation of the spectra F and Φ due to the interaction between the random wave field and nonuniform current-depth effects. As previously some kinematic constraints have to be satisfied. A generalization of (6.29) is

$$\sin\theta = \frac{\tanh(\mid \mathbf{k} \mid h)}{\tanh(\mid \mathbf{k}_0 \mid h_0)} \frac{\sin\theta_0}{[1 - (\mathbf{U} \cdot \mathbf{k}/\omega)]^2}, \tag{6.40}$$

and with $\mid \mathbf{k}_0 \mid h_0 = \infty$

$$\sin\theta = \tanh(\mid \mathbf{k} \mid h) \frac{\sin\theta_0}{[1 - (\mathbf{U} \cdot \mathbf{k}/\omega)]^2}. \tag{6.41}$$

Hence the following condition has to be satisfied

$$[1-(\mathbf{U}\cdot\mathbf{k}/\omega)] \geq [\tanh(\mid \mathbf{k} \mid h) \mid \sin\theta_0 \mid]^{\frac{1}{2}}. \tag{6.42}$$

Herein also when $\mid \sin\theta \mid = 1$ the corresponding spectral component is totally reflected.

The denominator of (6.39) must be positive. It means that

$$\frac{\partial W}{\partial \mid \mathbf{k} \mid} = \mathbf{C}_g \cdot \frac{\mathbf{k}}{\mid \mathbf{k} \mid} = \frac{\partial \sigma}{\partial \mid \mathbf{k} \mid} + \mathbf{U} \cdot \frac{\mathbf{k}}{\mid \mathbf{k} \mid} > 0, \tag{6.43}$$

where $\mathbf{C}_g = \mathbf{U} + \frac{\partial \sigma}{\partial |\mathbf{k}|}\frac{\mathbf{k}}{|\mathbf{k}|}$ is the group velocity in the fixed frame of reference. For wave propagating against the current, it means that the intrinsic group velocity must be opposite in direction and larger in magnitude relative to the current component in the direction of wave propagation. In the limit when $\frac{\partial \sigma}{\partial |\mathbf{k}|} = -\mathbf{U} \cdot \frac{\mathbf{k}}{|\mathbf{k}|}$, the corresponding spectral component can no longer propagate and the local spectral magnitude Φ becomes infinite (blocking phenomenon). In the vicinity of the blocking point the amplification of the waves may become very large.

Note that Grue and Palm (1986) considered a two-dimensional wave spectrum on a current.

Lavranov (1998b) simulated the occurrence of freak waves on the Agulhas current within the framework of Hamilton equations (6.7) and (6.9) coupled with the conservation of wave action (6.22). He considered a steady horizontal shear current of the form, $\mathbf{U} = [U(x,y), V(x,y), 0]$. Herein we have considered the current flowing in the y-direction while Lavrenov chose the x-axis for the direction of the current. Lavrenov considered the following inverse problem: what are the initial conditions (conditions corresponding to $\mathbf{U} = 0$) which produce rays focusing at a fixed point? For that he integrated numerically the Hamilton equations. From the wave spectrum transformation given by

$$S(\omega,\theta) = \frac{\mid \mathbf{k} \mid}{\mid \mathbf{k}_0 \mid} \frac{\frac{\partial W}{\partial |\mathbf{k}_0|}}{\frac{\partial W}{\partial |\mathbf{k}|}} \frac{\omega - \mathbf{U}\cdot\mathbf{k}}{\omega} S_0(\omega,\theta_0), \tag{6.44}$$

with

$$F(\mathbf{k}) = \frac{1}{\mid \mathbf{k} \mid} \frac{\partial W}{\partial \mid \mathbf{k} \mid} S(\omega,\theta), \tag{6.45}$$

the spatial distribution of the wave energy or mean wave height on the Agulhas current can be computed. The initial spectrum S_0 is

$$S_0(\omega,\theta) = \frac{8H_0}{\pi^2}\sin^2(\theta_0)(\frac{\omega_m}{\omega})^{-5}\omega^{-1}\exp[-1.2(\frac{\omega_m}{\omega})^5], \tag{6.46}$$

where H_0 is the mean height of the swell propagating on still water, and ω_m the peak frequency. Note that the initial value θ_0 is a function of values of ω and θ at the point (x,y). Lavrenov found that the maximum value of H/H_0 could reach 2.19.

6.3 Dispersion enhancement of transient wave packets

When short wave groups propagate in the front of longer wave groups that have larger group velocities they will be overtaken, and large-amplitude waves can occur at some geometrical point owing to the superposition of all the waves at the same location. Afterwards, the long waves will be in front of the short waves, and the amplitude of the wave train will decrease. It is obvious, that a significant focusing of the wave energy can occur only if all the quasi-monochromatic groups merge at a certain point. As it is well known, wind waves are not uniform in space and time. They correspond to wave groups with variable amplitude and frequency. This means that specific locations of transient wave packets should sometimes occur, leading to the formation of freak waves. This scenario can explain why the freak wave phenomenon is a rare event with short life time.

To illustrate the dispersive focusing phenomenon of water waves let us consider the simple case of unidirectional propagation in inhomogeneous media. The spatio-temporal evolution of the frequency is given by (6.9)

$$\frac{\partial \omega}{\partial t} + C_g(\omega)\frac{\partial \omega}{\partial x} = 0. \tag{6.47}$$

The group velocity is calculated using the relation (6.35) where ω herein is equal to σ the intrinsic frequency. For the sake of simplicity the depth h is assumed constant.

The initial condition is given by $\omega(x,0) = \omega_0(x)$, and the analytical solution of the nonlinear hyperbolic equation (6.47) is

$$\omega(x,t) = \omega_0(\xi) \quad \text{on} \quad x = \xi + C_g(\omega_0(\xi))t. \tag{6.48}$$

The characteristics curves are straight lines in the (x,t) plane since ω remains constant along these curves. The shape of the kinematic wave varies with space and time, and its slope is

$$\frac{\partial \omega}{\partial x} = \frac{\frac{\mathrm{d}\omega_0}{\mathrm{d}\xi}}{1 + t\frac{\mathrm{d}C_g}{\mathrm{d}\xi}}. \tag{6.49}$$

The case $\mathrm{d}C_g/\mathrm{d}\xi < 0$ (or $\mathrm{d}C_g/\mathrm{d}x < 0$ at $t = 0$) corresponds to long waves behind shorter waves and the increase of the slope of the kinematic wave up to infinity is associated with the phenomenon of long waves overtaking shorter waves. The merging of several wave groups with different frequencies at the same point and time (wave focusing) occurs at $t_f = 1/\max(-\mathrm{d}C_g/\mathrm{d}x)$. For deep water waves, all of the wave packets will meet at the same point, x_f, at time $t = t_f$ if $\omega(x,t) = g(t - t_f)/2(x - x_f)$.

The wave amplitude, a, satisfies the following equation

$$\frac{\partial a^2}{\partial t} + \frac{\partial}{\partial x}(C_g a^2) = 0. \tag{6.50}$$

This equation which corresponds to the conservation of wave energy can be obtained from (6.15) for $\mathbf{U} = 0$ and its solution is found explicitly by

$$a(x,t) = \frac{a_0(\xi)}{\sqrt{1 + t\frac{\mathrm{d}C_g}{\mathrm{d}\xi}}}, \tag{6.51}$$

where $a_0(\xi)$ is an initial distribution of the wave amplitude in space. Focusing points are singular points where the amplitude becomes infinite and behaves like $(t_f - t)^{-1/2}$.

Taking into account that wind waves evolve into frequency and amplitude modulated wave groups and that the kinematic approach predicts infinite wave height at caustics, the probability of freak wave occurrence should be very high. The kinematic approach assumes slow variations of the amplitude and frequency along the wave packet, and this approximation fails in the vicinity of focusing points. This theory is useful to emphasize that freak wave can be due to the dispersive nature of water waves, but provides unrealistic amplitudes of freak waves.

Now let us consider the Cauchy problem for singular initial data, say for instance a delta-function of intensity Q. We consider that the surface elevation of the wave packet is of the form

$$\eta(x,t) = \int_{-\infty}^{\infty} \eta(k) \exp(\mathrm{i}(kx - \omega t))\mathrm{d}k, \tag{6.52}$$

$$\eta(k) = \frac{1}{2}\int_{-\infty}^{\infty} \eta(x,0) \exp(-\mathrm{i}kx)\mathrm{d}x. \tag{6.53}$$

The frequency ω is given by (6.35). The singular delta-function disturbance (freak wave model) transforms into a smooth wave field whose asymptotic expression when t and x become large can be obtained through the stationary-phase method

$$\eta(x,t) = Q\sqrt{\frac{C_g}{2\pi x \mid \frac{\mathrm{d}C_g}{\mathrm{d}k} \mid}} \exp(\mathrm{i}(kx - \omega t - \frac{\pi}{4})), \tag{6.54}$$

where the group velocity C_g (and also wave frequency and wave number) can be calculated for instance from the self similar solution $C_g = (x - x_f)/(t - t_f)$ corresponding to the previous optimal focusing. Due to invariance of the Fourier integral to the signs of the coordinate, x, and time, t, the smooth modulated wave field given by (6.54) evolves to a singular solution. In the vicinity of the leading wave ($k \to 0$), expression (6.54) is not valid since the wavelength becomes comparable to the distance to the source and should be replaced with

$$\eta(x,t) = Q(\frac{2}{cth^2})^{\frac{1}{3}}\mathrm{Ai}[(\frac{2}{cth^2})^{\frac{1}{3}}(x - ct)], \tag{6.55}$$

where the long-wave approximation of the dispersion relation has been used

$$\omega = c(1 - \frac{k^2k^2}{6}), \tag{6.56}$$

$$c = \sqrt{gh}, \tag{6.57}$$

and Ai denotes the Airy-function. The amplitude of the leading wave decreases as $t^{-1/3}$, and its length increases as $t^{1/3}$.

It has been shown that a delta-function disturbance may evolve in a smooth wave field, and owing to spatial and temporal invariance we can claim that the initial smooth wave field like (6.54) and (6.55) with inverted coordinate and time will form a freak wave. These solutions demonstrate obviously which wave packets can generate a freak wave.

Generally speaking, the singular solutions of the linear problem have only mathematical interest. A Gaussian pulse is a more realistic model for freak waves. So, consider in the long-wave approximation the following Gaussian pulse as initial condition

$$\eta(x,0) = a_0 \exp(-K^2x^2), \tag{6.58}$$

where a_0 and K^{-1} are the amplitude and the width respectively. Then the integral (6.52) can be calculated exactly in an explicit form, and the wave field is

$$\eta(x,t) = \frac{a_0}{K(\frac{h^2ct}{2})^{\frac{1}{3}}} \exp\left[\frac{1}{2h^2ctK^2}(x - ct + \frac{6}{77h^2ctK^4})\right] \mathrm{Ai}\left[\frac{x - ct + \frac{9}{77h^2ctK^4}}{(h^2ct/2)^{\frac{1}{3}}}\right]. \tag{6.59}$$

Inverting coordinate and times, this wave packet evolves into a Gaussian pulse (6.58), and then again disperse according to (6.59). Figure 5 shows the freak wave formation in a dispersive wave packet in shallow water. Similar solutions can be found for wave groups with Gaussian envelope in deep water.

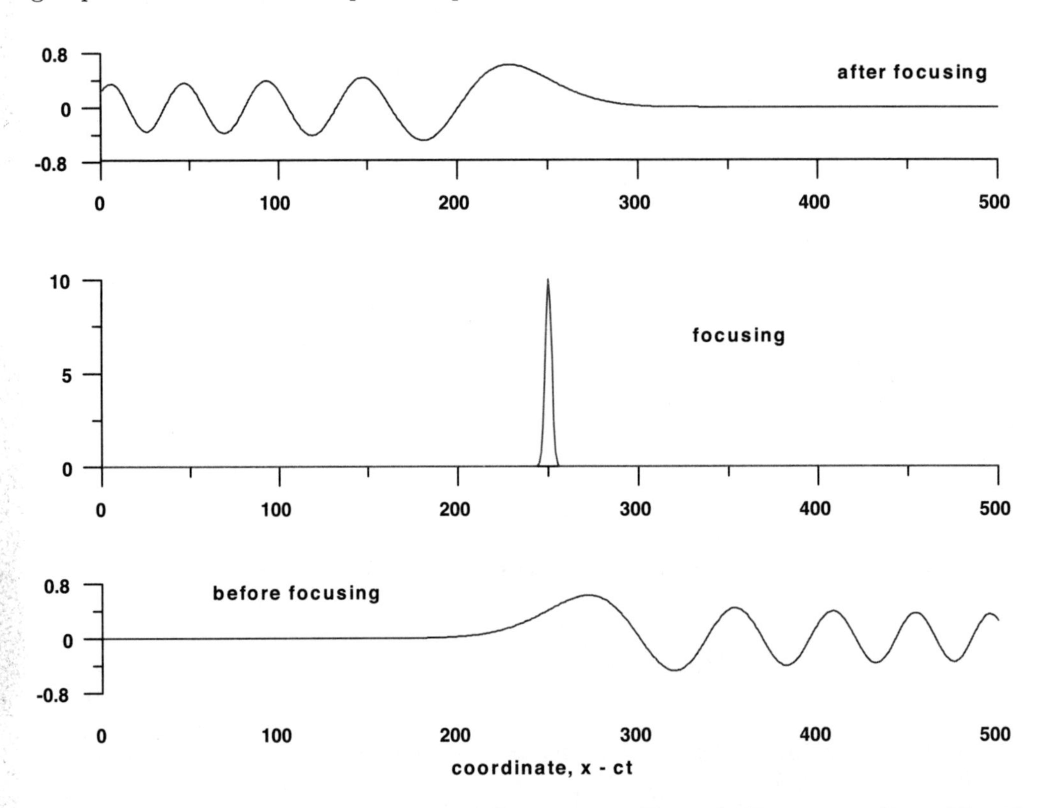

Figure 5. Formation of a freak wave of Gaussian profile in shallow water. From Kharif and Pelinovsky (2003).

7 Nonlinear approaches of the problem

From section 6 devoted to linear theories one may conclude that the main mechanisms generating freak waves are related with wave focusing of frequency modulated wave packets (concerning both dispersive and spatial focusing) and wave blocking on opposite currents. Both mechanisms are sensitive to the spectrum width of the wind waves field. In particular, the focusing mechanism requires a wide energetic spectrum with a specific phase distribution; meanwhile the wave-current interaction phenomenon is effective when the spectrum is very narrow. Due to the fact that freak waves occur more often than linear statistics predicts it seems natural to study their dynamics within the framework of nonlinear theories. The effect of nonlinearity is twofold: (i) it can destroy the phase coherence between spectral components of the wave field, washing out caustics and focuses that decreases the amplitude of the extreme events (ii) it produces instability of the wave field leading to the formation of anomalous high waves. Note that Peregrine and Smith (1979) and Peregrine (1983) considered nonlinear effects on waves near caustics. The second important ingredient is randomness of the wind wave field that also acts on the phase coherence of deterministic transient wave groups. All these aspects will be analyzed in this section where nonlinearity is taken into account. Weakly nonlinear theories as well as fully nonlinear models will be considered.

7.1 Weakly nonlinear freak wave packets in deep and intermediate depths

The one-dimensional nonlinear Schrödinger equation. The phenomenon of freak waves can be studied from the point of view of the simplest nonlinear equation governing the evolution of slowly modulated and weakly nonlinear deep water gravity waves in one space dimension: the Nonlinear Schrödinger equation (NLS equation)

$$\mathrm{i}(\frac{\partial a}{\partial t} + C_g\frac{\partial a}{\partial x}) = \frac{\omega_0}{8k_0^2}\frac{\partial^2 a}{\partial x^2} + \frac{\omega_0 k_0^2}{2} \mid a \mid^2 a, \tag{7.1}$$

where the surface elevation, $\eta(x,t)$ is given by

$$\eta(x,t) = \mathrm{Re}[a(x,t)\exp(\mathrm{i}k_0x - \mathrm{i}\omega_0 t)], \tag{7.2}$$

where k_0, $\omega_0 = \omega(k_0)$ and $C_g = \frac{\partial\omega}{\partial k}(k_0)$ are the wave number, frequency of the carrier wave and the group velocity. The complex amplitude a, is a slowly varying function of time and space. The NLS equation (7.1) is a universal equation that plays an important role in the understanding of nonlinear dynamics of water waves.

It is well known that initially-uniform wave trains suffer from modulational instability known as the Benjamin-Feir instability. Benjamin and Feir (1967) showed the progressive distortion of initially-uniform water waves as they propagate downstream from a wavemaker. Modulation is due to energy exchange between the fundamental harmonic with two neighboring waves (sidebands). This result is also in agreement with somewhat earlier analysis by Lighthill (1965), Zakharov (1966) and Zakharov (1968). The main features of this instability is described by the NLS equation (7.1). A uniform train of amplitude a_0 is unstable to Benjamin-Feir instability (BF instability or modulational instability) corresponding to long wave disturbances of wave number Δk, when the following condition is satisfied

$$\frac{\Delta k}{k_0} < 2\sqrt{2}k_0 a_0. \tag{7.3}$$

The maximum of instability occurs at $\Delta k/k_0 = 2k_0 a_0$, with the maximum growth rate equal to $\omega_0(k_0 a_0)^2/2$. The nonlinear stage of the BF instability has been thoroughly investigated analytically, numerically and experimentally (see e.g. the paper by Dias and Kharif (1999)).

Figure 6 describes the formation of strongly modulated wave groups due to the BF instability simulated numerically. In this case the chosen modulational instability is not the most unstable. One can observe in the first stage of the evolution the formation of a breather and in the second stage the occurrence of two groups corresponding to the excitation of the most unstable perturbation. The value of the initial wave steepness is 0.03. The characteristic timescale of appearance and disappearance of the wave groups is of order $[\omega_0(k_0 a_0)^2]^{-1}$. This phenomenon is the Fermi-Pasta-Ulam recurrence. Such a behavior can be due to the excitation of breather solutions of the nonlinear Schrödinger equation. Using a fully nonlinear irrotational flow solver, Henderson *et al.* (1999) computed the time evolution of a uniform wave train with a small modulation which grows. For very long runs (over thousands of periods) they observed the formation of short groups of steep waves whose envelopes fit with one of the analytical solutions of the nonlinear Schrödinger equation, namely the isolated Ma-soliton. The Ma-soliton (or Ma-breather) is time-periodic. Furthermore, the nonlinear Schrödinger equation admits also explicit solutions which are space-periodic: the Akhmediev-breather. When the period of the Ma-soliton or the wavelength of the Akhmediev-soliton become infinite one obtains the so-called algebraic breather (or Peregrine-soliton) given in a system of coordinates moving with the group velocity by

$$a(x,t) = a_0 \exp(\mathrm{i}\omega_0 t)\left[1 - \frac{4(1+2\mathrm{i}\omega_0 t)}{1+16k_0^2x^2+4\omega_0^2t^2}\right]. \tag{7.4}$$

This algebraic breather is shown in figure 7 in dimensionless coordinates a/a_0, $a_0k_0^2x$ and $a_0^2k_0^2\omega_0 t$. The maximal height of this wave (from trough to crest) exceeds the value 3. The Ma-, Akhmediev- and Peregrine- breathers can be considered as simple analytical models of freak waves because they satisfied the criterion (4.6) for the height of freak waves. Breather solutions describe simplified dynamics of modulationally unstable wave packets. Osborne *et al.* (2000) and Calini and Schober (2002) gave more detailed analysis of freak wave events during the nonlinear stage of modulational instability by using the inverse scattering approach (so-called homoclinic orbits).

Hence, the nonlinear instability of a weakly modulated wave packet in deep water may generate a short-lived anomalous high wave. This mechanism is different from the previous ones presented in section 6.

The wave spectrum of the modulated wave train may present many harmonics contained in a relatively narrow band for applicability of the nonlinear Schrödinger equation. In this case, the wave field is assumed to be a superposition of different spectral components propagating with different velocities depending on the wave number and amplitude

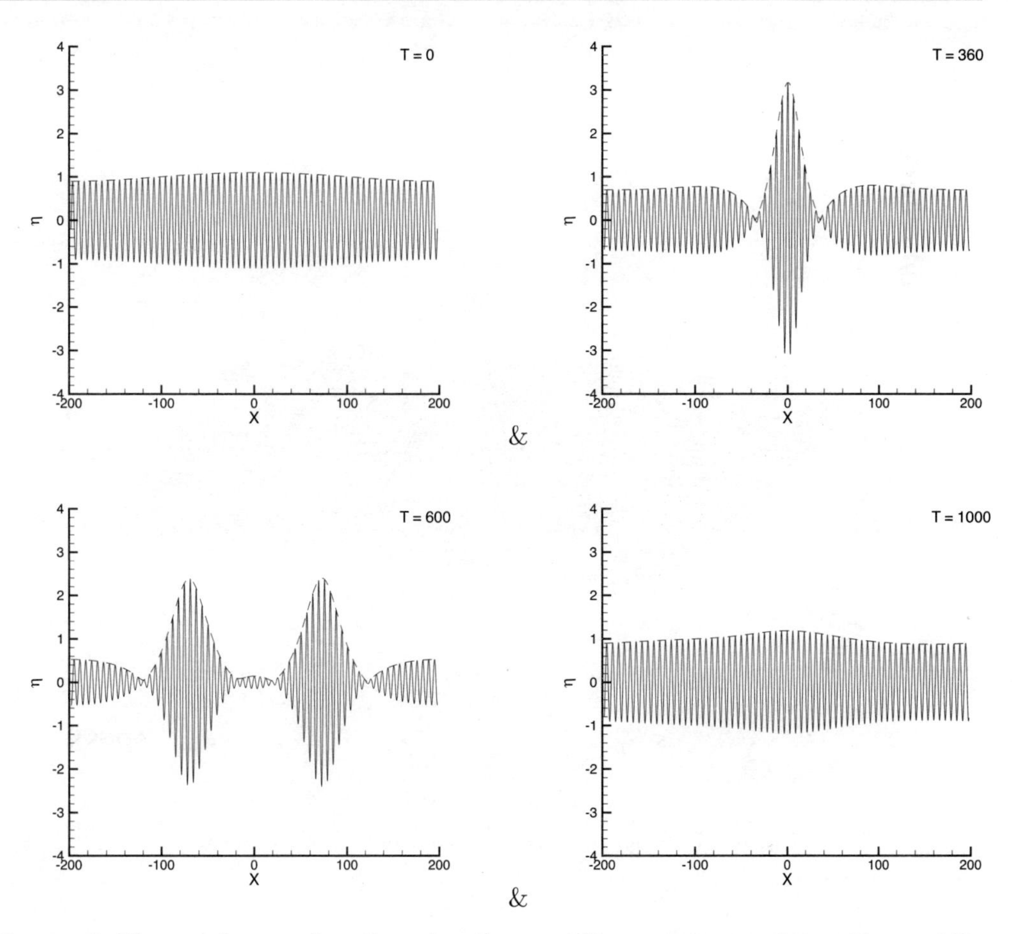

Figure 6. The envelope and surface elevation at different instants of time $T = \omega_0 t/2$, corresponding to a Stokes wave train of steepness $a_0 k_0 = 0.03$ amplified by an unstable perturbation corresponding to $\Delta k = a_0 k_0$.

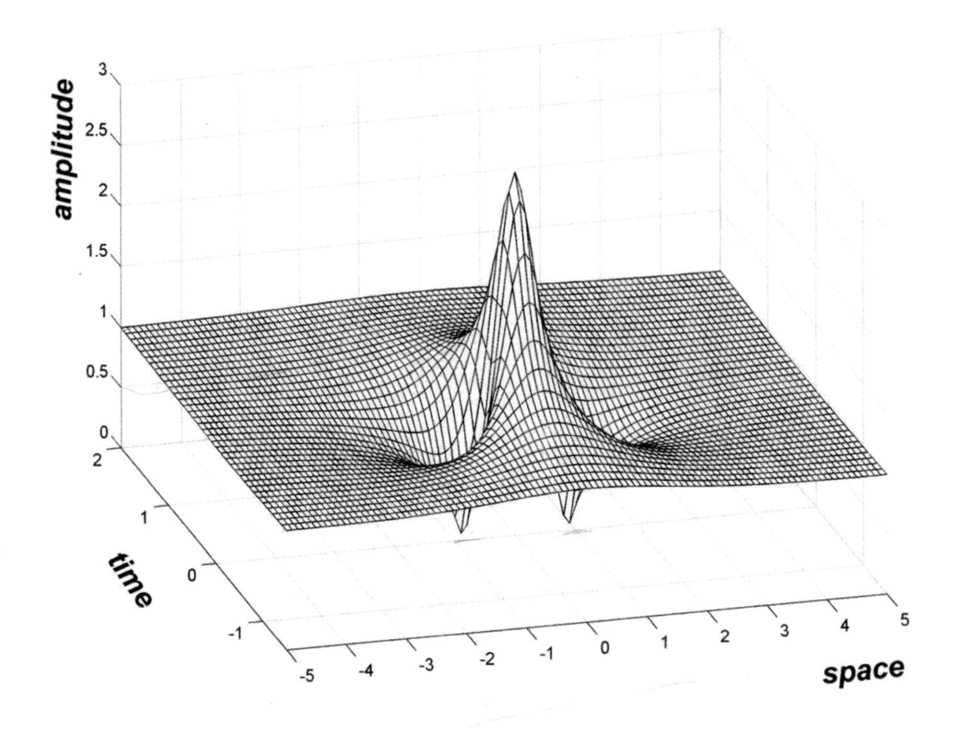

Figure 7. Algebraic breather as a model of freak wave.

as well. As a result, the focusing process is possible for specific phase relations between harmonics. Formally, this process can be analyzed by using the generalized kinematics equations (6.47) and (6.50) with the dispersion relation of water waves depending on the wave amplitude. It means that now both equations are coupled via the nonlinear dispersion relation. In this case the system of equations becomes elliptic and does not provide a simple interpretation in terms of caustics as for hyperbolic systems.

Equation (7.1) for a can be transform to the self-focusing nonlinear Schrödinger equation

$$\mathrm{i}\frac{\partial q}{\partial T} + \frac{\partial^2 q}{\partial X^2} + 2 \mid q \mid^2 q = 0, \tag{7.5}$$

by the transformation

$$T = \frac{1}{2}\omega_0 t, \qquad X = 2k_0 x - \omega_0 t, \qquad q = \frac{1}{\sqrt{2}}k_0 a^*, \tag{7.6}$$

where $*$ denotes complex conjugate.

Due to invariance of equation (7.5) under the transformation $T \rightarrow -T$, $\mathrm{i} \rightarrow -\mathrm{i}$, the simplest way to find the nonlinear wave packet evolving to a giant wave can be suggested: Take the expected shape of the freak wave as an initial condition for equation (7.5) and consider the resulting wave field as an initial condition that gives the freak wave under the previous invariant transformation. The initial value problem of equation (7.5) can be solved by the inverse scattering transform (IST). Generally, the scattering data include both continuous and discrete spectra: the continuous spectrum corresponds to dispersive wave packets which spread over large regions of space decaying as $T^{-1/2}$. In the case of no discrete spectrum (7.5) admits for large times the following solution corresponding to a phase modulated wave

$$q(X,T) = \frac{Q}{\sqrt{T}} \exp[\mathrm{i}(\frac{X^2}{4T} + 2Q^2 \ln T + \theta)], \tag{7.7}$$

where Q and θ are functions of X/T. Note that when Q and θ are real constants (7.7) is an exact solution of equation (7.5). Replacing T with $T_c - T$ and i with $-i$, equation (7.7) describes the transformation of an initial modulated wave train into a delta function at $T = T_c$, formally representing the freak wave.

Rather to consider a freak wave with an infinite amplitude corresponding to a singularity that develops at the focus, we can consider more realistic profiles for freak waves such as $q(X) = q_{fr}\mathrm{sech}(X/L)$, where the index fr refers to freak wave and L is the characteristic length of the wave. Satsuma and Yajima (1974) used the IST to solve equation (7.5) for sech initial profile. The number of solitons, N, depends on the mass of the freak wave, $M = \pi q_{fr} L$, trough the relation

$$N = I[\frac{M}{\pi} + \frac{1}{2}], \tag{7.8}$$

where I is the integer function.

When $M < \pi/2$ there is no soliton generation and the wave evolves like the solution (7.7) for large X and T. Using the previous transformation, the wave focuses in a freak wave and then disperses. When the amplitude of the freak wave satisfies $\pi/2 < M < 2\pi/3$ it is found that a freak wave can be generated by a dispersive wave packet plus a single soliton as shown by Kharif *et al.* (2001).

The previous examples have demonstrated that the mechanism of wave focusing valid in linear theory applies in nonlinear theory as well.

Within the framework of the NLS equation (7.5), we consider now the freak wave formation due modulational instabilities. It is well known that a simple solution of the NLS equation (7.5) depending on time only is

$$q(T) = q_0 \exp(2\mathrm{i} q_0^2 T), \tag{7.9}$$

where q_0 is a constant. The corresponding expression of a is

$$a(t) = a_0 \exp(-\frac{\mathrm{i}}{2}\omega_0 k_0^2 a_0^2 t). \tag{7.10}$$

This solution represents the fundamental component of the Stokes wave. We consider a perturbation of (7.9) and express it in the following form

$$q(X,T) = q_0(1 + p(X,T))\exp(2\mathrm{i}q_0^2 T), \tag{7.11}$$

where p is a complex to be determined. Substituting this expression in equation (7.5) and ignoring the products of $p(X,T)$ we obtain

$$\mathrm{i}\frac{\partial p}{\partial T} + \frac{\partial^2 p}{\partial X^2} + 2q_0^2(p + p^*) = 0. \tag{7.12}$$

We seek a solution for p of the form

$$p(X,T) = p_1 \exp(\Omega T + \mathrm{i}lX) + p_2 \exp(\Omega^* T - \mathrm{i}lX), \tag{7.13}$$

where p_1 and p_2 are complex constants, l is a real wavenumber and Ω is the growth rate when it is real and positive. Substituting (7.13) in equation (7.12) gives the following system of coupled equations

$$(\mathrm{i}\Omega - l^2)p_1 + 2q_0^2(p_1 + p_2^*) = 0, \tag{7.14}$$

$$(\mathrm{i}\Omega^* - l^2)p_2 + 2q_0^2(p_1^* + p_2) = 0. \tag{7.15}$$

Taking the complex conjugate of (7.15) we obtain finally

$$(\mathrm{i}\Omega - l^2 + 2q_0^2)p_1 + 2q_0^2 p_2^* = 0, \tag{7.16}$$

$$2q_0^2 p_1 + (-\mathrm{i}\Omega - l^2 + 2q_0^2)p_2^* = 0. \tag{7.17}$$

A necessary and sufficient condition for non trivial solutions is the vanishing of the determinant

$$\Omega^2 = l^2(4q_0^2 - l^2). \tag{7.18}$$

The growth rate Ω is real and positive or purely imaginary whether $l^2 < 4q_0^2$ or $l^2 > 4q_0^2$. Hence the Benjamin-Feir instability occurs when the following criterion is satisfied

$$l^2 < 2a_0^2 k_0^2. \tag{7.19}$$

The range of instability is

$$0 < l < l_c = \sqrt{2}a_0 k_0. \tag{7.20}$$

The maximum of instability corresponds to $l = a_0 k_0$ with a rate of growth given by

$$Re(\Omega_{\max}) = a_0^2 k_0^2. \tag{7.21}$$

In figure 8 are shown snapshots numerically computed from equation (7.5), for a cycle of recurrence (Fermi-Pasta-Ulam recurrence), of the envelope and free surface elevation of an initial perturbed Stokes wave given by

$$q(X,0) = q_0(1 + \alpha\cos(lX)), \tag{7.22}$$

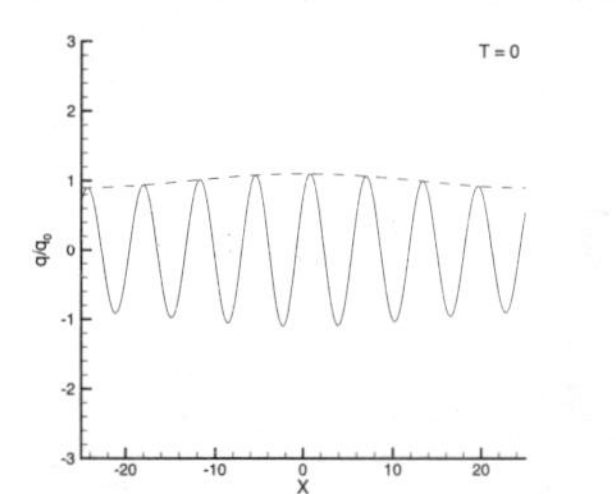

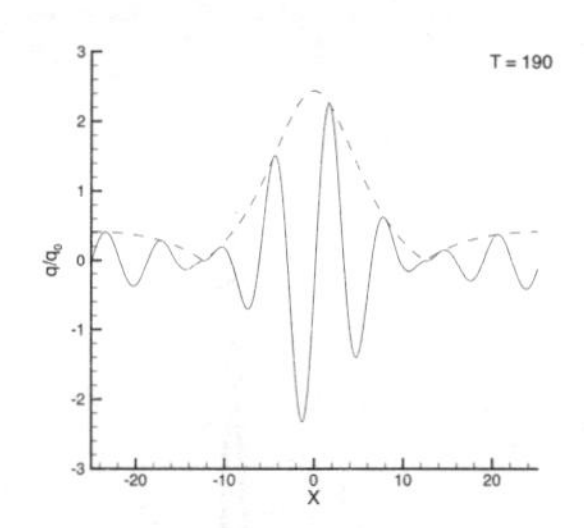

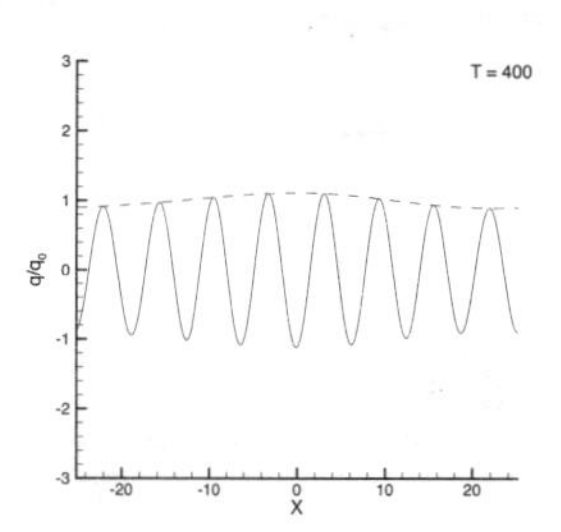

Figure 8. The envelope $\mid q \mid$ and elevation η at different instants of time, corresponding to a Stokes wave train of steepness $a_0k_0 = 0.125$ amplified by its most unstable perturbation.

where $q_0 = \sqrt{2}/16$, $l = 1/8$ and $\alpha = 0.1$. This case corresponds to the most unstable perturbation of a Stokes wave of wave steepness $a_0k_0 = 1/8$.
As it is well known, the width of the unstable sidebands is $O(\epsilon)$ with $\epsilon = a_0k_0$. The characteristic time scale of the instability is $O(\epsilon^{-2})$. Dispersion and nonlinearity are in balance and the characteristic time of existence of the freak wave is close to that of the modulational instability. It means that for small wave steepness this characteristic time scale is large and one can assume that the focusing mechanism should dominate the water wave dynamics. With increase of the wave steepness both modulational and focusing mechanism compete. In the case of no specific phase modulation of the wave packet, the nonlinear mechanism of the freak wave phenomenon due to Benjamin-Feir instability should dominate the dynamics of the wave field (see figure 8). But if a specific spatial distribution of the waves occurs (due to some external effect) the phase modulation can cardinally modify the modulational instability. The effect of the quadratic phase modulation on the nonlinear evolution of the modulational instability is investigated numerically on the basis of (7.5). The initial condition is

$$q(X,0) = q_0(1 + \alpha \cos(lX)) \exp(\mathrm{i}\frac{X^2}{D^2}), \tag{7.23}$$

where $q_0 = 0.043$, $l = 1/28$, $\alpha = 0.1$, and D varies greatly. In figure 9 are displayed the envelopes corresponding to maximal amplitude for different values of the parameter of the phase modulation D. The maxima occur at different instants of time. The effect of the phase modulation on the initial envelope is twofold: (i) it increases the wave amplitude and (ii) it reduces the period of occurrence of the freak wave. As a result, the phase modulation of the amplitude-modulated envelope leads to the formation of a more energetic wave impulse for a shorter period of time.
For $D = 28$, the long-term evolution of the nonlinear dynamics of the wave field is displayed in figure 10. The phase modulation generates a complex behavior of the wave envelope with several peaks and holes that can be considered as a group of freak waves. The time evolution of the maximal wave amplitude is shown in figure 11a. The very large amplitude peaks occur several times, and their amplitudes decrease with time. Then the process becomes more or less stationary and peaks with amplitudes about 0.12 appear regularly.

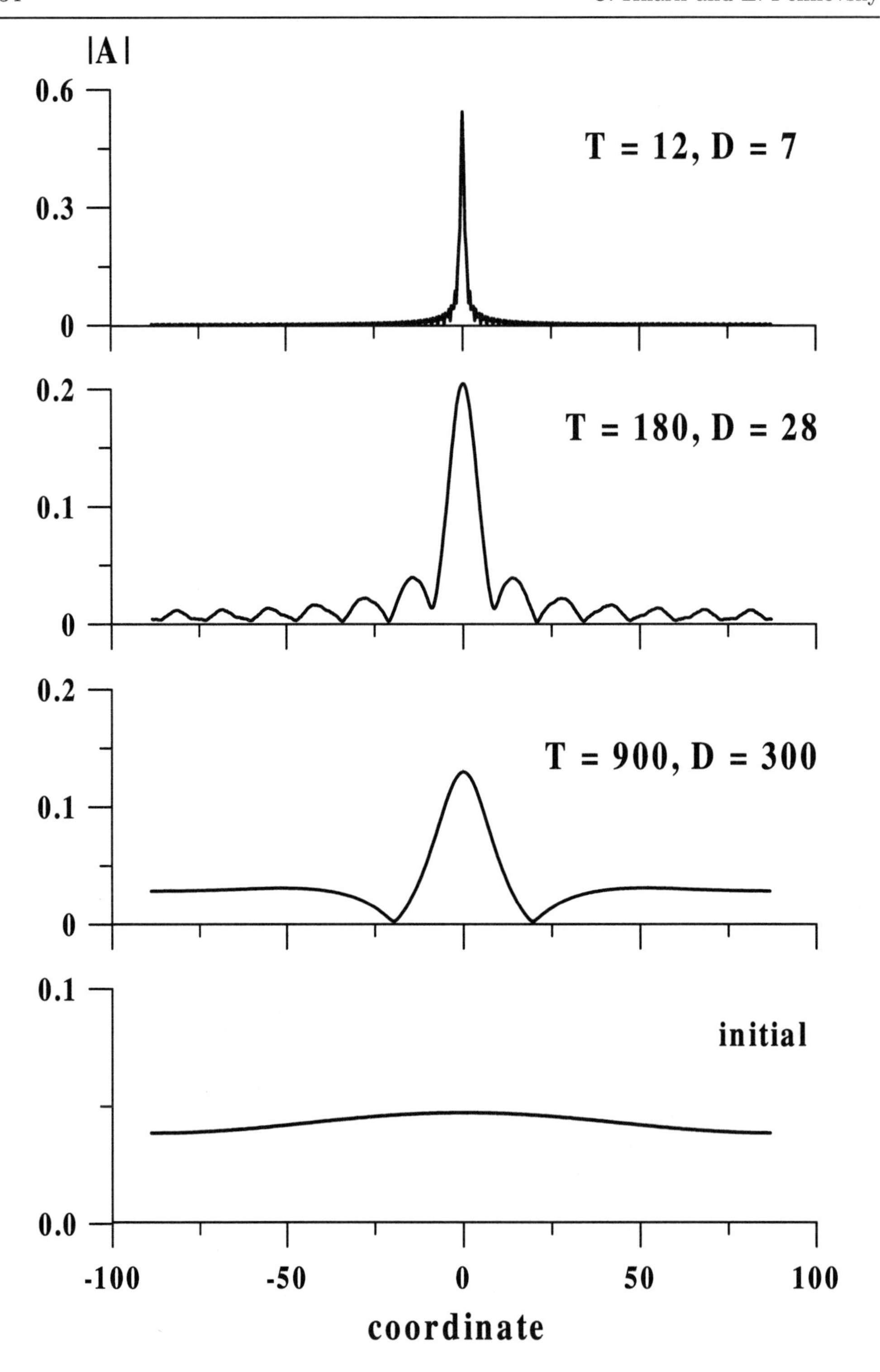

Figure 9. Development of the modulational instability for different values of D. Maximal amplitudes are provided. From Kharif *et al.* (2001).

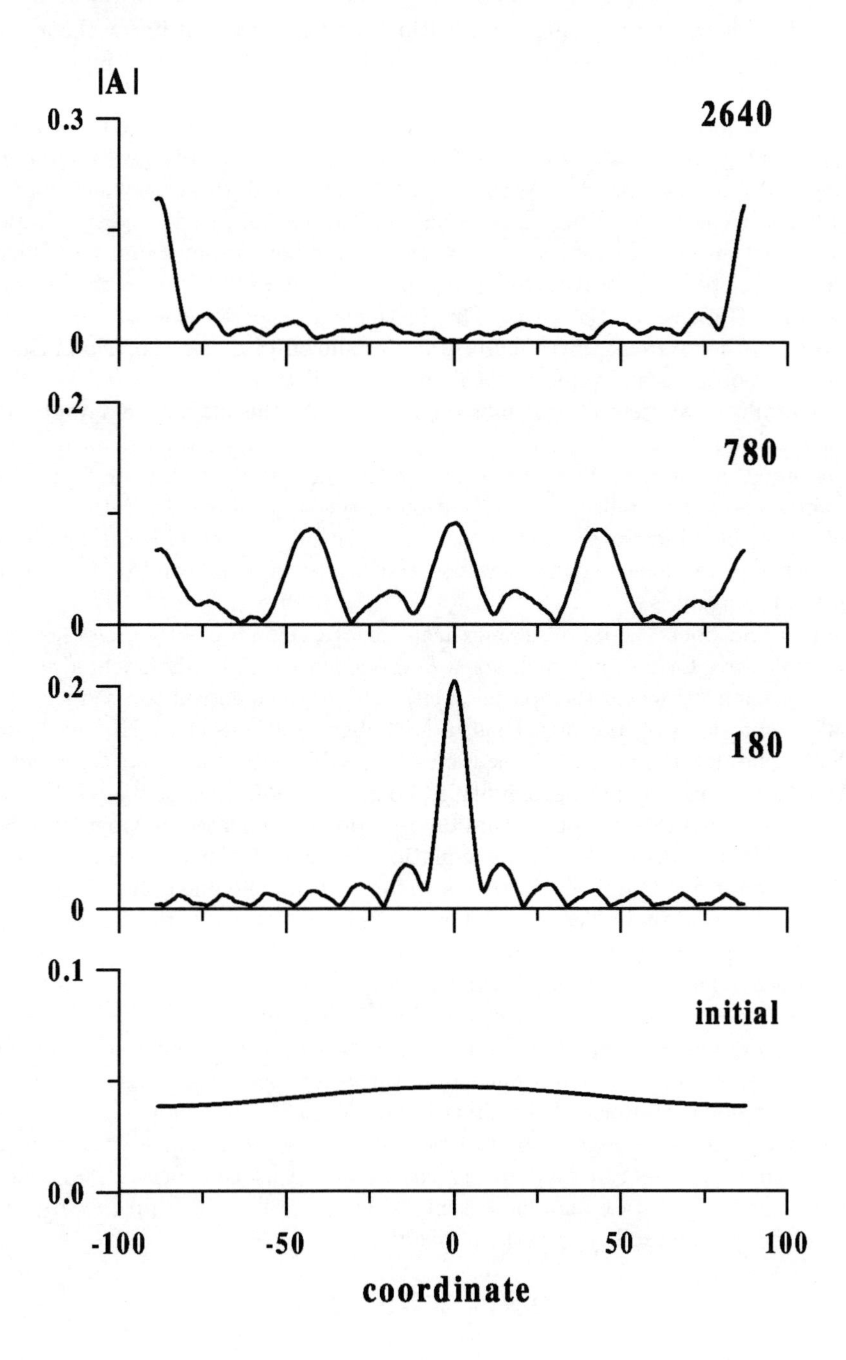

Figure 10. Development of the modulational instability for $D = 28$. From Kharif *et al.* (2001).

Figure 11b shows the time evolution within the framework of linear theory. In this case, the process of generation of significant peaks is almost periodic. Similar behavior is found for wide ranges of variation of the phase index D. Physically, the role of nonlinearity in the wave field dynamics can be explained as follows. The quadratic phase modulation in equation (7.23) corresponds to a linear variation of the wavenumber (wave frequency) with distance (fetch). As shown in subsection 6.3, this case which corresponds to short wave trains propagating ahead long wave trains can provide spatio-temporal focusing. At focusing points the slope of the wave number becomes infinite. During the defocusing the slope of the wave number changes its sign and decreases, the longer waves are now in front of the shorter waves. Due to the spatial periodicity of the domain, one can expect to observe freak waves many times as shown in figure 11b. Nonlinear effects produce the smoothing and uniformity of the phase distribution. Herein, the role of the classical modulational instability is more significant. At this stage, the amplitude of the freak wave is lower than three times the amplitude of the unperturbed wave train.

Therefore, the effect of phase modulation on an initial weakly modulated Stokes wave train leads to a significant amplification of the freak wave generation. The phase modulation of the wind wave field can be due to specific meteorological conditions and the link between the observed freak waves and heavy weather conditions is very often cited in the literature.

Calini and Schober (2002) observed that a chaotic regime greatly increases the likelihood of freak wave formation which are well modelled by higher homoclinic solutions of the NLS equation for which the spatial excitations have coalesced to produce a wave of maximal amplitude. Very recently, Islas and Schober (2005) used the NLS equation (7.5) to correlate the development of freak waves in oceanic sea states characterized by the JONSWAP spectrum with the proximity to homoclinic solutions of the NLS equation.

For random disturbances the situation is more complicated. Alber and Saffman (1978) and Alber (1978) derived an equation describing the evolution of the envelope of a random wave train. Their analysis started from the one-dimensional nonlinear Schrödinger equation or its equivalent for the finite depth case, the Davey-Stewartson system, rather than from the full equations of motion. Using a more general approximate equation, the Zakharov equation, Crawford, Saffman and Yuen (1980) investigated also the evolution of a random inhomogeneous field of nonlinear deep water gravity waves. Following Alber and Saffman (1978), they investigated the stability of a narrow-band homogeneous spectrum to inhomogeneous perturbations in the limiting cases of the one- and two-dimensional nonlinear Schrödinger equations. Using a more realistic spectrum they obtained results that agree qualitatively with those of Alber and Saffman, namely they found that the effect of randomness characterized by the spectral bandwidth, σ_s, is to reduce the growth rate and the extent of the instability. Within the framework of equation (7.5), the instability growth rate is given by

$$\Omega = l\sqrt{4q_0^2 - l^2} - 2\Sigma_s, \tag{7.24}$$

where $\Sigma_s = \sigma_s/(2k_0)$ and $q_0^2 = k_0^2\overline{a_0^2}$ ($\overline{a_0^2}$ is the mean square wave amplitude or twice the variance of the surface elevation). The parameter Σ_s is the normalized bandwidth of the

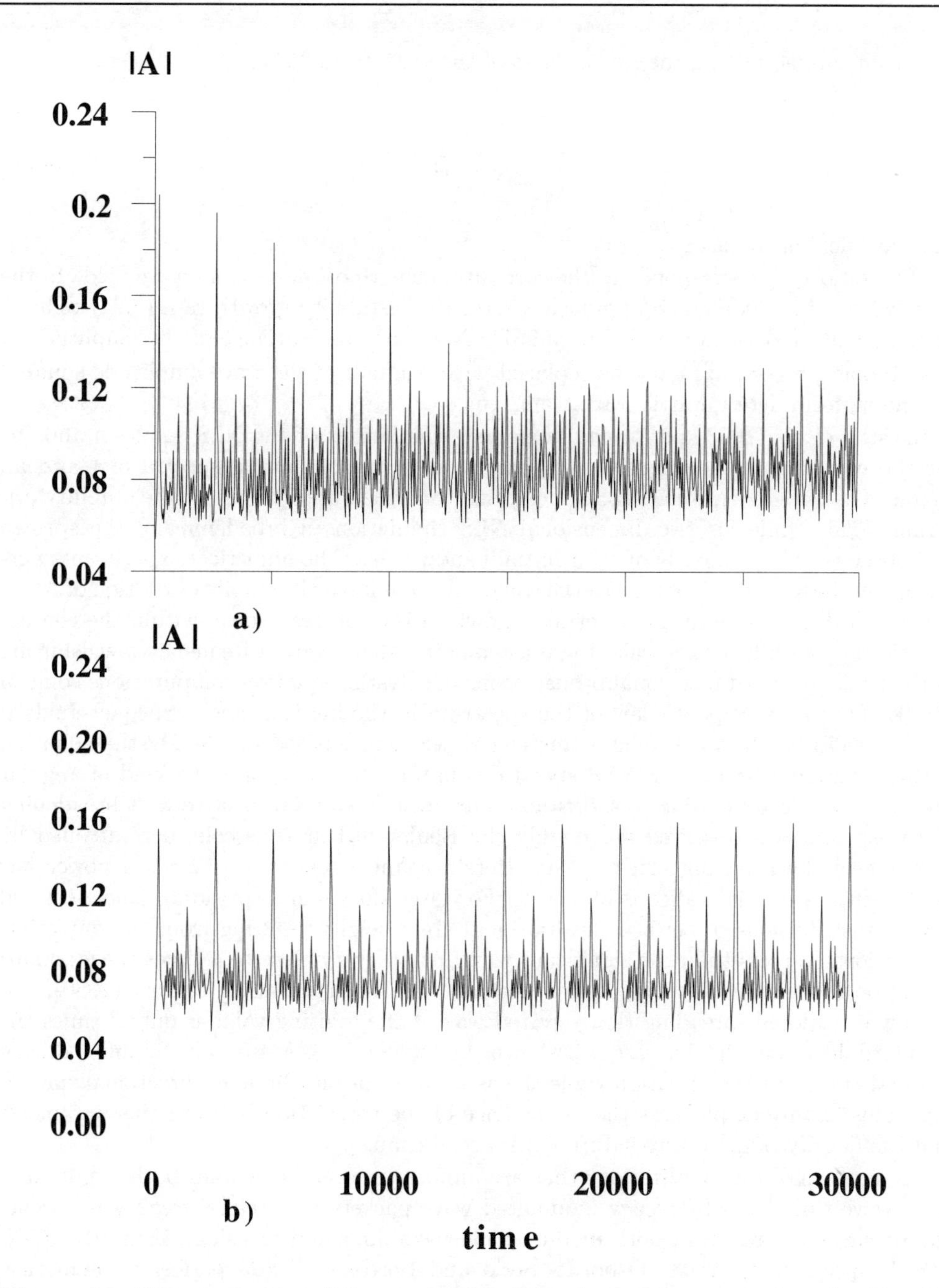

Figure 11. Time evolution of the maximal amplitude in the wave packets for $D = 28$: (a) nonlinear case and (b) linear case. From Kharif *et al.* (2001).

spectrum, characterizing the randomness of the wave field. When

$$\Sigma_s > q_0 \quad \text{or} \tag{7.25}$$

$$\frac{\sigma_s}{2k_0} \geq k_0\sqrt{\overline{a_0^2}}, \tag{7.26}$$

the wave field is stable.

The ratio $1/\sigma_s$ corresponds to the correlation length of the random wave field. In the limit where the randomness approaches zero, the instability growth rate (7.24) reduces to the result of Benjamin and Feir (1967). Note that the variance of the amplitude of the random process, $\overline{a_0^2}$, must be replaced with one-half of the wave amplitude squared for the uniform deterministic wave train, $\frac{1}{2}a_0^2$.

Dysthe *et al.* (2003) performed numerical simulations of the NLS equation and the Dysthe equation (a modified NLS equation) with random initial spectra of Gaussian shape. The one-dimensional NLS simulations confirmed approximately the stability criterion (7.26) while for two-dimensional NLS simulations a broadening of the spectra was observed independently of their initial bandwidths. The numerical experiments performed in the framework of the Dysthe equation invalidated the stability criterion of Alber since regardless to the initial spectral bandwidth the spectra evolve, within the characteristic Benjamin-Feir timescale, to an asymmetric shape with a frequency-downshifting of the peak. In two-dimensional propagation, the Dysthe equation computations confirm the $|\mathbf{k}|^{-5/2}$ (or ω^{-4}) power law of the spectrum in the inertial range. Socquet-Juglard *et al.* (2005) conducted similar numerical experiments, based on the Dysthe equation with now a truncated JONSWAP spectra as initial conditions and two kind of angular distributions corresponding to short- and long-crested waves respectively. A broadening of the spectra was observed on roughly the Benjamin-Feir timescale, accompanied by a frequency-downshifting of the peak. Herein again a tendency towards a power law $|\mathbf{k}|^{-5/2}$ in the inertial range is observed. For large directional spreading (short-crested waves) distributions of surface elevation and crest height resulting from the numerical simulations and less than the significant wave height (very close to 4 times the standard deviation), fit very well the theoretical second order distributions of Tayfun (1980). For narrow directional spreading (long-crested waves) the limiting value is only 3 times the standard deviation. In this case it is shown that the kurtosis is strongly amplified during the first stage of the simulation while this is not the case for broader directional spreading. This feature emphasizes the importance of the correlation between this parameter and the density of large waves during a spectral change.

In situ, wind wave realizations that are uniform in average contain both almost uniform wave trains and frequency modulated wave packets. Therefore, freak wave events can occur as the result of both modulational instability and focusing. Using the JONSWAP spectrum Onorato, Osborne, Serio and Bertone (2001) performed numerical experiments to study freak wave generation and its statistics. In particular, it was shown that if the spectrum is narrow (increasing the value of the enhancement coefficient in the JONSWAP spectrum) the probability of freak wave occurrence is increased. This increase can be explained by the effect of the modulational instability in addition to the wave focusing.

The NLS equation can be derived for basins of arbitrary depth. For finite depth, the coefficients of equation (7.1) depend on the parameter kh, where h is the water depth. For 2D motions Whitham (1967) and Benjamin (1967) found that instability could occur only if $kh > 1.363$. In shallow water, uniform wave trains are stable and the focusing phenomenon can be the main mechanism responsible for the formation of freak waves. Due to weak dispersion in shallow water, the coherence between spectral components becomes strong, leading to the formation of solitons and quasi-shock waves. In this case another model than the NLS equation is required that will be described in subsection 7.3.

The two-dimensional nonlinear Schrödinger equation. For 3D motions in deep water, the two-dimensional NLS equation is

$$\mathrm{i}(\frac{\partial a}{\partial t} + C_g \frac{\partial a}{\partial x}) = \frac{\omega_0}{8k_0^2}\frac{\partial^2 a}{\partial x^2} - \frac{\omega_0}{4k_0^2}\frac{\partial^2 a}{\partial y^2} + \frac{\omega_0 k_0^2}{2} \mid a \mid^2 a. \tag{7.27}$$

The leading order displacement of the free surface of the water, $\eta(x, y, t)$, is related to a by the expression

$$\eta(x, y, t) = \mathrm{Re}[(a(x, y, t)\exp(\mathrm{i}k_0 x - \mathrm{i}\omega_0 t)]. \tag{7.28}$$

It is important to note that the two-dimensional NLS equation (7.27) is principally anisotropic, and modulations of the wave packets in the longitudinal and transverse directions behave differently, in particular modulations in the transverse direction are stable. The study of the Benjamin-Feir instability of the Stokes waves of steepness $a_0 k_0$, developed previously can be easily extended to three-dimensional perturbations. Within the framework of equation (7.29) below, which corresponds to (7.27) rewritten in dimensionless form as

$$\mathrm{i}\frac{\partial q}{\partial T} + \frac{\partial^2 q}{\partial X^2} - \frac{\partial^2 q}{\partial Y^2} + 2 \mid q \mid^2 q = 0, \tag{7.29}$$

with $Y = \sqrt{2}k_0 y$.

We seek a solution for the perturbation of the form

$$p(X, T) = p_1 \exp(\Omega T + \mathrm{i}lX + \mathrm{i}mY) + p_2 \exp(\Omega^* T - \mathrm{i}lX - \mathrm{i}mY), \tag{7.30}$$

where p_1 and p_2 are complex constants, l and m are the longitudinal and transversal wave numbers respectively and, Ω the rate of growth when real and positive. The corresponding dispersion relation is

$$\Omega^2 = (4q_0^2 - l^2 + m^2)(l^2 - m^2). \tag{7.31}$$

This shows that the wave train is unstable within the region bounded by the pair of straight lines

$$l = \pm m,$$

and the pairs of hyperbolas

$$l^2 - m^2 = 4q_0^2.$$

Maximum of instability occurs along the curves

$$l^2 - m^2 = 2q_0^2,$$

and the maximum growth rate is

$$Re(\Omega) = a_0^2 k_0^2 = 2q_0^2.$$

The stability diagram, in the (l, m)-plane, is shown in figure 12 where the wave numbers l and m have been normalized by $\sqrt{2}a_0 k_0$.

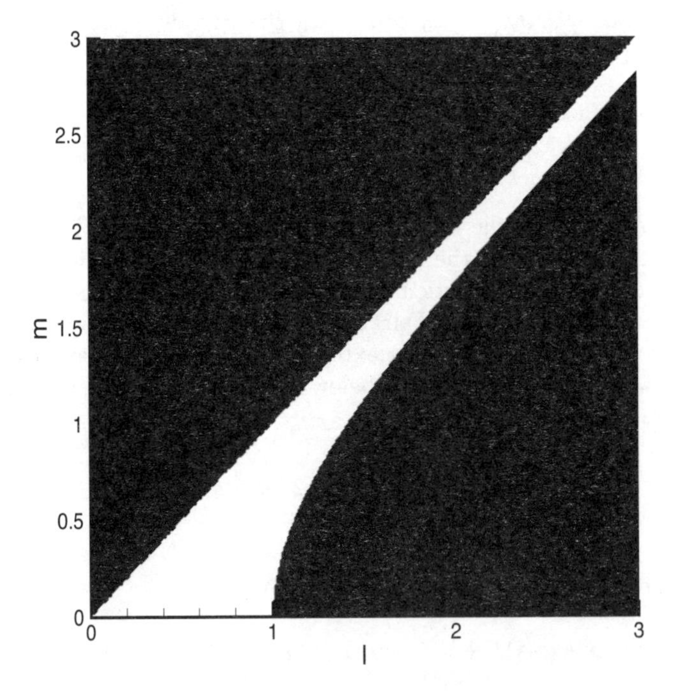

Figure 12. Diagram of instability of the 3D Benjamin-Feir instability in the plane of wave numbers. The unstable region corresponds to the white area.

The most important observation is that the instability region is unbounded in the perturbation wave vector plane. This is inconvenient of the two-dimensional NLS equation (7.27), which is absent for more relevant models discussed below.

Therefore, one can expect that both mechanisms of freak wave generation, modulational instability and wave focusing, should work for two-dimensional wave trains in deep water. The wave focusing mechanism may be produced by dispersive focusing (due to the dispersive nature of water waves) and geometrical focusing (due to waves propagating in different directions) as well. Hence, the phenomenon of freak wave generation appears to be richer for three-dimensional motions.

A numerical simulation of equation (7.29) has been performed with an initial condition corresponding to a Stokes wave train of steepness a_0k_0 perturbed in both longitudinal and transverse directions

$$q(X,Y,0) = q_0(1 + \epsilon_1 \cos(lX) + \epsilon_2 \cos(mY)). \tag{7.32}$$

Figure 13 shows several snapshots of the wave evolution computed within the framework of the two-dimensional NLS equation (7.29). In the first stage, the wave field is subjected

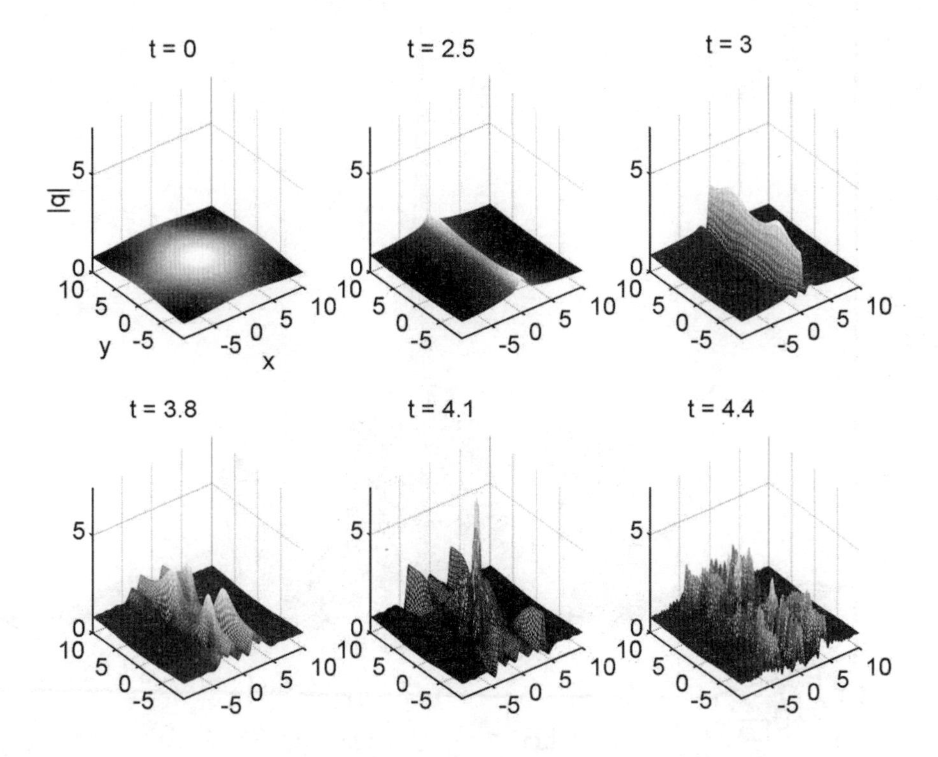

Figure 13. Evolution of a perturbed plane wave computed from equations (7.31) and (7.32), with $(l,m) = (0.3, 0.3)$ and $\epsilon_1 = \epsilon_2 = 0.10$. The amplitude has been renormalized in such a way that $q_0 = 1$. From Slunyaev *et al.* (2002).

to the Benjamin-Feir instability and one may see a growing quasi-one-dimensional wave crest (figure 13, $T \simeq 2.5$). Then the transverse instability is excited ($T \simeq 3$) leading first to the formation of several isolated peaks ($T \simeq 3.8$) followed by the merging of four peaks whose coupling gives rise to the giant wave ($T \simeq 4.1$), depicted in figure 14. Its amplitude exceeds seven times that of the initial wave. In numerical experiments by Osborne *et al.* (2000) the freak wave height reached the value 42m (the maximum crest of 29m is accompanied by a minimum 13m deep), the wavelength and steepness were about 450m and 0.33, respectively. The time dependence of the maximum of the envelope is plotted in figure 15. It is clearly observed, that while the growth rate of the Benjamin-Feir instability is reflected by a smoothly climbing curve with saturation, the coupling of several peaks happens rapidly and produces a large wave amplification. The disappearance of the giant wave is also a rapid process.

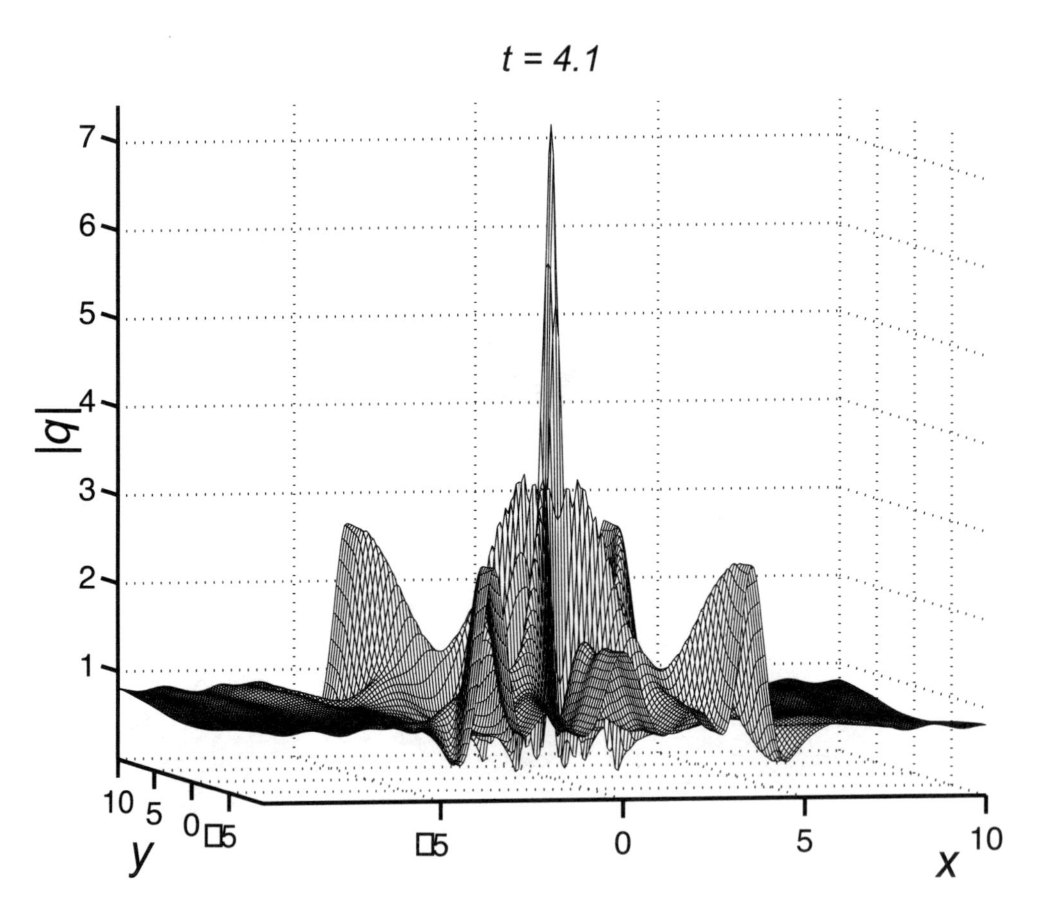

Figure 14. Huge wave generated from the weakly modulated plane wave shown in figure 13. From Slunyaev *et al.* (2002).

The phenomenon described in figures 13-15 is extremely sensitive to weak random perturbations. Nevertheless this does not cancel the rate of growth of the Benjamin-Feir instability at all. This result agrees with the conclusion of previous theoretical, numerical and experimental works. In figure 16 is considered the case of the initial

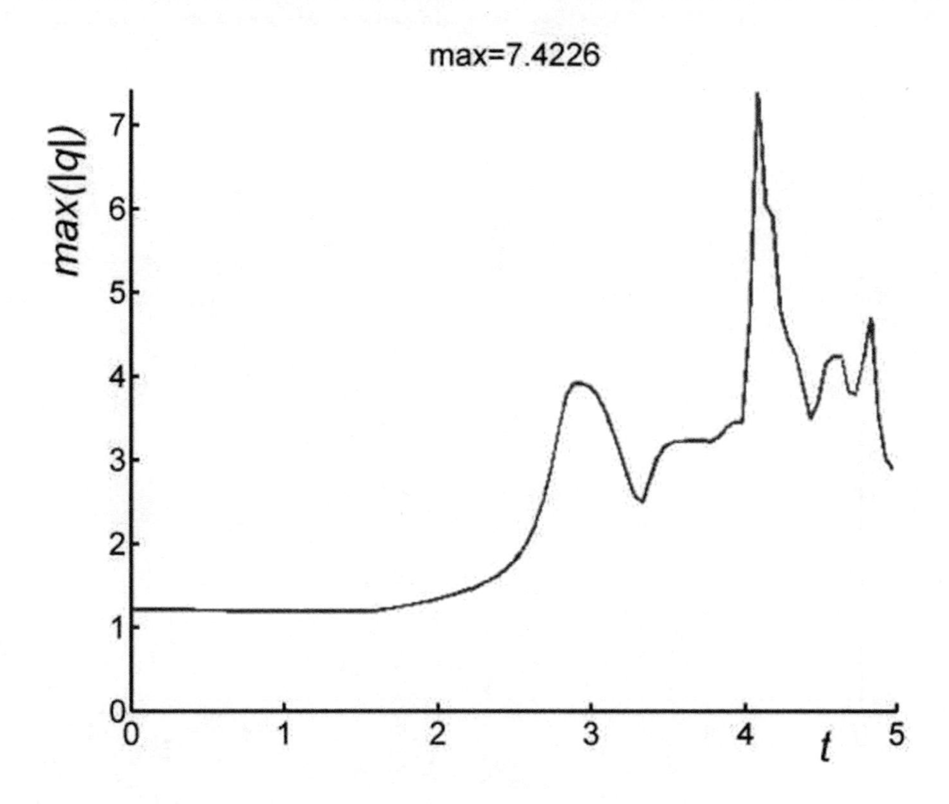

Figure 15. Time evolution of the maximum of the envelope corresponding to the numerical simulation described in figures 13. From Slunyaev *et al.* (2002).

wave field corresponding to figures 13-15 perturbed with a noise whose amplitude in the spectral space is 1% of the amplitude of the deterministic perturbation given by (7.32), and a spectral width five times wider. The time evolution of the maximum of the envelope of the wave field plotted in figure 16 shows that the coherent evolution leading to the merging of several peaks is destroyed. When the growth rate of the Benjamin-Feir instability reaches its maximum, the wave field presents a random behavior.

Now let us consider the dispersive focusing mechanism. For the nonlinear case, the optimal phase modulation may be found with the help of the invariance of the nonlinear Schrödinger equation with respect to the sign of the coordinates and time (changing q with its complex conjugate q^*). The expected freak wave is assumed to be of the following form

$$q_G(X,Y,T) = \frac{q_0}{(G_X G_Y)^{\frac{1}{4}}} \mathrm{e}^{\left(-\frac{l^2X^2}{G_X} - \frac{m^2Y^2}{G_Y}\right)} \mathrm{e}^{\mathrm{i}\left(\frac{4TX^2l^4}{G_X} - \frac{4TY^2m^4}{G_Y} - \mathrm{atan}(4Tl^2)/2 + \mathrm{atan}(4Tm^2)/2\right)} \tag{7.33}$$

where $G_X = 1 + 16T^2l^4$ and $G_Y = 1 + 16T^2m^4$.

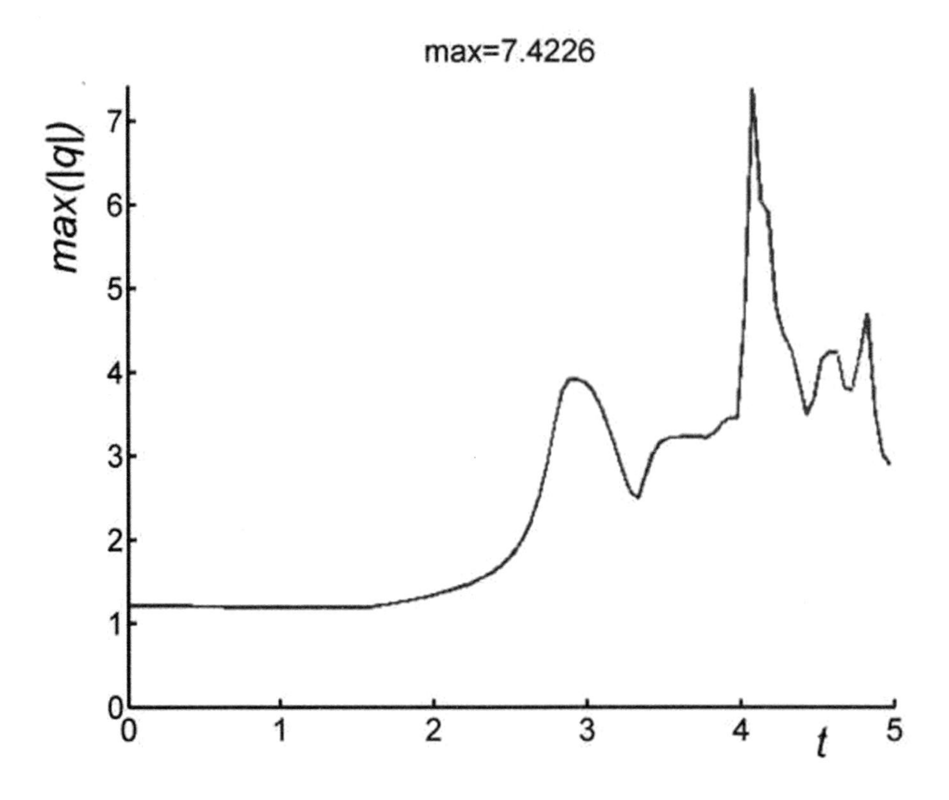

Figure 16. Same as figure 15, with a weak noise added to the initial condition (7.32). From Slunyaev *et al.* (2002).

Expression (7.33) is a two-dimensional Gaussian impulse, solution of the linearized two-dimensional nonlinear Schrödinger equation. For $T = 0$, this profile which is not modulated at this instant of time, presents a maximum of amplitude, q_0. It disperses when T is increased and, becomes modulated according to the quadratic terms of the phase, with a decreasing amplitude. Hence, if we choose as initial condition $q_G(X, Y, -T_\infty)$, this wave will reach its maximum amplification when $T = 0$. The solution (7.33) corresponds to two kinds of focusing: one is due to the phase modulation and the second is a geometrical focusing due to merging of many waves, propagating in different directions. The question is: What is the effect of the nonlinearity on the focusing phenomenon which is optimal in the linear case. Expression (7.33) at $T = -T_\infty$ is used as initial condition for the two-dimensional NLS equation (7.31). This nonlinear numerical simulation corresponds to the non-optimal case, compared with the optimal case. Figure 17 displays a comparison of the optimal focusing and non-optimal focusing. The time evolution of the maximum of the envelope within the framework of the linear and nonlinear theory are plotted in figure 17a and figure 17b respectively. One can observe that the amplification is smaller in figure 17b, but still large. The non-optimal phase modulation does not modify the dynamics of the wave field: A rapid growth of the amplitude is observed which is however less important than for the linear wave packet. This is due to the fact that not all the components of the wave train converge to one point at once.

In order to demonstrate how robust the dispersive focusing phenomenon to random wave field is, a numerical simulation similar to that corresponding to figure 16 has been performed. Mixed deterministic and random components are taken as initial wave field. The deterministic wave is the same as the initial condition used in figure 17 while the random wave is given by a broad Gaussian spectral distribution function and, an amplitude five times greater than the maximum of the deterministic modulated wave field. Hence, the deterministic field is completely hidden by the random waves in the first stage of wave evolution as shown in figure 18. But after a rapid growth a huge wave occurs in the field (see figure 18). It is clearly shown that even an intense noise cannot prevent the focusing mechanism.

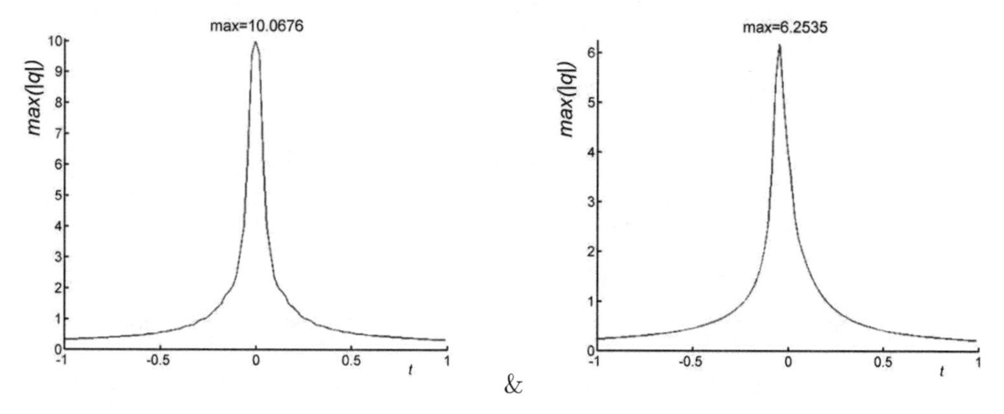

Figure 17. Time evolution of the maximum of the envelope: (a) optimal focusing corresponding to the linear theory (b) non-optimal focusing corresponding to the nonlinear theory of the two-dimensional NLS equation. From Slunyaev *et al.* (2002).

The Davey-Stewartson system. For finite depth the effect of the mean flow generated by modulated waves becomes important. The weakly nonlinear dynamics of three-dimensional wave trains in water of finite depth may be described by the Davey-Stewartson system, given in dimensionless form by

$$\mathrm{i}\frac{\partial q}{\partial T} + \frac{\partial^2 q}{\partial X^2} - \frac{\partial^2 q}{\partial Y^2} + 2q(\mid q\mid^2 - p) = 0, \tag{7.34}$$

$$s_1\frac{\partial^2 p}{\partial X^2} + s_2\frac{\partial^2 p}{\partial Y^2} = \frac{\partial^2}{\partial X^2}\mid q\mid^2, \tag{7.35}$$

where

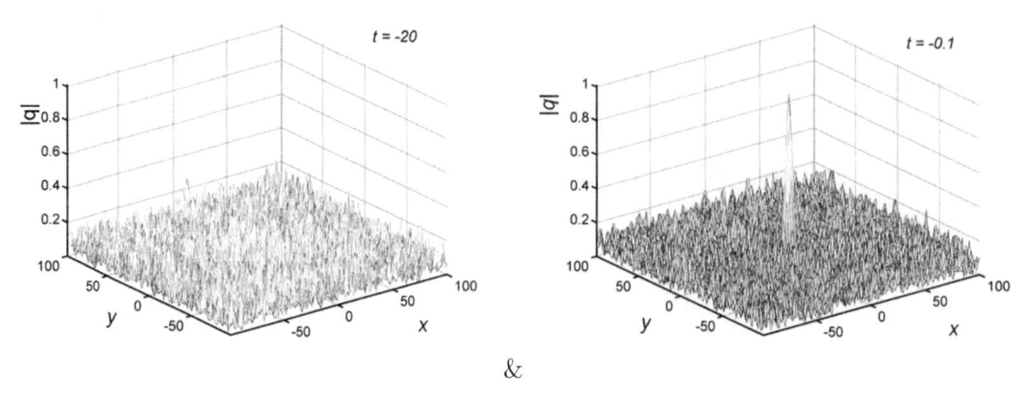

&

Figure 18. Formation of a huge wave from three-dimensional random wave field. From Slunyaev *et al.* (2002)

$$s_1 = S_1 \frac{N_1}{N_2^2}, \qquad s_2 = S_2 \frac{N_1}{N_2^2} \frac{D_1}{D_2}, \qquad D_1 = -\frac{4k}{c_{ph}} \frac{\partial c_g}{\partial k}, \qquad D_2 = \frac{4c_g}{c_{ph}},$$

$$N_1 = \tfrac{1}{8}(1 + 9\sigma^{-2} - 13(1-\sigma^2) - 2\sigma^4), \qquad N_2 = 1 + \frac{c_g}{2c_{ph}}(1-\sigma^2),$$

$$S_1 = \frac{kh}{\sigma} - \frac{c_g^2}{c_{ph}^2}, \qquad S_2 = \frac{kh}{\sigma}, \qquad \sigma = \tanh kh.$$

c_g and c_{ph} are the group and phase velocities, and the free surface elevation is given by

$$\eta(x,y,t) = \frac{c_{ph}^2}{g\sqrt{2N_1}}[q(X,Y,T)\exp(\mathrm{i}\omega T - \mathrm{i}kX) + c.c.] + ...,$$

$$X = \frac{2k}{\sqrt{D_1}}(x - c_g t), \qquad Y = \frac{2k}{\sqrt{D_2}} y, \qquad T = \tfrac{1}{2}\omega t.$$

The mean flow induced by the waves is described by a function $p(X,Y,T)$. The coefficients N_1 and N_2 correspond to nonlinear effects due to the fundamental wave and the interaction with the mean flow, respectively. D_1 and D_2 are the coefficients of longitudinal and transverse dispersion, and g is the acceleration due to gravity. Note that the coefficients of the Davey-Stewartson system depend on the dispersive parameter kh only.

For two-dimensional free surface flows, the Davey-Stewartson system system reduces to the one-dimensional NLS equation (7.5), focused for deep water and defocused for shallow water. For three-dimensional free surface flows in deep water the mean flow is negligible, and the Davey-Stewartson system reduces to the two-dimensional NLS equation (7.31). As shown by Anker and Freeman (1978), the Davey-Stewartson system becomes integrable for shallow water.

For moderate depth $kh = 1$, a numerical simulation of the focusing of a Gaussian impulse using the system (7.34)-(7.35) has been performed. The results are shown in figure 19 for optimal focusing and non optimal focusing. In the case of non optimal focusing, one can see the occurrence of two maxima in the wave envelope corresponding to focusing events in the longitudinal and transverse directions respectively. Note that the amplification is much smaller than in the optimal case.

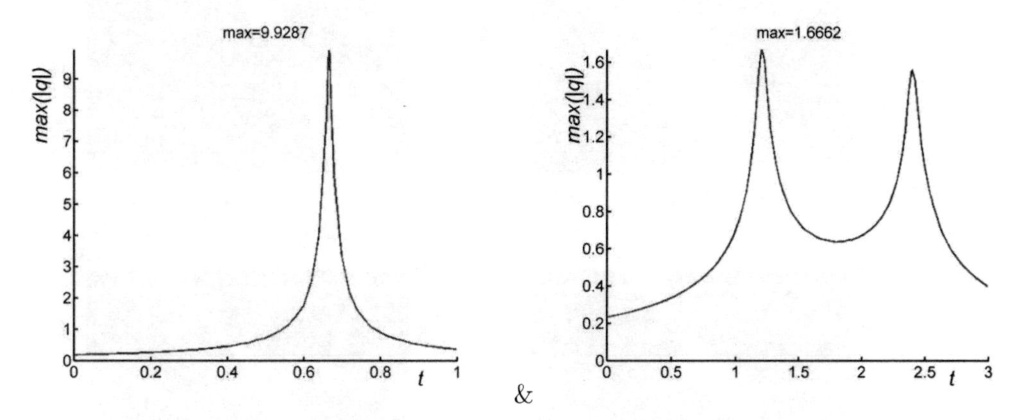

Figure 19. Time evolution of the maximum of the envelope for finite depth ($kh = 1$): (a) optimal focusing corresponding to the linear theory (b) non-optimal focusing corresponding to the nonlinear theory of the Davey-Stewartson system. From Slunyaev *et al.* (2002).

One of the main features of the dynamics of three-dimensional water wave flows is the existence of modulational instability for any depth. Figure 20 displays the stability diagrams of the Stokes wave in the plane of the longitudinal and transverse wave numbers for several values of kh. When depth decreases instability domains shrink and growth rates decrease. Hence, both mechanisms of freak wave phenomenon due to modulational and dispersive focusing should work for three-dimensional wave packets in finite depth and shallow water as well. A detailed analysis of freak wave scenarios within the framework of the Davey-Stewartson system can be found in Slunyaev *et al.* (2002).

7.2 Extended nonlinear models for freak waves

The freak wave phenomenon was discussed in the above subsection within the framework of weakly nonlinear models. In fact, freak wave events are large amplitude and short-lived phenomena. Hence, approximations of weak nonlinearity and narrow-banded spectrum do not correspond exactly to real data. For instance, the NLS equation (and the DS system) being invariant under transformations which invert coordinates presents symmetrical waveforms in the evolution of the solution (if it is initially symmetric). Laboratory experiments show asymmetric profiles of the wave envelope for large amplitudes: see, for instance, Shemer *et al.* (1998). Weakly nonlinear models predict also incorrect values of the Benjamin-Feir instability for short-scale modulations with regard to fully nonlinear computations given by Longuet-Higgins (see Dysthe, 1979): a wave cannot be considered as a weakly nonlinear wave if its steepness is approximately greater

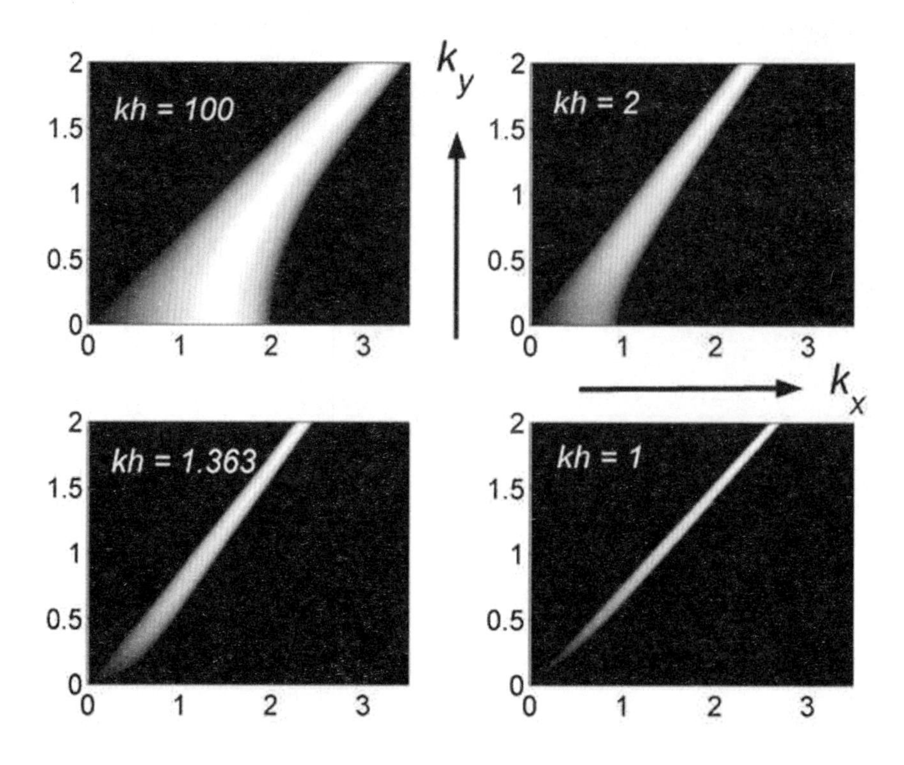

Figure 20. Stability diagrams in the plane of envelope wave numbers. From Kharif and Pelinovsky (2003).

than 0.10. For deep water Dysthe (1979) derived a modified nonlinear Schrödinger equation to fourth order in wave steepness. One of the main results at this order is precisely the influence of the wave-induced mean flow. This extended NLS equation is called now the Dysthe equation. Spatial version of the two-dimensional Dysthe equation are discussed by Lo and Mei (1985) and Kit and Shemer (2002). Extensions of NLS equations result in a significant modification of the diagrams of Benjamin-Feir instability, namely their reduction. Hence Stokes wave trains will be stable with respect to short-scale disturbances according to the predictions of rigorous theories. Modified NLS equations describe better the results of laboratory experiments (see Shemer *et al.* (1998), Ablowitz *et al.* (2000, 2001)). Onorato, Osborne, Serio and Damiani (2001) investigated the predictions of the statistical characteristics of freak waves within the framework of the NLS equation and the Dysthe equation. They found that the NLS equation overestimate the maximum of the wave amplitudes compared with the Dysthe equation.

Later, Trulsen and Dysthe (1996, 1997) extended the Dysthe equation by including higher order linear dispersive terms (up to the fifth derivative of the wave amplitude) that describe broader bandwidth water waves. Trulsen *et al.* (2000) noticing the importance of linear dispersion improved the Dysthe equation with exact linear dispersion. For numerical simulations of ocean waves, one must emphasize that this model does not suffer from energy leakage. The extended Dysthe equation is written as follows

$$\mathrm{i}(\frac{\partial a}{\partial T} + L(\partial_X, \partial_Y)a) = \frac{\omega_0 k_0^2}{2} \mid a \mid^2 a - \frac{\mathrm{i}\omega_0 k_0}{4}(6 \mid a \mid^2 \frac{\partial a}{\partial X} + a^2 \frac{\partial a^*}{\partial X}) + k_0 a \frac{\partial \Phi}{\partial X}(z=0), \tag{7.36}$$

where $\Phi(X, Y, T)$ is the induced mean flow satisfying the Laplace equation

$$\Delta\Phi = 0, \tag{7.37}$$

with the boundary conditions on the sea surface ($z = 0$)

$$\frac{\partial \Phi}{\partial z} = \frac{\omega_0}{2} \frac{\partial \mid a \mid^2}{\partial X}. \tag{7.38}$$

It vanishes far from the sea surface (in deep water).

The pseudo-differential operator L can be derived from the Fourier integral with the kernel $\omega(k) - \omega(k_0)$

$$L(\partial_X, \partial_Y) = \mathrm{i}\omega_0[(1 - \frac{\mathrm{i}}{k_0}\frac{\partial}{\partial X})^2 - \frac{1}{k_0^2}\frac{\partial^2}{\partial Y^2}]^{\frac{1}{4}} - \mathrm{i}\omega_0. \tag{7.39}$$

Dysthe *et al.* (2003) showed that additional terms to the NLS equation transform the initial Gaussian shape of the spectrum in an asymmetric profile with a steepening of the low-frequency side and a downshift of the spectral peak. Ablowitz *et al.* (2000, 2001) considered experimentally and numerically the long-time dynamics of the modulational instability of deep water waves. Their numerical simulations using a perturbed NLS equation (high-order NLS similar to the Dysthe equation) were found to be consistent with the laboratory experiments and support the conjecture that for periodic boundary conditions long-time evolution of modulated Stokes wave trains is chaotic. This sensitivity to the initial condition is associated with homoclinic manifold of the unperturbed NLS equation. Later Calini and Schober (2002) suggested that homoclinic chaos may increase the likelihood of freak wave formation. The main result herein is the chaotic dynamics of the wave field even for regular initial conditions: so freak waves appear and disappear randomly. Note that due to the presence of homoclinic structures the calculations are influenced by infinitesimal errors which makes computationally the NLS equation chaotic in certain range of parameter space despite that it is integrable (see Herbst and Ablowitz (1989) and Ablowitz, Schober and Herbst (1993)). Trulsen (2001) used the extended Dysthe equation to simulate numerically the New Year Wave event.

The two-dimensional Dysthe equation and its extensions have been used to study random wave field evolutions. Such computations need a large number of realizations to obtain the probability density function. As a result, an anisotropy in the wave field behavior is predicted. In particular the spectrum develops asymmetrically with a downshift of the spectral peak and an angular widening: a power-law behavior $k^{-2.5}$ above the peak is observed, see Dysthe *et al.* (2003). It is generally assumed that water wave fields are Gaussian in average only: kurtosis oscillates around 3 (value for Gaussian distribution function) which becomes sometimes very high, see Onorato, Osborne and Serio (2002). Large values of the kurtosis correspond to large tails of the distribution function, providing higher probability of freak wave occurrence.

A more general approximate model free of the assumption of narrow-band spectra is the Zakharov equation. Using the Hamiltonian formalism, Zakharov (1968) derived an integral equation for nonlinear deep water waves. Firstly, the standard equations of the motion are transformed as follows

$$\mathrm{i}\frac{\partial b(\mathbf{k})}{\partial t} = \frac{\delta H}{\delta b^*(\mathbf{k})}, \tag{7.40}$$

where $\mathbf{k}$ is the wave vector, $b(\mathbf{k})$ is the complex amplitude and $H(b(\mathbf{k}), b^*(\mathbf{k}))$ is the Hamiltonian. The asterisk means complex conjugate. In the spectral space the complex Fourier amplitudes $b(\mathbf{k})$ are expressed by means of integral power series in Fourier amplitudes of the elevation of the free surface and the velocity potential at the free surface. Then the Hamiltonian H is developed in terms of an integral power series in the complex amplitudes $b(\mathbf{k})$

$$H = H_0 + \sum_{n=4}^{\infty} H_n. \tag{7.41}$$

$H_0 = \int \omega(\mathbf{k}) b(\mathbf{k}) b^*(\mathbf{k}) \mathrm{d}\mathbf{k}$ is a quadratic term, H_n with $n > 3$ are of power n in $b(\mathbf{k})$ responsible of nonlinear effects, and $\omega(\mathbf{k}) = (g \mid \mathbf{k} \mid)^{\frac{1}{2}}$ is the linear dispersion relation for gravity waves. Herein the surface tension is neglected. By taking into account H_0, H_4 and H_5 in the truncated Hamiltonian we arrive at the so-called five-wave reduced Zakharov equation, derived by Krasitskii (1994):

$$\begin{aligned}
\mathrm{i}\tfrac{\partial b_0}{\partial t} &= \omega_0 b_0 \\
&+ \textstyle\int V_{0123} b_1^* b_2 b_3 \delta_{0+1-2-3} \mathrm{d}\mathbf{k}_1 \mathrm{d}\mathbf{k}_2 \mathrm{d}\mathbf{k}_3 \\
&+ \textstyle\int W_{01234} b_1^* b_2 b_3 b_4 \delta_{0+1-2-3-4} \mathrm{d}\mathbf{k}_1 \mathrm{d}\mathbf{k}_2 \mathrm{d}\mathbf{k}_3 \mathrm{d}\mathbf{k}_4 \\
&+ \frac{3}{2} \int W_{43210} b_1^* b_2 b_3 b_4 \delta_{0+1+2-3-4} \mathrm{d}\mathbf{k}_1 \mathrm{d}\mathbf{k}_2 \mathrm{d}\mathbf{k}_3 \mathrm{d}\mathbf{k}_4,
\end{aligned} \tag{7.42}$$

where

$$b_0 = b(\mathbf{k}),\ b_j = b(\mathbf{k}_j),\ \omega_0 = \omega(\mathbf{k}),$$

$$V_{0123} = V(\mathbf{k}, \mathbf{k}_1, \mathbf{k}_2, \mathbf{k}_3),\ W_{01234} = W(\mathbf{k}, \mathbf{k}_1, \mathbf{k}_2, \mathbf{k}_3, \mathbf{k}_4),\ W_{43210} = W(\mathbf{k}_4, \mathbf{k}_3, \mathbf{k}_2, \mathbf{k}_1, \mathbf{k}),$$
$$\delta_{0+1-2-3} = \delta(\mathbf{k} + \mathbf{k}_1 - \mathbf{k}_2 - \mathbf{k}_3).$$

V and W are kernels satisfying symmetry conditions. This equation describes four-wave and five-wave resonant interactions. Stassnie and Shemer (1984) extended the derivation, including five-wave interaction, to water waves on finite depth. Note nevertheless that their kernels lack these symmetry properties. In the previous sections, the nonlinear Schrödinger equation and the Davey-Stewartson equation can only describe modulational interactions of Benjamin-Feir type, between four waves. They are unable to take into account five-wave interaction processes. Within the framework of four-wave interaction, equation (7.42) reduces to the NLS equation under the assumption of narrow-band gravity waves. In finite depth one can recover the DS system. While the Zakharov equation is the most sophisticated approximate model for the spatio-temporal evolution of water waves it has not the ease of use of the NLS equation.

Annenkov and Badulin (2001) observed in the frequency spectrum of a 20 minutes wave recording at the 'Draupner' platform, the specific component peculiar to five-wave interactions. This component corresponds to class II instabilities phase-locked to the dominant component of the spectrum. Physically this instability is known to generate water wave horseshoe patterns frequently observed on the sea surface. More details can be found in the review paper by Dias and Kharif (1999). In order to have a better understanding of the role of this kind of resonance in the formation of rogue waves they performed numerical simulations using equation (7.42) in which modulational and five-wave interactions are both taken into account. They showed that cooperative effects of these interactions may be responsible for the occurrence of rogue waves and emphasized the role of oblique waves in this process.

The Zakharov equation should be a powerful model to study the formation of freak waves which are not only due to modulational instabilities. Unfortunately, up to now only very few authors used this equation to investigate freak waves. Using the Zakharov equation, Janssen (2003) showed numerically that four-wave interactions play an important role in the evolution of the spectrum of surface waves. In this paper two kinds of four-waves interactions are concerned: (i) for short times, $O(\epsilon^{-2})$, it is the Benjamin-Feir instability which operates and this kind of interaction is called improperly nonresonant four-wave interaction (ii) for longer times, $O(\epsilon^{-4})$ the resonance function of the Boltzmann equation tends towards a delta function and it is the resonant four-wave interaction which prevails ($\epsilon = \sqrt{k_0^2 < \eta^2 >}$ is a measure of the nonlinearity of the wave train). It is found that the theory of homogeneous four-wave interaction, extended to include effects of nonresonant transfer, compares favorably with the statistics computed from the numerical simulations: In particular spectrum shape and kurtosis, even for waves with a narrow spectrum and large steepness. Jansen claimed that these extreme conditions are favorable for the occurrence of freak waves.

7.3 Weakly nonlinear freak waves in shallow water

For shallow water, the ratio between nonlinearity and dispersion is usually high and the generation of higher harmonics becomes more important. The simplified model of unidirectional waves in shallow water taking into account weak nonlinearity and dispersion is the Korteweg-de Vries equation (KdV equation)

$$\frac{\partial \eta}{\partial t} + c(1 + \frac{3\eta}{2h})\frac{\partial \eta}{\partial x} + \frac{ch^2}{6}\frac{\partial^3 \eta}{\partial x^3} = 0, \tag{7.43}$$

where $c = \sqrt{gh}$ is the wave speed of linear long waves. The KdV equation which is the most simple model describing the spatio-temporal evolution of weakly nonlinear and weakly dispersive waves in shallow water was derived in 1895 by Korteweg and de Vries. The corresponding Cauchy problem admits an exact solution found by using the inverse scattering transform (IST). The solutions of (7.43) are stable (see figure 20), and therefore, the nonlinear mechanism of freak wave formation due to modulational instability does not operate in shallow water. If the initial wave field corresponds to a weakly modulated wave train, its shape is modified during its propagation, but the wave amplitude does not vary significantly as shown by Kit *et al.* (2000). Hence, two-dimensional freak waves in shallow water occur only through the focusing-dispersive mechanism. In order to demonstrate that, one considers the solution of the Cauchy problem for the initial singular data corresponding to a delta-function. Using the IST the initial delta-function transforms into a soliton (solitary wave) and an oscillating dispersive tail, located in space according to their speed values: The soliton moves with a large speed and is in front of the wave train. Due to invariance of the KdV equation with respect to the reversal of time and coordinate, this wave field when inverted in space should converge to the initial disturbance in finite time and, then again transform into a soliton and a dispersive tail. It means that there is no principal limitation for the formation of abnormal waves of large amplitude. Therefore, the wave focusing mechanism is applicable in nonlinear cases also, but the wave field is more complicated, including soliton and amplitude-frequency modulated wave packets. This dynamics was investigated in details by Pelinovsky *et al.* (2000) and Kharif, Pelinovsky and Talipova (2000) within the framework of equation (7.43). The disturbance described by the delta-function is formally outside the scope of the KdV equation (weak nonlinearity, long wave). Let us consider a more realistic form of the freak wave

$$\eta_f = \eta_0 \mathrm{sech}^2(\frac{x}{l}), \tag{7.44}$$

with arbitrary amplitude η_0 and width l. In general, the number, N, of solitary waves generated from the initial condition given by (7.44) is

$$N = I[\sqrt{\frac{3}{2}U_r + \frac{1}{4}} - \frac{1}{2}] + 1, \tag{7.45}$$

where I is the integer function, and the Ursell parameter

$$U_r = \frac{\eta_0 l^2}{h^3} \tag{7.46}$$

is the ratio of two small parameters: nonlinearity, η_0/h, to dispersion, h^2/l^2. The amplitudes of the generated solitons are

$$\eta_n = \frac{4\eta_0}{3U_r}[\sqrt{\frac{3U_r}{2} + \frac{1}{4}} - (n - \frac{1}{2})]^2, \tag{7.47}$$

where $n = 1, 2, ..., N$. If $U_r \gg 1$, the number of generated solitons is large, and the amplitude of the highest soliton is $2\eta_0$, exceeding the initial amplitude. This means that the expected freak wave is not an abnormal wave in the group of solitons, and is not interesting from the point of view of freak wave phenomenon. If $U_r \ll 1$, only one solitary wave is generated and its amplitude is

$$\eta_1 = 3\eta_0 U_r. \tag{7.48}$$

According to the definition of the freak wave, its amplitude should be at least more than twice the significant height of the background field (for wind waves). This means that the freak wave parameters have to fulfill the following condition

$$U_r < \frac{1}{6}. \tag{7.49}$$

So the freak wave is almost a linear wave despite its large amplitude.

In dimensionless variables

$$\zeta = \frac{\eta}{h}, \qquad y = \frac{x - ct}{h}, \qquad \tau = \frac{ct}{h}, \tag{7.50}$$

equation (7.43) becomes

$$\frac{\partial \zeta}{\partial \tau} + \frac{3}{2}\zeta\frac{\partial \zeta}{\partial y} + \frac{1}{6}\frac{\partial^3 \zeta}{\partial y^3} = 0. \tag{7.51}$$

Equation (7.51) is numerically integrated with the initial condition corresponding to a Gaussian impulse

$$\zeta(y, 0) = a_0 \exp[-(\frac{x - 5L/6}{d})^2]. \tag{7.52}$$

The amplitude is a_0 and the characteristic width d.

Figure 21 shows the evolution of the impulse at different instants of time. First, the impulse disturbance quickly turns into a dispersive train, and then for $\tau = 140$ the soliton with an amplitude equal to 0.1 is formed. This is more visible for $\tau = 360$, where the frontal part of the wave train is shown. Short-scale waves moving with speeds lower than the long wave limit propagate to the left because equation (7.51) is written in a system of reference moving with the speed of long waves in shallow water. Then the wave profile displayed in figure 21 for $\tau = 140$ is overturned (see figure 22), and it is used as initial condition for equation (7.51). The wave transformation of this initial wave is shown in figure 22 for different times. As expected, the wave train evolves to the Gaussian impulse (7.52) at time $\tau = 140$. It is important to emphasize that the Gaussian form is reconstructed almost exactly. This is due to the stability of the solution of the KdV equation. Then, in the evolution process, the Gaussian impulse is transformed again into a wave train with decreasing amplitude.

The time evolution of the maximum of amplitude of the wave train is plotted in figure 23. One can observe that the amplification is significant in the vicinity of the focusing. In particular, the period of time when the wave amplitude exceeds the initial value twice

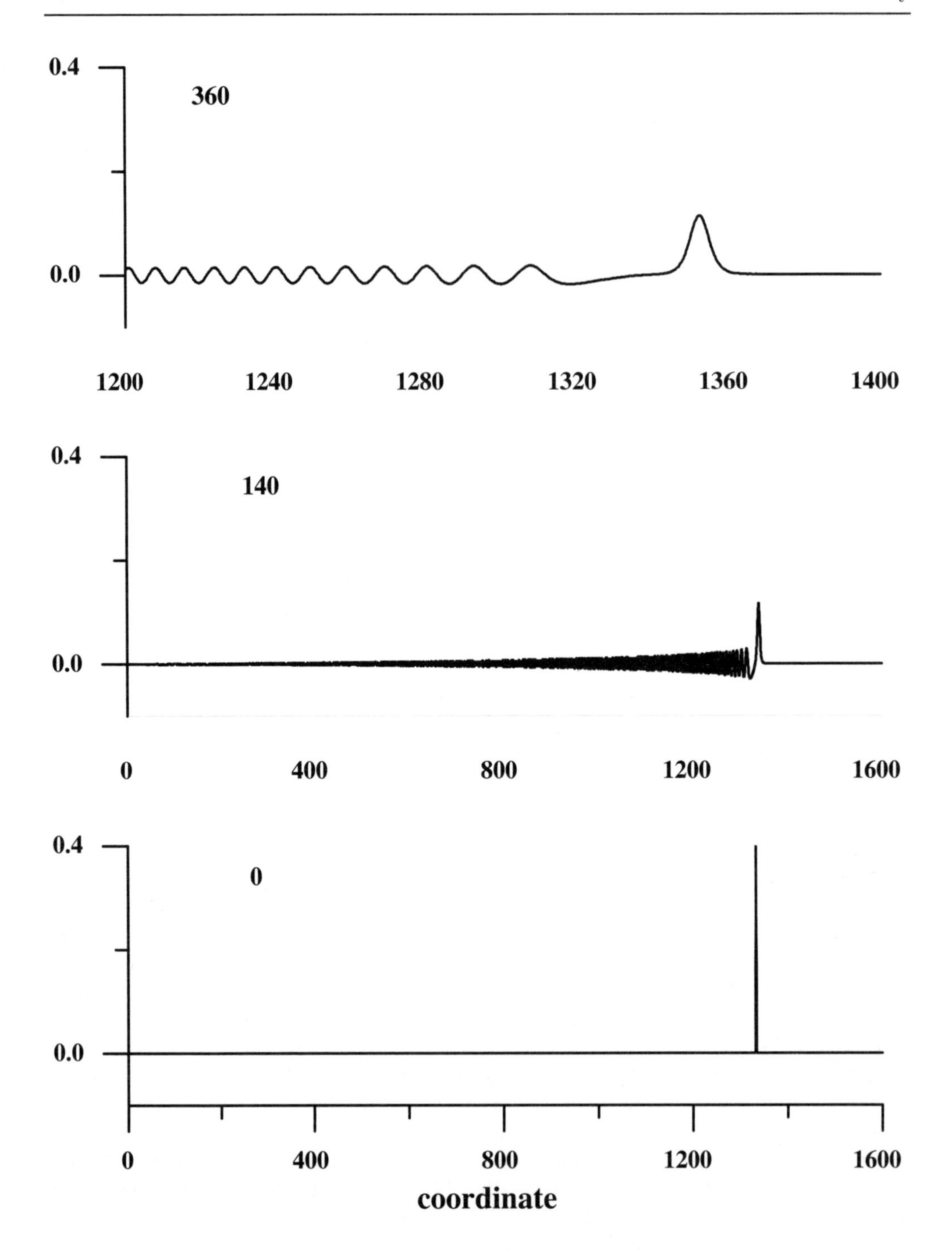

Figure 21. Evolution of an initial impulse into soliton and dispersive wave train at several instants of time within the framework of the KdV equation. From Pelinovsky *et al.* (2000).

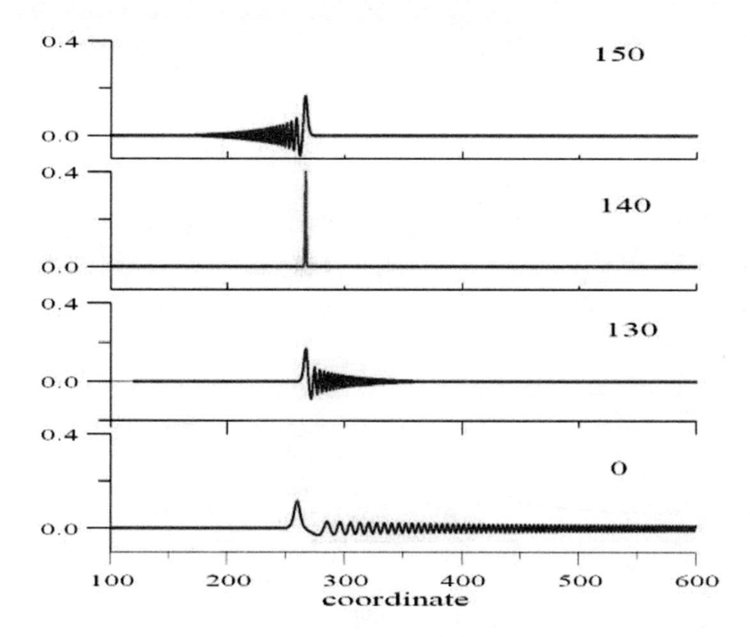

Figure 22. Freak wave generation from nonlinear dispersive wave train within the framework of the KdV equation. From Pelinovsky *et al.* (2000).

is approximately $\Delta\tau = 5$ and it is much smaller than the recurrence period $\tau = 280$. This example demonstrates that the freak wave occurs and disappears suddenly.

What is the effect of randomness on the dynamics of wave trains evolving to freak waves? Randomness should destroy the phase coherence among the components of the wave field and possibly prevent the generation of freak waves. Figure 24 shows numerical realizations of a random wave field at different instants of time, with an averaged amplitude about the value 0.10. Due to nonlinearity and dispersion, spectral components in the random process interact with each other, and sometimes the superposition can also lead to the occurrence of a freak wave: see for example the crest with the amplitude value close to 0.25 in figure 24 at $\tau = 100$. Physically, the freak wave generation in a random wave field is the same as in deterministic wave fields. Herein the random wave field is rather considered as a perturbation background for the mechanism of freak wave generation. To find the wave fields whose evolution leads to the freak wave, one uses again the previous scheme: solve the KdV equation with the initial condition corresponding to the expected freak wave, and then solve the reverse problem. In our case, we should consider the freak wave (Gaussian impulse) on the background of the random perturbations shown in figure 24. Figure 25 at $\tau = 0$ corresponds to the Gaussian impulse plus the random wave field. One can observe that for long times the initial Gaussian impulse is completely hidden in the random wave field.

Now consider the wave form at $\tau = 140$ as an initial condition for the KdV equation.

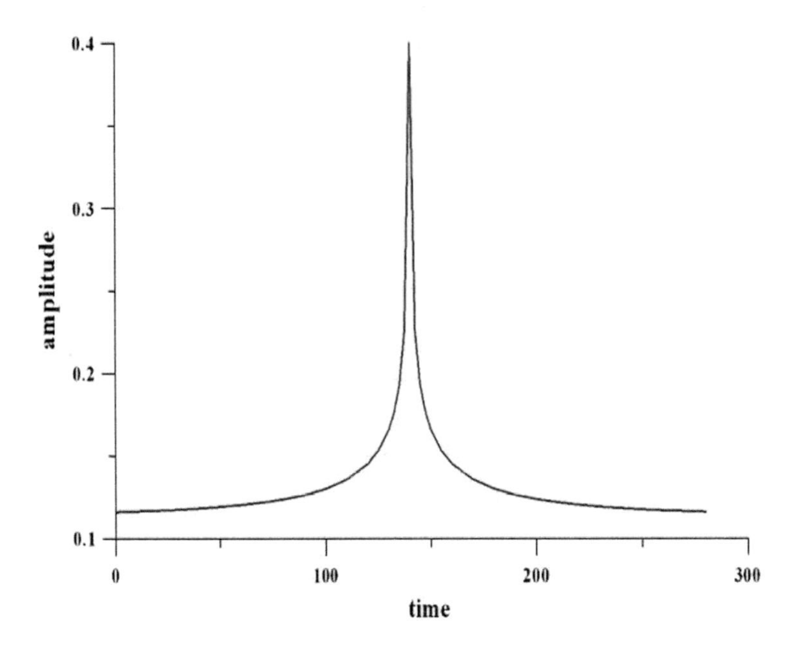

Figure 23. Time evolution of the amplitude of the peak of the wave field during the occurrence and disappearance of the freak wave. From Pelinovsky *et al.* (2000).

The result of the corresponding numerical simulation is given in figure 26. As expected, the wave field turns into the freak wave, and the wave of large amplitude is visible in the random wave field only during a short period of time. This example points out that the wave field, involving the deterministic frequency modulated wave train and random components, can produce a freak wave for optimal condition.

The present analysis shows that wave focusing should be the main mechanism responsible of freak waves in shallow water. Recently, Onorato *et al.* (2003) considering the nonlinear interactions of waves with a doubled-peaked power spectrum in shallow water, derived two coupled NLS equations and showed that uniform wave trains can be unstable to small perturbations. This suggests that in the case of a doubled-peaked spectrum in shallow water the modulational instability may lead to the freak wave generation as for deep water.

For three-dimensional free surface water wave motions, the simplest model is the Kadomtsev-Petviashvili equation (KP equation)

$$\frac{\partial}{\partial x}[\frac{\partial \eta}{\partial t} + c(1 + \frac{3\eta}{2h})\frac{\partial \eta}{\partial t} + \frac{ch^2}{6}\frac{\partial^3 \eta}{\partial x^3}] = -\frac{c}{2}\frac{\partial^2 \eta}{\partial y^2}, \tag{7.53}$$

where y is the transverse horizontal coordinate.

Equation (7.53) is integrable as the KdV and NLS equations. In particular, the two-soliton solution can be written explicitly (in frame of reference moving with the long-wave

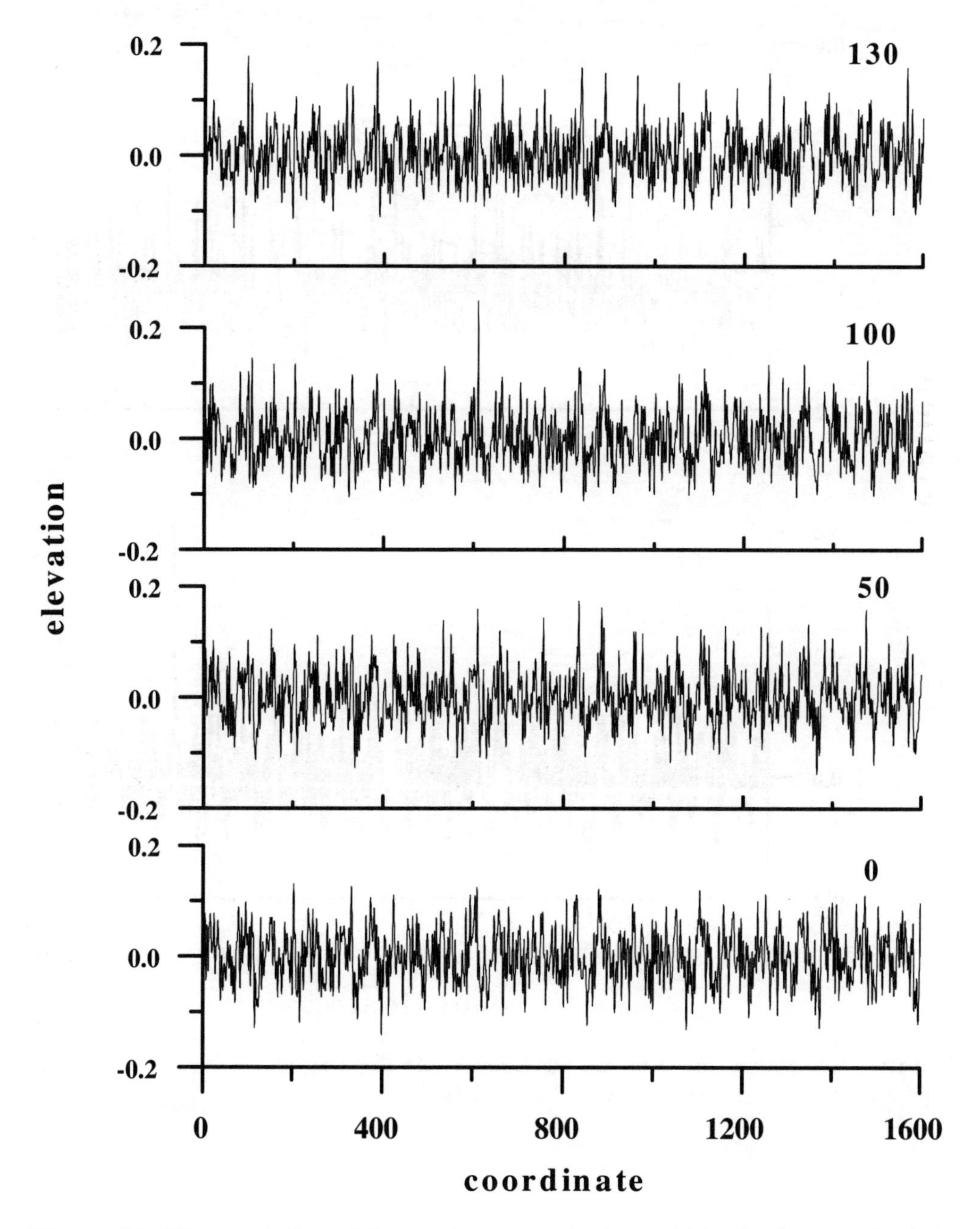

Figure 24. Time evolution of the random perturbations within the framework of the KdV equation. From Pelinovsky *et al.* (2000).

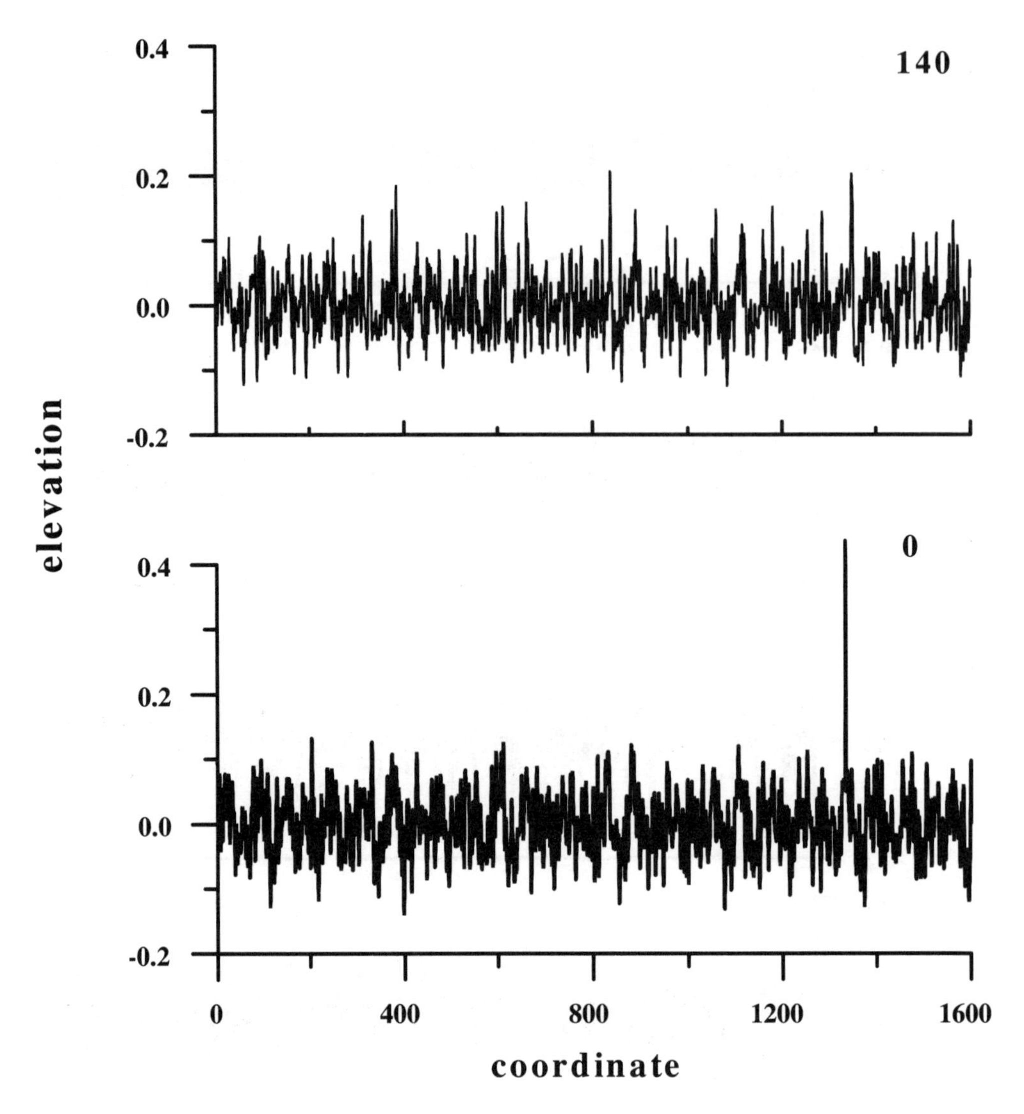

Figure 25. Disappearance of the large-amplitude impulse on the background of the random wave field. From Pelinovsky *et al.* (2000).

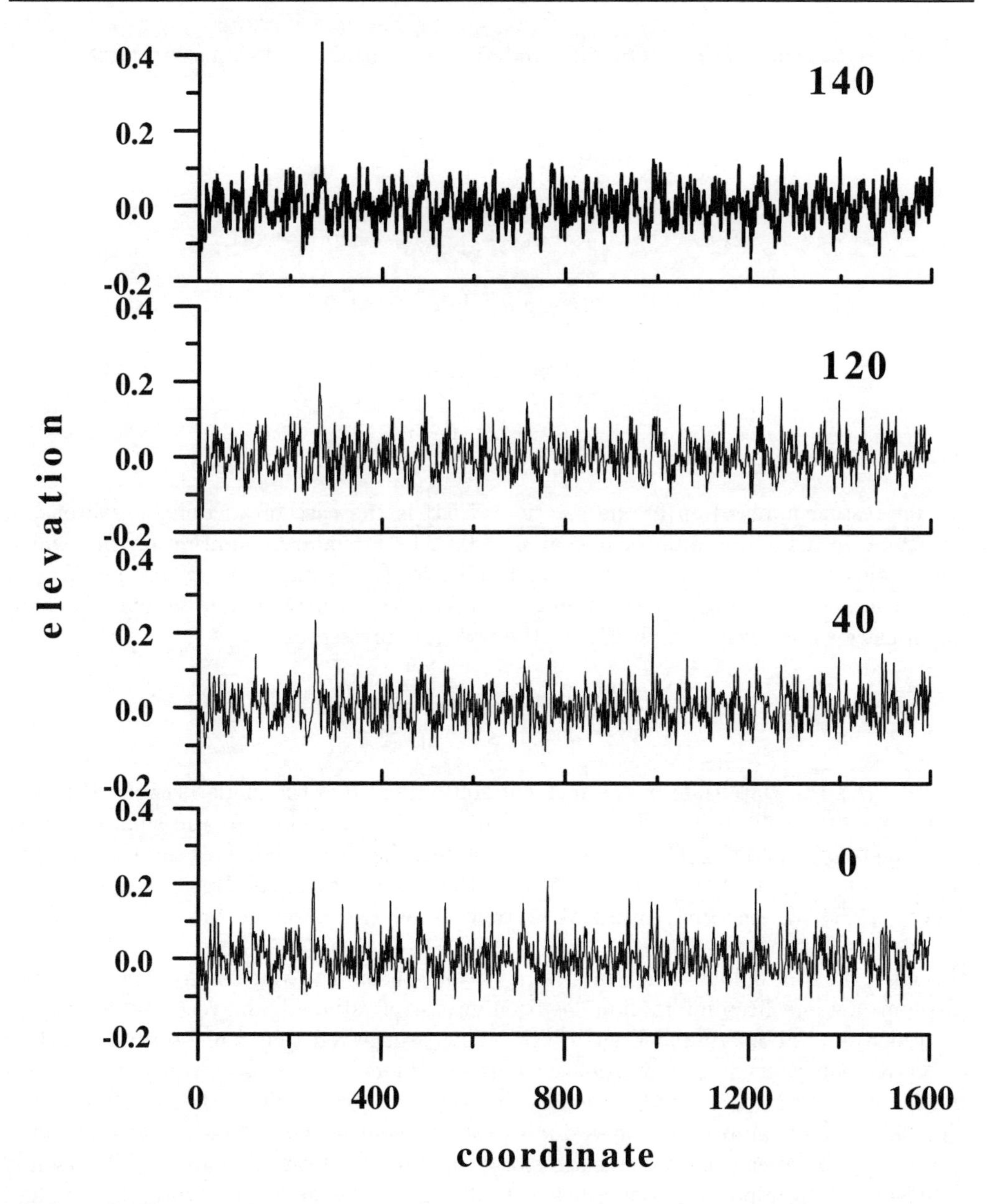

Figure 26. Freak wave occurrence from random wave field. From Pelinovsky *et al.* (2000).

speed), see Satsuma (1976), Onkuma and Wadati (1983) and Pelinovsky (1996)

$$\eta(x,y,t) = \frac{4h^3}{4}\frac{\partial^2 \ln F}{\partial x^2}, \tag{7.54}$$

with

$$F(x,y,t) = 1 + \exp\zeta_1 + \exp\zeta_2 + d\exp(\zeta_1+\zeta_2),$$

$$\zeta_i = k_i x - p_i y - V_i t, \quad i = 1,2,$$

$$V_i = c(k_i^2 + 3p_i^2), \quad i = 1,2,$$

$$d = \frac{(k_1+k_2)^2-(p_1-p_2)^2}{(k_1-k_2)^2-(p_1-p_2)^2}.$$

An interesting application of the solution (7.54) is the case of an oblique soliton approaching to a vertical wall located at $y = 0$. The boundary condition on the wall is automatically satisfied when the soliton amplitudes (proportional to k) and speeds (V) are equal, and $p_1 = -p_2$ (corresponding to wave reflection). Omitting the mathematical manipulations, the wave amplitude on the wall is expressed as

$$\frac{H}{H_0} = \frac{4}{1+\sqrt{1-\frac{3H_0}{4h\tan^2\theta}}}, \tag{7.55}$$

where H_0 is the amplitude of the incident soliton and θ is the angle between the crest of the incident soliton and the y-axis. For small angles, close to the square root of the nonlinear parameter, H_0/h, the wave amplification factor becomes significant (close to 4). Using a perturbation method within the framework of the Boussinesq equation, Miles (1977) was the first to find the strong amplification of the soliton amplitude in the vicinity of the wall. For very small angle, the solution (7.55) becomes irrelevant. In this case, the soliton interaction induces a virtual soliton, the Mach stem. The resonance phenomenon in soliton interaction has been intensively studied, and recently Peterson *et al.* (2003) and Soomere and Engelbrecht (2005) suggested that multi-soliton solutions of the KP equation can explain the occurrence of three-dimensional extreme wave events in shallow water (not only in the vicinity of the vertical wall). Three-dimensional surface elevation of large-amplitude waves due to the two-soliton interaction for several values of the angle between soliton crests are plotted in figure 27 from the paper by Peterson *et al.* (2003). It is important to emphasize that these waves have an infinite life-time and propagate with constant speed.

7.4 The fully nonlinear equations

We have discussed the freak wave phenomenon within the framework of weakly nonlinear models. In fact freak waves are really huge waves which should be considered as strongly nonlinear waves. Studies about freak waves within the framework of fully nonlinear and dispersive models are very few. For fully nonlinear cases, the governing

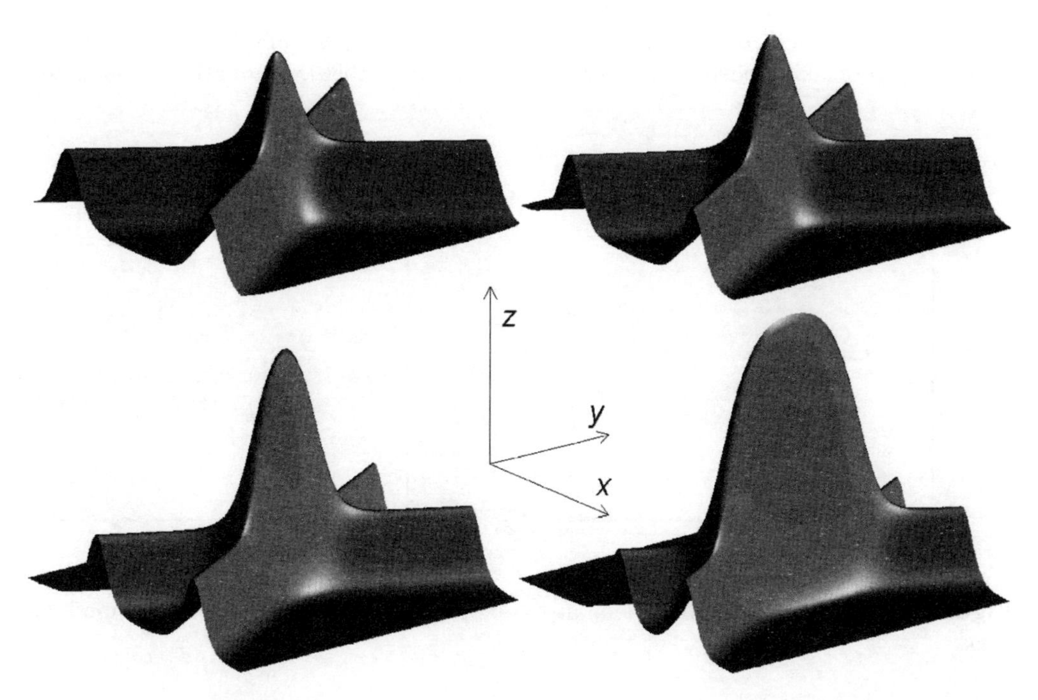

Figure 27. Large-amplitude waves due to soliton interactions in shallow water. From Peterson *et al.* (2003).

equations, given in section 5, are the Laplace equation (5.1), the dynamic boundary condition (5.4), the kinematic boundary condition (5.3) and the bottom condition (5.2) with $h(x, y) = const.$ For infinite depth the bottom condition becomes $\nabla\phi \to 0$ as $z- \to \infty$.

Generally direct numerical simulations of freak waves requires large spatial and temporal domains including a very large number of nodes and time steps.

Henderson, Peregrine and Dold (1999) investigated numerically the time evolution of a two-dimensional uniform wave train with a small growing modulation. They performed numerical experiments: it was observed that the energy can focus into a short group of steep waves called steep wave event (SWE). Details about the numerical code they used to study water wave modulations can be found in the paper by Dold (1992). They found that the breather solutions of the NLS equation fit numerical SWE rather well. These SWE's can be considered as freak wave events. So, the freak wave mechanism related with the Benjamin-Feir instability is confirmed in fully nonlinear computations. Figure 28 shows the water wave profile, at the maximum of modulation of an initially Stokes wave of wave steepness 0.10 disturbed by its most unstable perturbation. One can observe the formation of the huge wave due to modulational instability. The profile, at the maximum of modulation, is similar to that computed by Dommermuth and Yue (1987) for an initial value of the wave steepness equals to 0.13. Note that for the latter steepness value the numerical method, breaks down after few modulation-demodulation cycles owing to possible local breaking phenomena , see Dias and Kharif (1999). Recently Clamond and

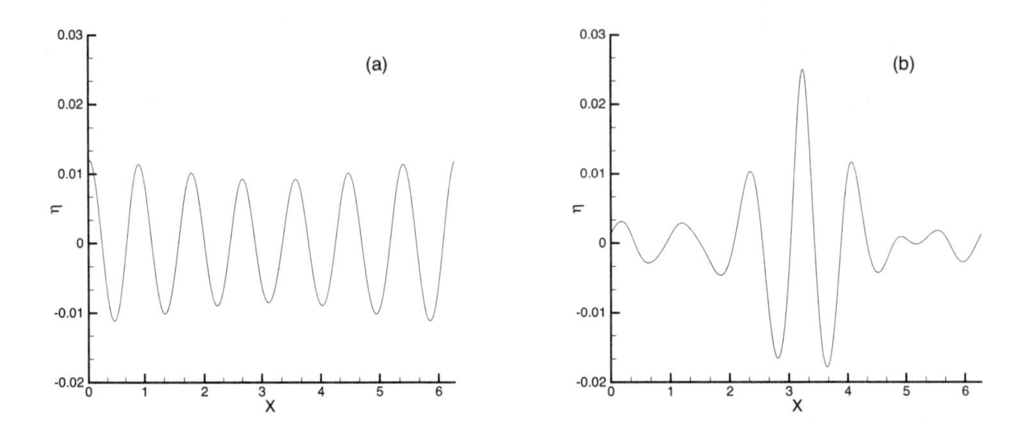

Figure 28. Water wave profiles (left) at $t = 0$, and (right) at the maximum of modulation.

Grue (2002) and Clamond *et al.* (2006) performed fully nonlinear numerical simulations of the long time evolution of a two-dimensional localized long wave packet and compared their computations with those resulting from the classical NLS equation and extended Dysthe equation. Their computations were obtained by using the fast computational method for nonlinear water waves developed previously by Clamond and Grue (2001) and it was shown by Clamond *et al.* (2006) that the results were consistent with those provided by the pseudo-spectral method of West, Brueckner and Janda (1987). The latter authors wrote the potential in a finite perturbation series up to a given order M, which is the order of approximation. Figure 29 shows the evolution of surface gravity wave packets presenting initially an envelope soliton by the fully nonlinear model developed by Clamond and Grue (2001). The initial profile corresponds to a Stokes wave of steepness 0.091 multiplied by a bell function and it is relevant for water waves. Clamond *et al.* (2006) found, as expected, that there is a good agreement with approximate models for relative short period of time, $O(T_0/\epsilon^2)$, for the NLS equation and $O(T_0/\epsilon^3)$ for the Dysthe equation, where T_0 is the fundamental wave period and ϵ is the wave steepness. For long periods of time it is found that approximate models should not necessarily provide correctly the nonlinear dynamics of the water waves and subsequently the probability of occurrence of freak waves. This conclusion should also apply for the Zakharov equation that is the more accurate among all the approximate models. The time evolution of the maximum crest amplitude is shown in figures 30–31 within the framework of different models. Of course, this is only one example of long-time evolution of isolated wave packets, but it emphasizes that full nonlinearity plays an important role for very long-time simulations.

Bateman, Swan and Taylor (2001) extended to three-dimensional motions the spectral wave model of Craig and Sulem (1993) to describe the water waves evolution of extreme waves due to the focusing of wave components involving spread of energy in both frequency and direction. The occurrence of freak waves would be due to direction spreading of waves. Using a JONSWAP spectrum, a three-dimensional simulation of the formation of a freak wave is presented.

Within the framework of fully nonlinear equations Brandini and Grilli (2001) investigated numerically the formation of rogue waves by using both spectral methods and a three-dimensional boundary element method. See also Ch. 4 of the volume.

The previous deterministic numerical experiments were performed to better understand the physical mechanisms that can contribute to the formation of abnormal waves. Mori and Yasuda (2001) considered exact equations numerically integrated by using the pseudo-spectral method of Dommermuth and Yue (1987) to investigate wave statistics. Unfortunately, they truncated the solution to fourth-order in wave steepness and this corresponds to the Zakharov equation (or Dysthe equation for narrow-band spectra). So their model does not capture the full nonlinearity contained in the equations. This limitation is all the more serious as the depth is shallow. As noted previously for long-time computation the water wave evolution can be very different from that given by fully nonlinear models. However, one believes that their conclusions remain valid when higher-order nonlinear interactions are considered, that is nonlinear terms should affect the long-term statistics of the waves. They emphasized that nonlinear interactions increase the occurrence probability of freak waves in deep water in comparison with the Rayleigh distribution.

Recently, Zakharov, Dyachenko and Vasilyev (2002) suggested a new method for numerical simulation of fully nonlinear water waves, based on combination of the conformal mapping and Fourier transform. The number of harmonics can be more than a dozen of thousands. Choosing the initial condition to be the superposition of a Stokes wave with wave steepness 0.1 and a weak Gaussian random wave field, they obtained the formation of a freak wave whose amplitude exceeds its initial level more than three times.

In ocean engineering the safe and economic design of any fixed or floating structures depend on a reliable estimate of the extreme wave loading. The knowledge of the underlying wave particle kinematics beneath the largest wave crests represents key information appropriate to the determination of the design loading. Johannessen and Swan (1997) showed that water particle kinematics is strongly dependent upon the nonlinear wave-wave interaction. They emphasized that if an appropriate description of a extreme wave event is to be achieved, both nonlinearity and unsteadiness of the wave motion must be incorporated. Bateman, Swan and Taylor (2003) computed velocities and accelerations beneath extreme three-dimensional ocean wave. They confirmed the result of Longuet-Higgins (1986) that the maximum magnitude of the vertical accelerations arises at some distance beneath the water surface.

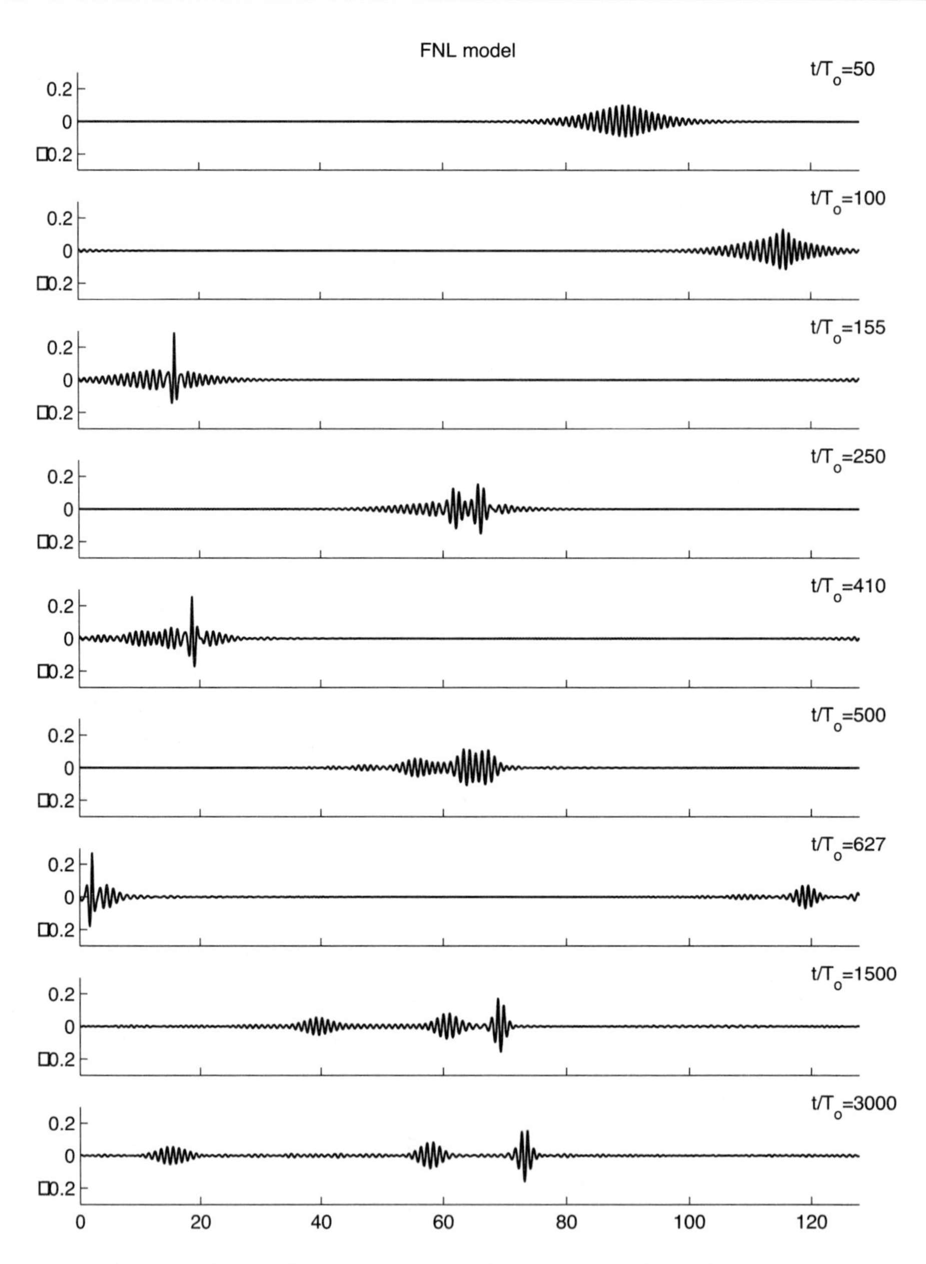

Figure 29. Wave packet evolution within the framework of the fully nonlinear equations. From Clamond *et al.* (2006).

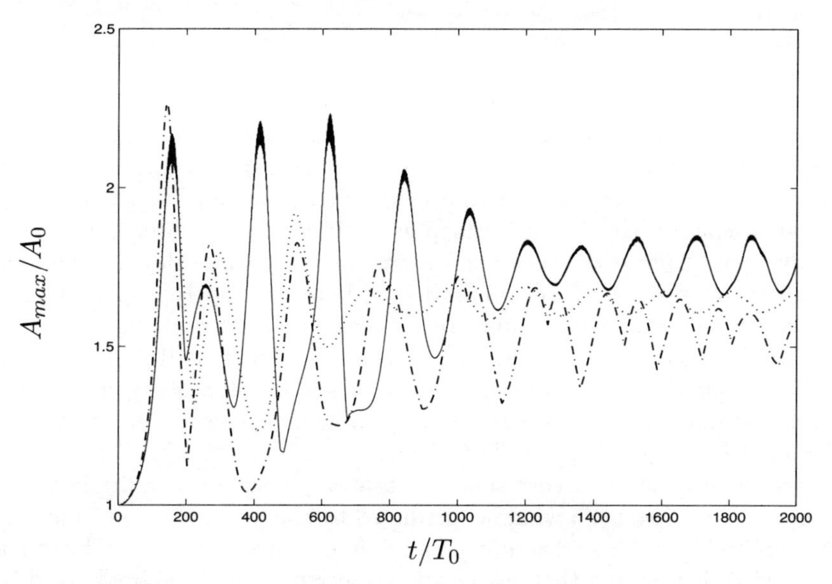

Figure 30. Time evolution of the maximum crest amplitude (normalized by initial amplitude) computed within the framework of several models. (-) CG, (...) extended Dysthe equation, (-.) NLS. From Clamond *et al.* (2006).

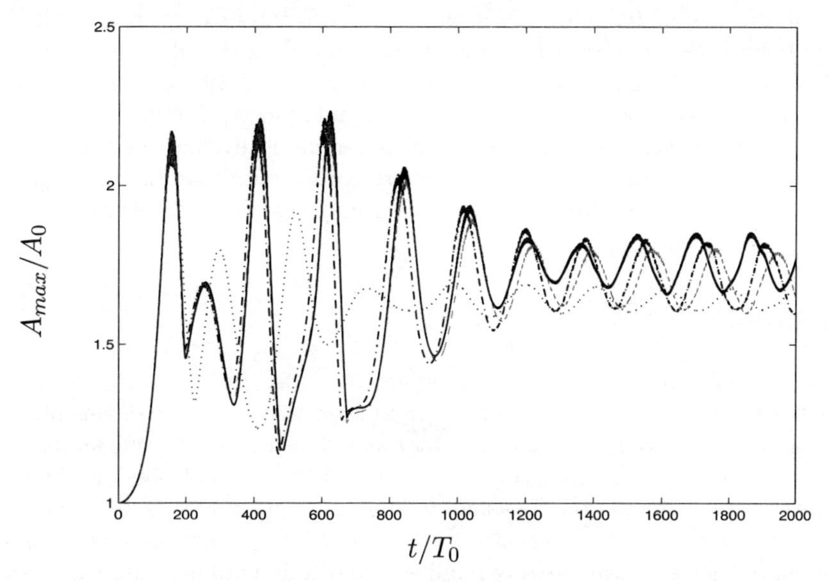

Figure 31. Time evolution of the maximum crest amplitude (normalized by initial amplitude) computed within the framework of several models. (-) CG, (- -) HOSM with $M = 4$, (-.-) HOSM with $M = 3$, (...) extended Dysthe equation. From Clamond *et al.* (2006).

8 Experiments

Many laboratory studies, cited in section 2, have concerned the focusing of transient surface water waves mechanically generated. Herein is briefly presented a unique experimental study of the effect of the wind on the focusing of transient wave groups developed by Giovanangeli *et al.* (2004a,b) in the large wind-wave tank of IRPHE (Marseille-Luminy). It is constituted of a closed loop wind tunnel located over a water tank 40m long, 1m deep and 2.6m wide. The wind tunnel over the water flow is 40m long, 3.2m wide and 1.6m high. The blower allows to produce wind speeds up to 14m/s and a computer-controlled wave maker submerged under the up stream beach can generate regular or random wave fields in frequency range from 0.5Hz to 2Hz. For each value of the mean wind speed from 0 to 8m/s, the water surface elevation η was measured at 1m fetch and at different fetches between 5m and 35m. The wave maker was driven by an analog electronic signal varying linearly from 1.3Hz to 0.8Hz in 10s with constant amplitude of the displacement corresponding to nearly constant amplitude of the initial wave group. When there is no wind, as predicted by the linear theory of wave focusing, the high-amplitude wave corresponding to the freak wave occurs at a fixed fetch. In presence of wind, it is shown that the location of occurrence of the freak wave is moved downstream and its amplitude is increased. It is also observed that high-amplitude waves are sustained in the group while it was not the case without wind since the freak wave event is followed by the defocusing of the wave group. The effect of the wind is to weaken the defocusing process. Furthermore, the experiments have emphasized the importance of the form drag associated with flow separation which can be described by the Jeffreys' sheltering theory. Hence, numerical simulations based on a BIEM method coupled with the Jeffreys' sheltering mechanism have been performed for different values of the wind speed by Kharif *et al.* (2005). Both experimental and numerical results are in quite qualitative agreement. It is shown experimentally and numerically that the effect of the wind increases the timescale of existence of the short group containing the rogue wave. This means that the Jeffreys sheltering phenomenon may sustain rogue waves by increasing their time of occurrence.

9 Conclusion

The freak wave phenomenon has been investigated intensively these last years, theoretically and experimentally. It is important to emphasize that existence of this phenomenon in the ocean has been confirmed in deep water as well as in coastal zone and now a lot of freak wave records are collected. All of the physical scenarios of possible extreme wave event (wave-current interaction, focusing, modulational instability, soliton-collision, etc) were, in fact, known but only recently were included in mathematical models of freak waves of various level (linear, weakly nonlinear and fully nonlinear models). Numerical simulations based on these models have been useful to better understand the different mechanisms yielding to the formation of huge waves. Effect of randomness on the freak wave generation was also analyzed. The random character of the wave field can be due for instance to wind or currents. Results of computations can provide the probability of freak wave occurrence for simplified models.

Most of laboratory experiments have been conducted mainly to study the focusing mechanism of the freak wave occurrence and wave-current interaction. To our knowledge the first experiment concerning the direct effect of the wind on the formation of freak waves has been perform very recently in IRPHE. Further experiments and development of models including wind action are needed to consider freak wave events inside storm zones.

Within the framework of the fully nonlinear equations, it would be useful to consider three-dimensional water wave fields in order to check the validity of approximate models for simplified conditions. At present, large three-dimensional numerical simulations of the fully nonlinear equations including variable bathymetry and current are out of reach. Up to now the applicability of the analytical and numerical results provided by the different models presented in the previous sections to real wind waves has not been checked. Despite all the approximations on which are based the different models, a significant progress in the understanding and modelling of the physics of the freak wave phenomenon has been realized.

Bibliography

M.J. Ablowitz, C.M. Schober and B.M. Herbst (1993). Numerical chaos, roundoff errors, and homoclinic manifolds. Phys. Rev. Lett. Vol. 71(17), 2683–7.

M.J. Ablowitz, J. Hammack, D. Henderson and C.M. Schober (2000). Modulated periodic Stokes waves in deep water. Phys. Rev. Lett. Vol. 84(5), 887–90.

M.J. Ablowitz, J. Hammack, D. Henderson and C.M. Schober (2001). Long-time dynamics of the modulational instability of deep water waves. Physica D Vol. 152-153, 416–33.

I.E. Alber (1978). The effects of randomness on the stability of two-dimensional surface wavetrain. Proc. Roy. Soc. Lond. A Vol. 363, 525–46.

I.E. Alber and P.G. Saffman (1978). Stability of random nonlinear water waves with finite bandwidth spectra, In TRW Defense and Space Syst. Gr. Rep. No 31326-6035-RU-00bv.

D. Anker and N.C. Freeman (1978). On the soliton solutions of the Davey-Stewartson equation for long wave. Proc. Roy. Soc. London A Vol. 360, 529–40.

S.Y. Annenkov and S.I. Badulin (2001). Multi-wave resonances and formation of high-amplitude waves in the ocean, In Rogues Waves 2000. Ed. by M. Olagnon and G.A. Athanassoulis, Brest, 205–13.

W.J.D. Bateman, C. Swan and P.H.Taylor (2001). On the efficient numerical simulation of directionally spread surface water waves. J. Comp. Phys. Vol. 174, 277–301.

W.J.D. Bateman, C. Swan and P.H.Taylor (2003). On the calculation of the water particle kinematics arising in a directionally spread wavefield. J. Comp. Phys. Vol. 186, 70–90.

T.B. Benjamin (1967). Instability of periodic wave trains in nonlinear dispersive systems. Proc. Roy. Soc. Lond. A Vol. 299, 59–75.

T.B. Benjamin and J.E. Feir (1967). The desintegration of wave trains on deep water. Part 1. Theory. J. Fluid Mech. Vol. 27, 417–30.

C. Brandini and S. Grilli (2001). Evolution of three-dimensional unsteady wave modulations. Rogues Waves, Brest 2000. Ed. by M. Olagnon and G.A. Athanassoulis, 275–82.

F.P. Bretherton and J.R. Garrett (1969). Wavetrains in inhomogeneous moving media. Proc. Roy. Soc. London A Vol. 302, 529–54.

U. Brinch-Nielsen and I.G. Jonsson (1986). Fourth-order evolution equations and stability analysis for Stokes waves on arbitrary water depth. Wave Motion. Vol. 8, 455–72.

M.G. Brown and A. Jensen (2001). Experiments on focusing unidirectional water waves. J. Geophys. Res. Vol. 106(C8), 16917–28.

A. Calini and C.M. Schober (2002). Homoclinic chaos increases the likelihood of rogue wave formation. Phys. Lett. A Vol. 298, 335–49.

D. Clamond and J. Grue (2001). A fast method for fully nonlinear water wave computations. J. Fluid Mech. Vol. 398, 45–60.

D. Clamond and J. Grue (2002). Interaction between envelope solitons as a model for freak wave formation. Pt 1: Long-time interaction. C.R. Mecanique Vol. 330, 575–80.

D. Clamond, M. Francius, J. Grue and C. Kharif (2006). Strong interaction between envelope solitary surface gravity waves. Eur. J. Mech. B/Fluids Vol. 5.

G. Clauss (2002). Dramas of the sea: Episodic waves and their impact on offshore structures. Appl. Ocean Res. Vol. 24, 147–61.

G. Contento, R. Codiglia and F. D'Este (2001). Nonlinear effects in 2D transient nonbreaking waves in a closed flume. Appl. Ocean Res. Vol. 23, 3–13.

W. Craig and C. Sulem (1993). Numerical simulation of gravity waves. J. Comput. Phys. Vol. 108, 73–83.

D.R. Crawford, P.G. saffman and H.C. Yuen (1980). Evolution of a random inhomogeneous field of nonlinear deep-water gravity waves. Wave Motion Vol. 2, 1–16.

F. Dias and C. Kharif (1999). Nonlinear gravity and capillary-gravity waves. Annu. Rev. Fluid Mech. Vol. 31, 301–46.

J.W. Dold (1992). An efficient surface-integral algorithm applied to unsteady gravity waves. J. Comp. Phys. Vol. 103, 90–115.

D.G. Dommermuth and D.K.P. Yue (1987). A high-order spectral method for the study of nonlinear gravity waves. J. Fluid Mech. Vol. 184, 267–88.

A.I. Dyachenko and V.E. Zakharov (2005). Modulational instability of Stokes wave → freak wave. J.E.T.P. Vol. 81(6), 318–22.

K.B. Dysthe (1979). Note on a modification to the nonlinear Schrödinger equation for application to deep water waves. Proc. Roy. Soc. London A Vol. 369, 105–14.

K.B. Dysthe (2001). Modelling a "Rogue Wave" - Speculations or a realistic possibility? Rogues Waves, Brest 2000. Ed. by M. Olagnon and G.A. Athanassoulis, 255–64.

K.B. Dysthe (2001). Refraction of gravity waves by weak current gradients. J. Fluid Mech. Vol. 442, 157-59.

K.B. Dysthe and K. Trulsen (1999). Note on breather type solutions of the NLS as a model for freak waves. Phys. Scripta Vol. T82, 48-52.

K.B. Dysthe, K. Trulsen, H.E. Krogstad and H. Socquet-Juglard (2003). Evolution of a narrow-band spectrum of random surface gravity waves. J. Fluid Mech. Vol. 478, 1–10.

J.P. Giovanangeli, C. Kharif and E. Pelinovsky (2004a). An experimental study on the effect of wind on the generation of freak waves, In Proc. of 1st European Geophysical Union 6, Natural Hazards, Abstract number EGU-A-07436, Nice 2004.

J.P. Giovanangeli, C. Kharif and E. Pelinovsky (2004b). Experimental study of the wind effect on the focusing of transient wave groups. Proc. Rogue waves 2004, Brest.

D.M. Graham (2000) NOAA vessel swamped by rogue wave. Oceanspace 284.

J. Grue and E. Palm (1985). Wave radiation and wave diffraction from a submerged body in a uniform current. J. Fluid Mech. Vol. 151, 257-78.

J. Grue and E. Palm (1986). The influence of a uniform current on slowly varying forces and displacements. Appl. Ocean Res. Vol. 8(4), 232-39.

S. Haver and O.J. Andersen (2000). Freak waves: Rare realizations of a typical population or typical realizations of a rare population? Proc. of 10th ISOPE Conference, 123–30.

K.L. Henderson, D.H. Peregrine and J.W. Dold (1999). Unsteady water wave modulations: Fully nonlinear solutions and comparison with the nonlinear Schrödinger equation. Wave Motion Vol. 29, 341-61.

B.M. Herbst and M.J. Ablowitz (1989). Numerically induced chaos in the nonlinear Schrödinger equation. Phys. Rev. Lett. Vol. 62(18), 2065–8.

A.L. Islas and C.M. Schober (2005). Predicting rogue waves in random oceanic sea states. Phys. Fluids Vol. 17, 031701-1–4.

P.A.E.M. Janssen (2003). Nonlinear four-wave interactions and freak waves. J. Phys. Oceanogr. Vol. 33, 863–84.

H. Jeffreys (1925). On the formation of wave by wind. Proc. Roy. Soc. A Vol. 107, 189–206.

T.B. Johannessen and C. Swan (1997). Nonlinear transient water waves - Pt. 1. A numerical method of computation with comparisons to 2D laboratory data. Appl. Ocean Res. Vol. 19, 293–308.

T.B. Johannessen and C. Swan (2001). A laboratory study of the focusing of transient and directionally spread surface water waves. Proc. Roy. Soc. Lond. A Vol. 457, 971–1006.

T.B. Johannessen and C. Swan (2003). On the nonlinear dynamics of wave groups produced by the focusing of surface-water waves. Proc. Roy. Soc. Lond. A Vol. 459, 1021–52.

C. Kharif (1990). Some aspects of the kinematics of short waves over longer gravity waves on deep water, In Water Wave kinematics. Ed. by A. Torum and O.T. Gudmestad, Kluwer Academic, Dordrecht, 265–79.

C. Kharif, E. Pelinovsky and T. Talipova (2000). Formation de vagues géantes en eau peu profonde. C. R. Acad. Sci. Paris, Ser. IIb Vol. 328(11), 801–7.

C. Kharif, E. Pelinovsky, T. Talipova and A. Slunyaev (2001). Focusing of nonlinear wave groups in deep water. J.E.T.P. Lett. Vol. 73(4), 170–5.

C. Kharif and E. Pelinovsky (2003). Physical mechanisms of the rogue wave phenomenon. Eur. J. Mech. B/Fluids Vol. 22, 603–34.

C. Kharif, J.P. Giovanangeli, J. Touboul and E. Pelinovsky (2005). Influence of the Jeffreys' sheltering mechanism on rogue waves, Workshop on Rogue Waves, Dec. 2005, Edinburgh.

E. Kit, L. Shemer, E. Pelinovsky, T. Talipova, O. Eitan and H. Jiao (2000). Nonlinear wave group evolution in shallow water. J. Waterways Port. Coast. Ocean Engng. Vol. 126(5), 221–8.

E. Kit and L. Shemer (2002). Spatial versions of the Zakharov and Dysthe equations for deep-water gravity waves. J. Fluid Mech. Vol. 450, 201–5.

V.P. Krasitskii (1994). On reduced equations in the Hamilitonian theory of weakly nonlinear surface waves. J. Fluid Mech. Vol. 272, 1–30.

I.V. Lavrenov (1998a). Mathematical modeling of the wind waves in spatially inhomogeneous Ocean. Hydrometeozdat, St Petersburg.

I.V. Lavrenov (1998b). The wave energy concentration at the Agulhas current off South Africa. Natural Hazards Vol. 17, 117–27.

G. Lawton (2001). Monsters of the deep (The perfect wave). New Scientist Vol. 170 (2297), 28–32.

M.J. Lighthill (1965). Contributions to the theory of waves in nonlinear dispersive systems. J. Inst. Math. Appl. 1, 269–306.

E. Lo and C.C. Mei (1985). A numerical study of water-wave modulation based on a high-order nonlinear Schrödinger equation. J. Fluid Mech. Vol. 150, 395-416.

M.S. Longuet-Higgins and R.W. Stewart (1961). The changes in amplitude of short gravity waves on steady non-uniform currents. J. Fluid Mech. Vol. 10, 529–49.

M.S. Longuet-Higgins and R.W. Stewart (1964). Radiation stress in water waves; a physical discussion, with applications. Deep Sea Res. Vol. 11, 529–62.

M.S. Longuet-Higgins (1986). Eulerian and Lagrangian aspects of surface waves. J. Fluid Mech. Vol. 173, 683–707.

J.K. Mallory (1974). Abnormal waves in the south-east coast of South Africa. Int. Hydrog. Rev. Vol. 51, 99–129.

S.R. Massel (1996). Ocean surface waves: Their physics and prediction. Word Scientific, Singapore.

J.W. Miles (1977). Resonantly interacting solitary waves. J. Fluid Mech. Vol. 79, 171–9.

N. Mori and Y. Yasuda (2001). Effects of high-order nonlinear interactions on unidirectional wave trains. Rogues Waves, Brest 2000. Ed. by M. Olagnon and G.A. Athanassoulis, 229–44.

N. Mori, P.C. Liu and Y. Yasuda (2002). Analysis of freak wave measurements in the sea of Japan. Ocean Engrg. Vol. 29, 1399–414.

K. Onkuma and M. Wadati (1983). The Kadomstev-Petviashvili equation, the trace method and the soliton resonance. J. Phys. Soc. Japan Vol. 52, 749–60.

M. Onorato, A.R. Osborne, M. Serio and T. Damiani (2001). Occurrence of freak waves from envelope equations in random ocean wave simulations. Rogues Waves 2000 edited by M. Olagnon and G.A. Athanassoulis (Brest, France), 181–91.

M. Onorato, A.R. Osborne, M. Serio and S. Bertone (2001). Freak waves in random oceanic sea states. Phys. Rev. Lett. Vol. 86(25), 5831–4.

M. Onorato, A.R. Osborne and M. Serio (2002). Extreme wave events in directional, random oceanic sea states. Phys. Fluids Vol. 14(4), L25–8.

M. Onorato, D. Ambrosi, A.R. Osborne and M. Serio (2003). Interaction of two quasi-monochromatic waves in shallow water. Phys. Fluids Vol. 15, 3871.

A.R. Osborne, M. Onorato and M. Serio (2000). The nonlinear dynamics of rogue waves and holes in deep-water gravity wave train. Phys. Lett. A Vol. 275, 386-93.

E. Pelinovsky (1996). Hydrodynamics of tsunami waves. Applied Physics Institute Press, Niznhy Novgorod, 1996.

E. Pelinovsky, T. Talipova and C. Kharif (2000). Nonlinear dispersive mechanism of the freak wave formation in shallow water. Physica D Vol. 147(1-2), 83-94.

D.H. Peregrine (1983). Wave jumps and caustics in the propagation of finite-amplitude water waves. J. Fluid Mech. Vol. 136, 435–52.

D.H. Peregrine and R. Smith (1979). Nonlinear effects upon waves near caustics. Philos. Trans. Roy. Soc. London A Vol. 292 10, 341–70.

P. Peterson, T. Soomere, J. Engelbrecht and E. van Groesen (2003). Soliton interaction as a possible model for extreme waves. Nonlinear Processes in Geophysics Vol. 10, 503–10.

S.E. Sand, N.E. Hansen, P. Klinting, O.T. Gudmestad and M.J. Steindorff (1990). Freak wave kinematics, In Water Wave kinematics. Ed. by A. Torum and O.T. Gudmestad, Kluwer Academic, Dordrecht, 535–49.

R.N. Sanderson (1974). The unusual waves off south east Africa. Marine Observer Vol. 44, 180–3.

J. Satsuma (1976). N-soliton solution of the two-dimensional Korteweg-de Vries equation. J. Phys. Soc. Japan Vol. 40, 286–90.

J. Satsuma and N. Yajima (1974). Initial value problems of one-dimensional self-modulation of nonlinear waves in dispersive media. Suppl. Prog. Theor. Phys. Vol. 55, 284–306.

L. Shemer, E. Kit, H. Jiao, O. Eitan (1998). Experiments on nonlinear wave groups in intermediate water depth. J. Waterways Port. Coast. Ocean Engng. Vol. 124, 320-7.

A. Slunyaev, C. Kharif, E. Pelinovsky and T. Talipova (2002). Nonlinear wave focusing on water of finite depth. Physica D Vol. 173(1-2), 77-96.

R. Smith (1976). Giant waves. J. Fluid Mech. Vol. 77, 417–31.

H. Socquet-Juglard, K.B. Dysthe, K. Trulsen and H.E. Krogstad (2005). Probability distributions of surface gravity waves during spectral changes, J. Fluid Mech. Vol. 542, 195-216.

T. Soomere and J. Engelbrecht (2005). Extreme elevations and slopes of interacting solitons in shallow water. Wave Motion Vol. 41, 179–92.

C.T. Stansberg (2001). Random waves in the laboratory - what is expected for the extremes?. In M. Olagon and G.A. Athanassoulis (Eds.), Rogue Waves 2000 (Brest, France), 289–301.

M. Stassnie and L. Shemer (1984). On modifications of the Zakharov equation for surface gravity waves. J. Fluid Mech. Vol. 143, 47–67.

H. Sturm (1974). Giant waves. Ocean Vol. 2(3), 98–101.

H.U. Sverdrup and W.H. Munk (1947). Wind, sea, and swell; theory of relations for forecasting. U.S. Navy Hydrographic Office, H.O. 601. y

M.A. Tayfun (1980). Narrow-band nonlinear sea waves. J. Geophys. Res. Vol. 85, 1548–52.

M.A. Tayfun, R.A. Dalrymple and C.Y. Yang (1976). Random wave-current interactions in water of varying depth Ocean Engng. 3, 403–20.

K. Trulsen (2001). Simulating the spatial evolution of a measured time series of a freak wave. Rogues Waves, Brest 2000. Ed. by M. Olagnon and G.A. Athanassoulis, 265–74.

K. Trulsen and K.B. Dysthe (1996). A modified nonlinear Schrödinger equation for broader bandwidth gravity waves on deep water. Wave Motion Vol. 24, 281–9.

K. Trulsen and K.B. Dysthe (1997). Freak waves - a three-dimensional wave simulation. Proc. 21st Symposium on Naval Hydrodynamics, 550–6.

K. Trulsen, I. Kliakhandler, K.B. Dysthe and M.G. Velarde (2000). On weakly nonlinear modulation of waves on deep water. Phys. Fluids Vol. 12, 2432–7.

B.J. West, K.A. Brueckner and R.S. Janda. A method of studying nonlinear random field of surface gravity waves by direct numerical simulation. J. Geophys. Res. Vol. 92, C 11, 11803-24.

B.S. White and B. Fornberg (1998). On the chance of freak waves at the sea. J. Fluid Mech. Vol. 255, 113–38.

G.B. Whitham (1965). A general approach to linear and non-linear dispersive waves using a Lagrangian. J. Fluid Mech. Vol. 22, 273–83.

G.B. Whitham (1967). Nonlinear dispersion of water waves J. Fluid Mech. Vol. 27, 399–412.

J. Willebrand (1975). Energy transport in a nonlinear and inhomogeneous random gravity wave field. J. Fluid Mech. Vol. 70, 113–26.

C.H. Wu and A. Yao (2004). Laboratory measurements of limiting freak waves on currents. J. Geophys. Res. Vol. 109, C12002, 1–18.

V.E. Zakharov (1966). Instability of waves in nonlinear dispersive media. J.E.T.P. Vol. 51, 1107–14.

V.E. Zakharov (1968). Stability of periodic waves of finite amplitude on the surface of deep water. J. Appl. Mech. Tech. Phys. Vol. 9, 190–94.

V.E. Zakharov, A.I. Dyachenko and O. Vasilyev (2002). New method for numerical simulation of nonstationary potential flow of incompressible fluid with a free surface. Eur. J. Mech.B/Fluids Vol. 21, 283–91.

Rapid computations of steep surface waves in three dimensions, and comparisons with experiments

John Grue

Mechanics Division, Department of Mathematics, University of Oslo, Norway

Abstract A novel fully nonlinear, rapid method for computations of ocean surface waves in three dimensions is outlined in this chapter. The essential step is to use Fourier transform to invert the integral equation over the ocean surface that solves the Laplace equation. This leads to a relation for the normal velocity of the free surface that is useful for iterations. This relation has a global contribution that is obtained by FFT and local contribution that is evaluated by rapidly converging integrals in the horizontal plane. The global part, evaluated by FFT, captures the most essential parts of the wave field. Together with an efficient time integration of the prognostic equations, where the linear part is integrated analytically and a time variable step size control is used for the nonlinear part, this results in a highly rapid computational strategy. Methods for efficient nonlinear wave generation and absorption are outlined. Conservation of various quantities of the wave field and convergence are discussed. Fully nonlinear computations of very large (rogue) wave events at sea, like the Camille and Drauper waves, are compared to laboratory measurements of the waves using Particle Image Velocimetry. The method is used to exemplify the generation, propagation and shoaling of very long wave phenomena like tsunamis, and class I and class II instabilities in water of infinite and finite depth. Both steady and oscillatory crescent wave patterns are predicted up to breaking. Competition between class I and class II instabilities is discussed.

1 Introduction

The previous three chapters have decribed processes in the ocean environment where the motion can be represented by their long wave features (ray theory), depth-integrated equations or approximate phase resolving models. From a spectral point of view, the long wave models capture the essential motion at small wavenumbers, while the approximate phase resolving models assume that the motion is narrow banded or slowly modulated. The outcome is efficient computational strategies for the wave motion on global scale - like the tsunami propagation in the world's oceans, or the formation of extreme waves due to the evolution in long wave fields.

A different strategy is to develop models that resolve the motion on a broad range spectrum, including the nonlinear interactions taking place between the motion on long scales (small wavenumbers) and short scales (high wavenumbers). The success of such a strategy depends on the efficiency of the model, where an essential task is to remove computational limitations. In nonlinear surface wave simulations a bottleneck is the solution of the Laplace equation, which needs to be obtained at each step of the time integration of the flow. Below we outline a method where this limitation has been overcome. The result is a method that in a fast and robust way simulates strongly nonlinear ocean surface waves in three dimensions. The method was derived by Clamond and Grue (2001,§6) for the case of infinite water depth and by Grue (2002) for the case of a finite water depth. The method and its full numerical implementation has recently been fully documented and tested out, see Fructus, Clamond, Grue and Kristiansen (2005a), Clamond, Fructus, Grue and Kristiansen (2005). The method has five essential parts:

1. A fast converging iterative solution procedure of the Laplace equation where the dominant, global part of the solution is obtained by Fast Fourier Transform, while the remaining part is highly nonlinear and consists of integrals with kernels that decay quickly in the space coordinate. Practical computations show that the explicit version of the method provides results with high accuracy, while any accuracy is obtained by a continued iteration, up to three times, when the method has converged. Moreover, computations of strongly nonlinear ocean surface wave phenomena can be performed on small PCs.
2. In the time stepping procedure, the linear part of the kinematic and dynamic boundary conditions are integrated analytically, working in spectral space. The remaining truly nonlinear part is integrated forward in time by using an RK-54 with variable step-size control.
3. Both partially and fully de-aliased computational schemes with no smoothing or regridding are successfully tested out.
4. Efficient fully nonlinear wave generation and wave damping procedures are derived and implemented.
5. The method is for the moment being generalized to include an arbitrary shape of the sea bottom that also can be moving. Preliminary results of the highly nonlinear generation of tsunamis are included.

The most essential steps of the method are described here for the case of a constant water depth h and a non-moving sea floor. We consider non-overturning waves. The method is based on application of potential theory. Let $\phi(\mathbf{x}, y, t)$ denote the velocity potential, and $\eta(\mathbf{x}, t)$, $\mathbf{x} = (x_1, x_2)$, y, and t denote surface elevation, horizontal coordinate, vertical coordinate and time, respectively. We introduce the potential function $\widetilde{\phi}(\mathbf{x}, t)$ at the free surface determined by $\widetilde{\phi}(\mathbf{x}, t) = \phi(\mathbf{x}, \eta(\mathbf{x}, t), t)$ (Zakharov, 1968). The kinematic and dynamic boundary conditions at $y = \eta(\mathbf{x}, t)$ give

$$\eta_t - V = \tilde{v}_D, \tag{1.1}$$

$$\widetilde{\phi}_t + g\eta + \frac{|\nabla_1\widetilde{\phi}|^2 - V^2 - 2V\,\nabla_1\eta\cdot\nabla_1\widetilde{\phi} + |\nabla_1\eta\times\nabla_1\widetilde{\phi}|^2}{2(1+|\nabla_1\eta|^2)} = -\tilde{p}_G - \tilde{p}_D. \tag{1.2}$$

In (1.1)–(1.2) a scaled normal velocity of the free surface, $V = \phi_n[1 + |\nabla_1\eta|^2]^{\frac{1}{2}}$ is introduced, where ϕ_n denotes the normal velocity and ∇_1 denotes the horizontal gradient. Further, $\tilde{p}_G$ denotes a pressure acting on the free surface corresponding to a pneumatic wave generator, and $\tilde{p}_D$ a pressure acting in the form of a damping (or dissipation). Both $\tilde{p}_G$ and $\tilde{p}_D$ are normalized by the density of the fluid. $\tilde{v}_D$ denotes a damping term in the kinematic equation, and corresponds a mass flux through the free surface. The functions $\tilde{p}_G$, $\tilde{p}_D$, $\tilde{v}_D$ are discussed in section 6.

2 Efficient solution of the Laplace equation

Equations (1.1)–(1.2) may be integrated in time once a relation between η, $\widetilde{\phi}$ and V is found. The latter is obtained from the solution of the Laplace equation when $\widetilde{\phi}$ and η are given on the free surface. By use of Green's theorem we obtain

$$\int\int_S \left(\frac{1}{r} + \frac{1}{r_1}\right) \frac{\partial \phi'}{\partial n'} \, \mathrm{d}S = 2\pi\widetilde{\phi} + \int\int_S \widetilde{\phi}' \frac{\partial}{\partial n'} \left(\frac{1}{r} + \frac{1}{r_1}\right) \mathrm{d}S, \tag{2.1}$$

where $\widetilde{\phi} = \widetilde{\phi}(\mathbf{x})$, $\widetilde{\phi}' = \widetilde{\phi}(\mathbf{x}')$, $r = [\mathbf{R}^2 + (y'-y)^2]^{\frac{1}{2}}$, $r_1 = [\mathbf{R}^2 + (y'+y+2h)^2]^{\frac{1}{2}}$, $\mathbf{R} = \mathbf{x}' - \mathbf{x}$ and S denotes the free surface. The normal vector $\mathbf{n}$ is pointing out of free surface.

We first derive the equations assuming infinite water depth, i.e. $h \to \infty$, and $1/r_1 \to 0$.

The element of the free surface is given by

$$\mathrm{d}S = [1 + (\nabla_1' \eta')^2]^{\frac{1}{2}} \, \mathrm{d}\mathbf{x} \tag{2.2}$$

where ∇_1 denotes the horizontal gradient. We now introduce

$$D = \frac{\eta' - \eta}{R}. \tag{2.3}$$

The quantity D denotes the difference in the elevation at the two points $\mathbf{x}'$ and $\mathbf{x}$, divided by the horizontal distance $R = |\mathbf{x}' - \mathbf{x}|$ between them. We have that

$$D \sim R^{-1} \qquad R \to \infty, \tag{2.4}$$

$$D \to \eta_R \qquad R \to 0 \tag{2.5}$$

The equation (2.1) then becomes

$$\int\int_{-\infty}^{\infty} \frac{V' \, \mathrm{d}\mathbf{x}'}{R(1+D^2)^{\frac{1}{2}}} = 2\pi\widetilde{\phi} + \int\int_{-\infty}^{\infty} (1+D^2)^{-\frac{3}{2}} \left(\frac{\mathbf{R}\cdot\nabla_1'\eta'}{R^3} - \frac{\eta'-\eta}{R^3}\right) \widetilde{\phi}' \, \mathrm{d}\mathbf{x}'. \tag{2.6}$$

First, we reorganize the integrand on the l.h.s. of (2.6) by writing

$$\frac{1}{R(1+D^2)^{\frac{1}{2}}} = \frac{1}{R} + \frac{1}{R}\left[(1+D^2)^{-\frac{1}{2}} - 1\right] \tag{2.7}$$

The first term, $1/R$ is kept on the l.h.s. of (2.6), while the second term, $\frac{1}{R}\left[(1+D^2)^{-\frac{1}{2}} - 1\right]$ is moved to the r.h.s.

In the next step we consider the integral on the r.h.s. of (2.6). Exploiting that

$$\frac{\mathbf{R}\cdot\nabla_1'\eta'}{R^3}-\frac{\eta'-\eta}{R^3}=-\nabla_1'\cdot\left[(\eta'-\eta)\nabla_1'\frac{1}{R}\right], \tag{2.8}$$

the last term in (2.6) may be rewritten, i.e.

$$\begin{aligned}\int\int_{-\infty}^{\infty}&\left(\frac{\mathbf{R}\cdot\nabla_1'\eta'}{R^3}-\frac{\eta'-\eta}{R^3}\right)\widetilde{\phi}'\,\mathrm{d}\mathbf{x}'\\ &=-\int\int_{-\infty}^{\infty}\widetilde{\phi}'\nabla_1'\cdot\left[(\eta'-\eta)\nabla_1'\frac{1}{R}\right]\,\mathrm{d}\mathbf{x}'\\ &=\int\int_{-\infty}^{\infty}(\eta'-\eta)\nabla_1'\widetilde{\phi}'\cdot\nabla_1'\frac{1}{R}\mathrm{d}\mathbf{x}',\end{aligned} \tag{2.9}$$

where we have used Gauss' theorem to obtain that

$$\int\int_{-\infty}^{\infty}\nabla_1'\cdot\left[\widetilde{\phi}'(\eta'-\eta)\nabla_1'\frac{1}{R}\right]\,\mathrm{d}\mathbf{x}'=0. \tag{2.10}$$

The modified and reorganized version of the equation reads

$$\begin{aligned}\int\int_{-\infty}^{\infty}\frac{V'}{R}\,\mathrm{d}\mathbf{x}'&=2\pi\widetilde{\phi}+\int\int_{-\infty}^{\infty}(\eta'-\eta)\nabla_1'\widetilde{\phi}'\cdot\nabla_1'\frac{1}{R}\,\mathrm{d}\mathbf{x}'\\ &-\int\int_{-\infty}^{\infty}\widetilde{\phi}'\left[(1+D^2)^{-\frac{3}{2}}-1\right]\nabla_1'\cdot\left[(\eta'-\eta)\nabla_1'\frac{1}{R}\right]\,\mathrm{d}\mathbf{x}'\\ &-\int\int_{-\infty}^{\infty}\frac{V'}{R}\left[(1+D^2)^{-\frac{1}{2}}-1\right]\,\mathrm{d}\mathbf{x}'.\end{aligned} \tag{2.11}$$

A decomposition $V=V_1+V_2+V_3+V_4$ is then introduced, where V_1, V_2, V_3, V_4 satisfy, respectively,

$$\int\int_{-\infty}^{\infty}\frac{V_1'}{R}\,\mathrm{d}\mathbf{x}'=2\pi\widetilde{\phi}, \tag{2.12}$$

$$\int\int_{-\infty}^{\infty}\frac{V_2'}{R}\,\mathrm{d}\mathbf{x}'=\int\int_{-\infty}^{\infty}(\eta'-\eta)\nabla_1'\widetilde{\phi}'\cdot\nabla_1'\frac{1}{R}\,\mathrm{d}\mathbf{x}', \tag{2.13}$$

$$\int\int_{-\infty}^{\infty}\frac{V_3'}{R}\,\mathrm{d}\mathbf{x}'=-\int\int_{-\infty}^{\infty}\widetilde{\phi}'\left[(1+D^2)^{-\frac{3}{2}}-1\right]\nabla_1'\cdot\left[(\eta'-\eta)\nabla_1'\frac{1}{R}\right]\,\mathrm{d}\mathbf{x}', \tag{2.14}$$

$$\int\int_{-\infty}^{\infty}\frac{V_4'}{R}\,\mathrm{d}\mathbf{x}'=-\int\int_{-\infty}^{\infty}\frac{V'}{R}\left[(1+D^2)^{-\frac{1}{2}}-1\right]\,\mathrm{d}\mathbf{x}'. \tag{2.15}$$

A Fourier transform is then applied to invert (solve) the equation(s) for V. For the left hand sides of (2.12)–(2.15) we get

$$\mathcal{F}\left\{\int\int_{-\infty}^{\infty}\frac{V_j'}{R}\,\mathrm{d}\mathbf{x}'\right\}=\frac{2\pi}{k}\int\int_{-\infty}^{\infty}V_j'\,e^{-\mathrm{i}\mathbf{k}\cdot\mathbf{x}'}\,\mathrm{d}\mathbf{x}'=\frac{2\pi\mathcal{F}\{V_j\}}{k},\quad j=1,...,4, \tag{2.16}$$

where $\mathcal{F}$ denotes Fourier transform, $\mathbf{k}$ denotes the wavenumber in Fourier space, and $k^2 = \mathbf{k} \cdot \mathbf{k}$. In (2.16) we have exploited that

$$\mathcal{F}\{1/R\} = \frac{2\pi}{k} e^{-\mathrm{i}\mathbf{k}\cdot\mathbf{x}'}. \tag{2.17}$$

The transformed equations (2.12)–(2.15) become

$$\mathcal{F}\{V_1\} = k\mathcal{F}\{\widetilde{\phi}\} \tag{2.18}$$
$$\mathcal{F}\{V_2\} = -k\mathcal{F}\{\eta V_1\} - \mathrm{i}\,\mathbf{k} \cdot \mathcal{F}\{\eta \nabla_1 \widetilde{\phi}\} \tag{2.19}$$
$$\mathcal{F}\{V_3\} = k\mathcal{F}\{T(\widetilde{\phi})\} \tag{2.20}$$
$$\mathcal{F}\{V_4\} = k\mathcal{F}\{N(V)\} \tag{2.21}$$

which defines V $(= V_1 + V_2 + V_3 + V_4)$ implicitly, but, as we shall see, is very useful for an iterative solution procedure. The functions $T(\widetilde{\phi})$ and $N(V)$ are given by

$$T(\widetilde{\phi}) = \frac{1}{2\pi} \int\!\!\int_{-\infty}^{\infty} \widetilde{\phi}' \, [1 - (1 + D^2)^{-3/2}] \, \nabla_1' \cdot \left[(\eta' - \eta) \, \nabla_1' \frac{1}{R} \right] \mathrm{d}\mathbf{x}', \tag{2.22}$$
$$N(V) = \frac{1}{2\pi} \int\!\!\int_{-\infty}^{\infty} \frac{V'}{R} [1 - (1 + D^2)^{-\frac{1}{2}}] \, \mathrm{d}\mathbf{x}'. \tag{2.23}$$

3 Successive approximations

The set of equations (2.18)–(2.21) is suitable for an iterative solution procedure, as mentioned. Applying the method of successive approximations, we obtain the linear approximation:

$$\mathcal{F}(V_1) = k\,\mathcal{F}(\widetilde{\phi}). \tag{3.1}$$

This is used in the analytical integration of the linear part of (1.1)–(1.2). This step is crucial for a rapid and robust time stepping method. An explicit quadratic approximation is given by

$$\mathcal{F}(V_2) = \mathcal{F}(V_1) - k\mathcal{F}(\eta V_1) - \mathrm{i}\mathbf{k} \cdot \mathcal{F}(\eta \nabla_1 \widetilde{\phi}). \tag{3.2}$$

Applying another analytical iteration, an explicit cubic approximation is obtained by

$$\mathcal{F}(V_3) = \mathcal{F}(V_2) + k\mathcal{F}[N(V_1)]. \tag{3.3}$$

In the quartic approximation, still explicit, we obtain

$$\mathcal{F}(V_4) = \mathcal{F}(V_3) + k\mathcal{F}[N(V_2) + T(\widetilde{\phi})]. \tag{3.4}$$

The full equation is expressed by

$$\mathcal{F}(V) = \mathcal{F}(V_2) + k\mathcal{F}[N(V) + T(\widetilde{\phi})]. \tag{3.5}$$

A continued iteration of (3.5) gives a solution of V to any desired accuracy.

An improved model is obtained by including the leading term in (2.23) in the global FFT part of the formulation, i.e.

$$N(V) = \frac{1}{2\pi}\int\int_{-\infty}^{\infty} \frac{V'D^2}{2R}\mathrm{d}\mathbf{x}' + \frac{1}{2\pi}\int\int_{-\infty}^{\infty} \frac{V'}{R}\left[1 - \tfrac{1}{2}D^2 - (1+D^2)^{-\frac{1}{2}}\right]\mathrm{d}\mathbf{x}' \qquad (3.6)$$

where the first term on the right is expressed by transform:

$$\frac{1}{2\pi}\int\int_{-\infty}^{\infty} \frac{V'D^2}{2R}\mathrm{d}\mathbf{x}' = -\tfrac{1}{2}\left[\eta^2\mathcal{F}^{-1}\{k\mathcal{F}\{V\}\} - 2\eta\mathcal{F}^{-1}\{k\mathcal{F}\{\eta V\}\} + \mathcal{F}^{-1}\{k\mathcal{F}\{\eta^2 V\}\}\right]. \qquad (3.7)$$

The model is documented and tested out in Fructus et al. (2005). The evaluation of the global part using Fast Fourier Transform (FFT) is very fast and provides a highly accurate solution. (Extraction of more convolution terms from N and T is found to lead to an unstable time integration.)

The local contributions to T and N have kernels that decay rapidly in the (x_1, x_2)-plane, with integrands $\widetilde{\phi}\eta^3 R^{-4}$ and $V\eta^4 R^{-5}$, respectively, and require evaluation over a very limited part of the horizontal plane. The integration region corresponds to quadrats with side one characteristic wavelength. In many practical applications it suffices to calculate only the FFT-part of V, which is very fast.

4 Effect of a finite depth

We here follow the derivations made in Grue (2002). The integrals in the equation (2.1) containing the effect of a finite water depth, i.e. those involving $1/r_1$, are considered.

In the first step, we invoke the contribution from the term $1/r_1$ on the l.h.s. of equation (2.1), which is expressed in a reorganized form by

$$\frac{1}{r_1} = \left(\frac{1}{R_1} - \frac{2h(\eta'+\eta)}{R_1^3}\right) - \left(\frac{1}{R_1} - \frac{2h(\eta'+\eta)}{R_1^3} - \frac{1}{r_1}\right), \qquad (4.1)$$

where $R_1^2 = R^2 + (2h)^2$. Thus we evaluate

$$\frac{1}{2\pi}\mathcal{F}\left(\int\int_{-\infty}^{\infty} V'\left(\frac{1}{R_1} - \frac{2h(\eta'+\eta)}{R_1^3}\right)\mathrm{d}\mathbf{x}'\right) = \frac{e_1}{k}\mathcal{F}(V) - e_1\mathcal{F}(\eta V) - \mathcal{F}(\eta\mathcal{F}^{-1}(e_1\mathcal{F}(V))), \qquad (4.2)$$

where $e_1 = e^{-2kh}$ is introduced. The contribution from the second term in (4.1) to the r.h.s. of the integral equation.

For the second integral on the r.h.s. of (2.1) we obtain

$$\frac{1}{2\pi}\mathcal{F}\left(\int\int_S \widetilde{\phi}'\frac{\partial}{\partial n'}\frac{1}{r_1}\mathrm{d}S\right) = -e_1\mathcal{F}(\widetilde{\phi}) + \mathcal{F}(\eta\mathcal{F}^{-1}(k\,e_1\mathcal{F}(\widetilde{\phi}))) - \frac{\mathrm{i}\mathbf{k}}{k}e_1\cdot\mathcal{F}(\eta\nabla_1\widetilde{\phi}) + \mathcal{F}(T_1(\widetilde{\phi})) \qquad (4.3)$$

where $T_1(\widetilde{\phi})$ is obtained below in (4.7). The Fourier transform of the resulting integral equation gives then

$$\begin{aligned}\mathcal{F}(V) &= kE_1\mathcal{F}(\widetilde{\phi}) - kE_1\mathcal{F}(\eta V_1) - \mathrm{i}\mathbf{k}\cdot\mathcal{F}(\eta\nabla_1\widetilde{\phi}) \\ &\quad + kC_1\left[\mathcal{F}(T(\widetilde{\phi}) + T_1(\widetilde{\phi})) + \mathcal{F}(N(V) + N_1(V))\right] \\ &\qquad + kC_1\left[e_1\mathcal{F}(\eta(V - V_1)) + \mathcal{F}(\eta\mathcal{F}^{-1}(e_1\mathcal{F}(V - V_1)))\right], \end{aligned} \tag{4.4}$$

where

$$\mathcal{F}(V_1) = kE_1\mathcal{F}(\widetilde{\phi}) = k\tanh(kh)\mathcal{F}(\widetilde{\phi}), \tag{4.5}$$

and $E_1 = (1-e_1)/(1+e_1)$, $C_1 = 1/(1+e_1)$. T_1 and N_1 are given by

$$N_1(V) = \frac{1}{2\pi}\int\!\!\int_{-\infty}^{\infty} V'\left(\frac{1}{R_1} - \frac{2(\eta'+\eta)h}{R_1^3} - \frac{1}{r_1}\right)\mathrm{d}\mathbf{x}', \tag{4.6}$$

$$\begin{aligned}T_1(\widetilde{\phi}) &= -\frac{1}{2\pi}\int\!\!\int_{-\infty}^{\infty}\widetilde{\phi}'\,\frac{12h^2\,(\eta'+\eta)}{R_1^5}\mathrm{d}\mathbf{x}' \\ &\quad + \frac{1}{2\pi}\int\!\!\int_{-\infty}^{\infty}\widetilde{\phi}'\,(\mathbf{R}\cdot\nabla_1\eta - (\eta+\eta') - 2h)\left(\frac{1}{r_1^3} - \frac{1}{R_1^3}\right)\mathrm{d}\mathbf{x}'. \end{aligned} \tag{4.7}$$

Equation (4.4) inverts the equation for the scaled normal velocity $V = \phi_n[1+|\nabla_1\eta|^2]^{\frac{1}{2}}$. This is used as input to the equations to integrate the surface elevation and the surface potential forward in time.

The integrals (4.6) and (4.7) have rapidly decaying kernels. For the former we have, to leading order, $N_1(V) \sim \int\!\int \mathcal{O}(V\eta^2 R_1^{-5})\mathrm{d}\,\mathbf{x}'$, while for the latter $T_1(\widetilde{\phi}) \sim \int\!\int \mathcal{O}(\widetilde{\phi}\eta^2 R_1^{-5})\mathrm{d}\,\mathbf{x}'$. This means that the contributions to the integrals are very local, and the integration may be truncated. In practice they are evaluated over one characteristic wavelength. (N_1 and T_1 may be expressed by sums of convolutions, contributing to a stable time integration. Potential unstable contributions from high wavenumbers are damped by the factor $e_1 = e^{-2kh}$.)

5 Time integration

The prognostic equations are integrated in Fourier space, i.e.

$$\mathcal{F}(\eta)_t - k\tanh(kh)\mathcal{F}(\widetilde{\phi}) = \mathcal{F}(H_1), \quad \mathcal{F}(\widetilde{\phi})_t + g\mathcal{F}(\eta) = \mathcal{F}(H_2) \tag{5.1}$$

where

$$H_1 = V - \mathcal{F}^{-1}[k\tanh(kh)\mathcal{F}(\widetilde{\phi})] + \tilde{v}_D, \tag{5.2}$$

$$H_2 = -\frac{|\nabla_1\widetilde{\phi}|^2 - V^2 - 2V\,\nabla_1\eta\cdot\nabla_1\widetilde{\phi} + |\nabla_1\eta\times\nabla_1\widetilde{\phi}|^2}{2(1+|\nabla_1\eta|^2)} - \tilde{p}_G - \tilde{p}_D. \tag{5.3}$$

Let us introduce $\omega = [gk\tanh(kh)]^{\frac{1}{2}}$, and

$$\hat{F} = \begin{bmatrix} k\mathcal{F}(\eta) \\ \frac{k\omega}{g}\mathcal{F}(\widetilde{\phi}) \end{bmatrix}, \quad \hat{N} = \begin{bmatrix} k\,\mathcal{F}(H_1) \\ \frac{k\omega}{g}\,\mathcal{F}(H_2) \end{bmatrix}, \quad A = \begin{bmatrix} 0 & -\omega \\ \omega & 0 \end{bmatrix}. \tag{5.4}$$

The prognostic equations become

$$\hat{F}_t + A\hat{F} = \hat{N} \tag{5.5}$$

with solution

$$\hat{F}(t) = \hat{F}(0)e^{-At} + \int_0^t e^{A(s-t)}\hat{N}\mathrm{d}s. \tag{5.6}$$

The integral is evaluated by using six stages fifth-order Runge-Kutta scheme. In a strategy to vary the time step in the computations, an embedded fourth-order scheme is used to select the size of the next time step. This means that the scheme adapts itself according to the nonlinearity of the wave motion that is involved: the wave step is reduced when nonlinearity is strong. In addition, it is an advantage to implement a stabilization technique which is derived for stiff problems in partial differential equations. This contributes to reduce the computation time. In an anti-aliasing strategy the number of evaluation nodes of the variables $(\eta, \widetilde{\phi}, V)$ are doubled, and the upper half of the Fourier components set to zero. This removes the effect of aliasing in convolution products up to cubic order.

6 Nonlinear wave generation and absorption

6.1 Generation

Pneumatic wave making has the advantage, when the applied pressure is sufficiently smooth, that no mathematical singularites are imposed on the wave field. This is different to the situation when a wave making object – like a physical moving wave paddle – intersects the free surface. In the latter case the flow at the paddle will always contain a singularity which is a complication in the computation of the otherwise smooth flow.

Following Clamond et al. (2005), we outline nonlinear pneumatic wave generation procedure which is based on the integration of the nonlinear equation (5.6), with the leading contribution determined from linear wave making theory. Multichromatic waves of frequencies σ_j, $j = 1, ..., J$, are generated by specifying the generating pressure by

$$\tilde{p}_G = \sum_{j=1}^{J} \sin(\sigma_j t + \delta_j)P_j(\mathbf{x}) \quad (t \geq 0^+), \tag{6.1}$$

and $\tilde{p}_G = 0$ for $(t < 0)$, where P_j are amplitudes and δ_j constant phase angles. In the case of long-crested waves, the functions P_j are chosen:

$$P_j = A_j e^{-x_1^2/2\lambda_j^2}, \tag{6.2}$$

with Fourier transform $\hat{P}_j = \sqrt{2\pi}A_j\lambda_j e^{-k_1^2\lambda_j^2/2}$. The parameters A_j and λ_j are taken from linear theory derived in Wehausen and Laitone (1960, section 21), giving

$$\lambda_j = \frac{1}{\kappa_j}, \quad A_j = ga_j\sqrt{\frac{e}{2\pi}}\left(1 + \frac{2\kappa_j h}{\sinh 2\kappa_j h}\right), \tag{6.3}$$

where κ_j are determined from $\sigma_j^2 = g\kappa_j \tanh \kappa_j h$. The constants a_j are the linear far field amplitudes of the waves.

Cylindrical waves are generated by defining

$$P_j = A_j e^{-r_j^2/4\lambda_j^2}, \quad A_j = \frac{ega_j}{\sqrt{8\pi}} \left(1 + \frac{2\kappa_j h}{\sinh 2\kappa_j h}\right), \tag{6.4}$$

where $r_j = |\mathbf{x} - \mathbf{x}_j|$ and $a_j/\sqrt{\kappa_j r_j}$ denote the wave amplitudes in the far field.

The form of an advancing pressure moving with speed U along the x_1 direction is given by

$$\tilde{p}_G = P(x_1 - Ut, x_2) - \bar{P}_0, \quad (t \geq 0^+), \tag{6.5}$$

and $\tilde{p}_G = 0$ for $t < 0$ where $P_0 = P(t = 0)$ and $\bar{P}_0$ denotes the spatial average of P_0. Functions of the form $P_0 = Ae^{-x_1^2/\lambda_1^2 - x_2^2/\lambda_2^2}$ are suitable pressure distributions.

6.2 Absorbing conditions

These are desired at the open boundaries (Romate, 1992; Givoli, 2004). A variant of the dynamic boundary condition (1.2), after taking the horizontal gradient, reads

$$\mathbf{U}_t + \nabla_1 T = 0, \tag{6.6}$$

where $\mathbf{U} = \nabla_1 \tilde{\phi} = \mathbf{V}_\mathrm{T}\sqrt{1 + |\nabla_1 \eta|^2}$ and $\mathbf{V}_\mathrm{T}$ denotes the fluid velocity tangential to the free surface. Further, $T = g\eta + [|\nabla_1 \tilde{\phi}|^2 - V^2 - 2V\,\nabla_1\eta \cdot \nabla_1 \tilde{\phi} + |\nabla_1 \eta \times \nabla_1 \tilde{\phi}|^2]/[2(1 + |\nabla_1 \eta|^2)]$ and $\tilde{p}_G = 0$. A term that damps the tangential fluid velocity is obtained by adding to (6.7) a term $\gamma\mathbf{U}$ (Clamond et al. 2005), i.e.

$$\mathbf{U}_t + \gamma\mathbf{U} + \nabla_1 T = 0, \tag{6.7}$$

where $\gamma(\mathbf{x})$ is a non-zero prescribed function in the damping region of the computational domain. This way of damping the wave field has been successfully tested out and has been found to be very efficient.

By an integration the function $\tilde{p}_D$ in (1.2) becomes

$$\tilde{p}_D = \nabla_1^{-1} \cdot [\gamma \nabla_1 \tilde{\phi}] - B(t) \tag{6.8}$$

where the inverse operator is defined by $\nabla_1^{-1} = (\nabla_1^2)^{-1}\nabla_1$ and $B(t)$ the Bernoulli constant. The inverse operator is evaluated by Fourier transform, i.e.

$$\mathcal{F}\left(\nabla_1^{-1} \cdot (\gamma \nabla_1 \tilde{\phi})\right) = \frac{\mathrm{i}\mathbf{k}}{-k^2} \cdot \mathcal{F}(\gamma \nabla_1 \tilde{\phi}). \tag{6.9}$$

Mass absorption. To absorb a mass flux at the boundary of the computational domain, a damping term of the form

$$\tilde{v}_D = -\Gamma\eta \tag{6.10}$$

is suitable in the kinematic condition (1.1). The function $\Gamma(\mathbf{x})$ is positive and nonzero in the regions where damping is required. See also Buchmann, Ferrant and Skourup (2000).

7 Convergence

7.1 Integration constants

In a paper by Benjamin and Olver (1982), on Hamiltonian structure, symmetries and conservation laws for water waves, they derived theorems for conserved quantities (they called them conserved densities). We shall not here go through the derivations, but merely state a few of the results. For the three-dimensional water wave problem, where the velocity potential satisfies the Laplacian in the fluid domain, the boundary conditions (1.1)–(1.2) at the free surface, and the bottom boundary condition of zero normal velocity at the flat sea floor, the following quantities, when there is no effect of surface tension, are conserved (Benjamin and Olver, 1982, theorem 5.3):

$$
\begin{aligned}
&T_1 = -\widetilde{\phi}\eta_{x_1}, \quad T_2 = -\widetilde{\phi}\eta_{x_2} \quad \text{(horizontal impulses)},\\
&T_3 = H = \tfrac{1}{2}\widetilde{\phi}V + \tfrac{1}{2}g\eta^2 \quad \text{(energy)},\\
&T_4 = \eta \quad \text{(mass)},\\
&T_5 = \widetilde{\phi} + gtT_4,\\
&T_6 = x_1\eta - tT_1, \quad T_7 = x_2\eta - tT_2,\\
&T_8 = \tfrac{1}{2}\eta^2 - tT_5 + \tfrac{1}{2}gt^2T_4,\\
&T_9 = (x_1\eta_{x_2} - x_2\eta_{x_1})\widetilde{\phi},\\
&T_{10} = (x_1 + \eta\eta_{x_1}\widetilde{\phi} + gtT_6 + \tfrac{1}{2}gt^2T_1,\\
&T_{11} = (x_2 + \eta\eta_{x_2}\widetilde{\phi} + gtT_7 + \tfrac{1}{2}gt^2T_2,\\
&T_{12} = (\eta - x_1\eta_{x_1} - x_2\eta_{x_2})\widetilde{\phi} + t(9gT_8 - 5H) + \tfrac{9}{2}gt^2T_5 - \tfrac{3}{2}g^2t^3T_4.
\end{aligned}
$$

These quantities should be conserved by the numerical integration of the equations.

7.2 Convergence test

Fructus et al. (2005a) performed several tests documenting the performance of the numerical scheme of the three-dimensional method. Particularly the quantity T_3 (the energy) and the phase of η were evaluated. Conservation of mass was enforced by prescribing in the scheme $\mathcal{F}(\eta)$ at $\mathbf{k} = 0$. One of the test cases included the marching of a progressive Stokes wave train. The waves with frequency $\omega_0 = \omega(k_0)$ and wave slope $(ak)_0$ were computed using the procedure of Fenton (1988). The computational domain included two wavelengths in each direction. In all computations, aliasing was avoided using a zero-padding technique in the computation of convolution products, see section 5. The discretization was $2N_1 \times 2N_2$ nodes. The Stokes wave train was marched forward during 1000 periods, and then integrated backwards the same time period. The accuracy of the mechanical energy, given by $E = \int T_3 \mathrm{d}\mathbf{x}$, was measured. The relative difference $|E - E_0|/E_0$ exhibited a maximum of $3 \cdot 10^{-5}$ during the whole simulation. The wave slope was $ak = 0.2985$. No trend, or steady increase or decrease, of the energy was observed. It is important to note that in the present scheme no smoothing is applied. This means that the energy is conserved. (Application of a smoothing technique imposes a small energy loss.)

Another quantity was also measured: the phase of the computed wave relative to the exact one, viz. $\epsilon_{\rm err} = 100|T - T_0|/T_0$, where $T = 2\pi/\omega$ denotes the computed and T_0 the exact period. Results from several discretizations are presented in table 1, indicating convergence with increasing resolution. In table 2 we explore convergence of the form

$$\text{Error}(N_1) = C_1 e^{-\sigma_1 N_1}. \tag{7.1}$$

From the table we estimate a value of $\sigma_1 \simeq 0.40 - 0.46$, indicating exponential convergence. Likewise, in table 3 we explore convergence of the form

$$\text{Error}(N_2) = C_2 e^{-\sigma_2 N_2}. \tag{7.2}$$

From the table we estimate a value of σ_2 in the range 0.06–0.12, indicating again exponential convergence.

$N_2 + N_2$	$N_1 + N_1 = 16$	32	64	256
16	4.1958	0.1509	0.0234	0.0238
32	4.2035	0.1650	0.0115	0.0116
64	4.2044	0.1752	0.0046	0.0058
128	4.2044	0.1747	0.0001	0.0001

Table 1. Relative error in percent of the wave period: $\epsilon_{\rm err} = 100|T - T_0|/T_0$ computed for a Stokes wave with $ak = 0.2985$.

$N_1 + N_1$	16		32		64
$\epsilon_{\rm err}$	4.204		0.1747		0.0001
$\ln \epsilon_{\rm err}$	1.436		-1.744		-9.210
σ_1		0.40		0.46	

Table 2. Estimate of σ_1 of eq. (7.1) vs. $N_1 + N_1$, with fixed $N_2 + N_2 = 128$.

$N_2 + N_2$	16		32		64		128
$\epsilon_{\rm err}$	0.0234		0.0115		0.0046		0.001
$\ln \epsilon_{\rm err}$	-3.755		-4.465		-5.382		-9.210
σ_2		0.09		0.06		0.12	

Table 3. Estimate of σ_2 of eq. (7.2) vs. $N_2 + N_2$, with fixed $N_1 + N_1 = 64$.

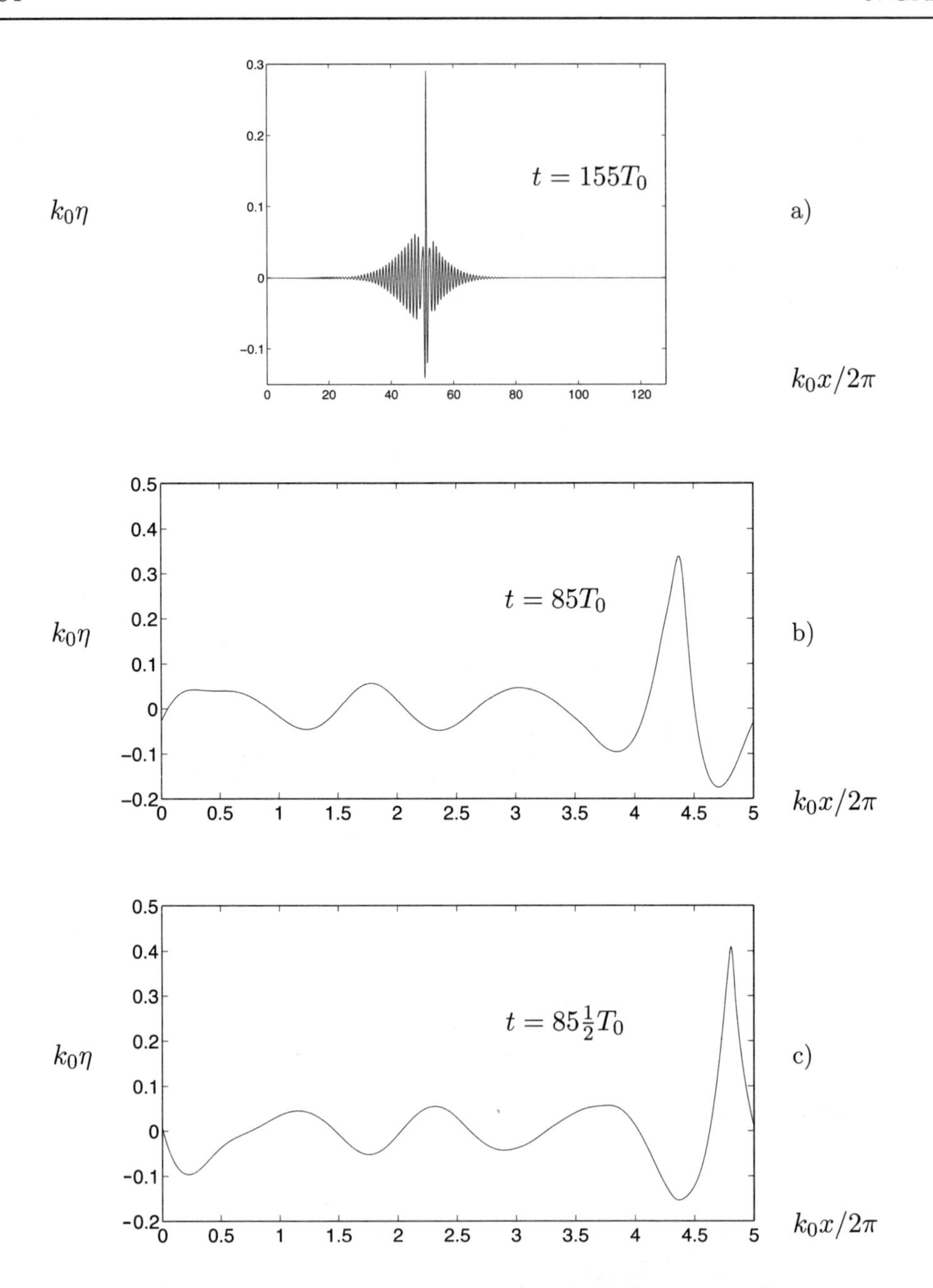

Figure 1. a) Large wave event after 155 periods of the evolution of a Stokes wave packet with initial wave slope $a_0k_0 = 0.091$ and envelope $\text{sech}[0.263\sqrt{2}a_0k_0^2x]$. a_0 initial wave amplitude. k_0 initial wavenumber. From Clamond and Grue (2002). b) and c) Wave elevation after 85 and $85\frac{1}{2}$ wave periods (T_0). Growth to a large wave event in a short wave tank. Periodic waves with $a_0k_0 = 0.1125$ perturbed at wavenumbers $0.8k_0, 1.2k_0$, with non-dimensional perturbation amplitude of 0.1.

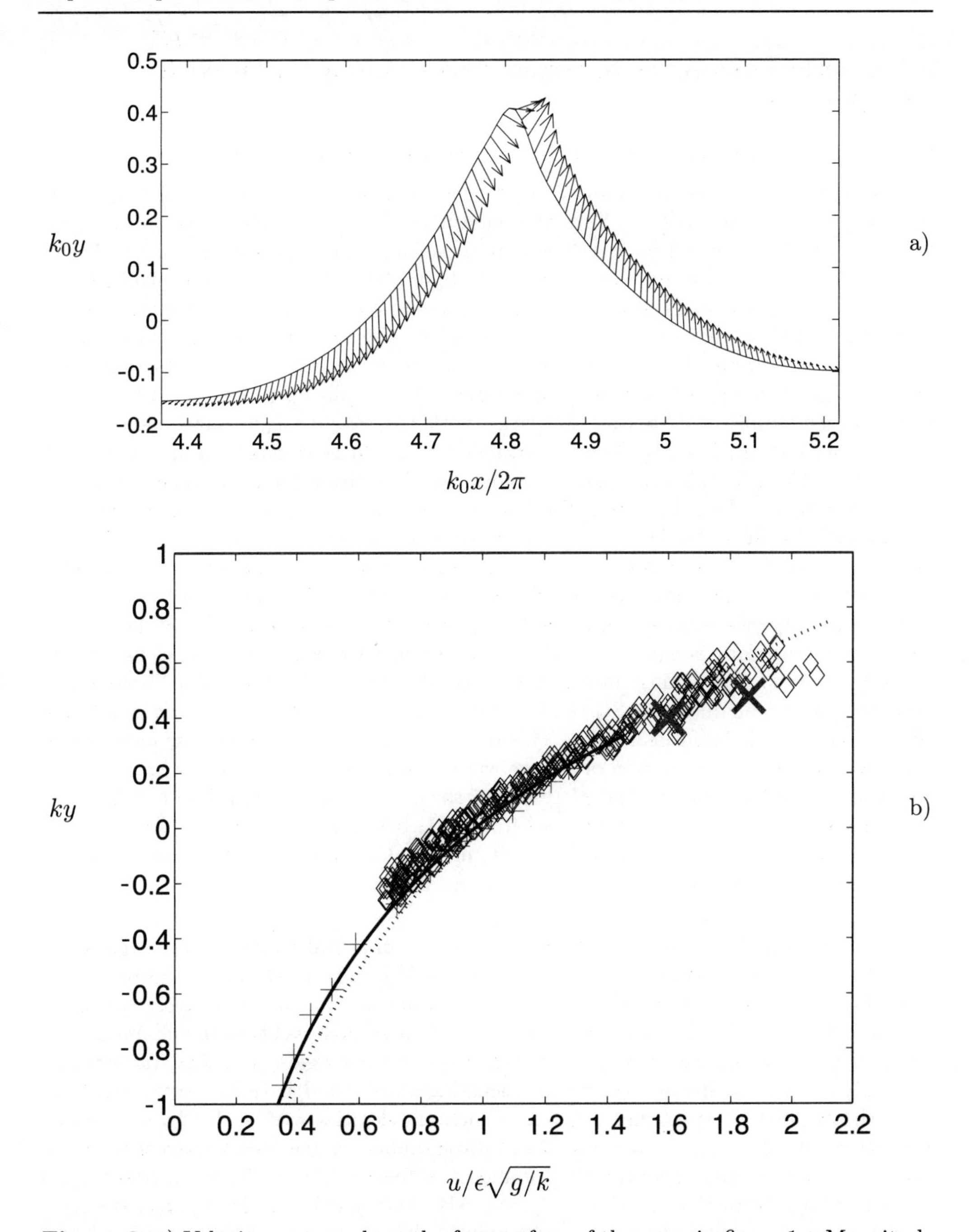

Figure 2. a) Velocity vectors along the free surface of the wave in figure 1c. Magnitude of velocity vectors indicated in plot b. b) Velocity profile below the crests of several steep wave events. Experiments (diamonds) (Grue et al. 2003; Grue and Jensen, 2006), experiments (+) (Baldock et al. 1996). Fully nonlinear calculations (solid line) (Clamond and Grue, 2002), $\exp(ky)$ (dots), Big $\times$ present model. ϵ and k are calulated from eq. (8.1).

8 Numerical examples of rogue waves. Comparison with experiments

8.1 Very steep wave events. Comparison with PIV-experiments

The method is useful to simulate the formation of large wave events of both short and long crested wave motion. Several simulations of the evolution of long wave trains that also are long crested have shown that these undergo a split-up into a number of group solitons (Clamond and Grue, 2002), see figure 29 in Ch. 3. There is strong interaction between the group solitons during the split-up process. The interactions introduce perturbations that trigger the side-band instability. This in turn causes nonlinear focussing within the group. This represents the key mechanism of the formation of the large events. The first example is from a simulation of the split-up process of a long wave group, which initially is a Stokes wave packet with wave slope $a_0k_0 = 0.091$ and envelope given by $\mathrm{sech}[0.263\sqrt{2}a_0k_0^2x]$. The first of totally three large events is taking place after 155 wave periods in this example, see figure 1a. The elevation at maximum is about three times higher than the initial. For later reference and comparison with experiments we characterize the event by the local wave slope ϵ and local wavenumber k which are defined in equation (8.1) below. These parameters become $\epsilon = 0.29$ and $k\eta_m = 0.34$ for the large event in figure 1a, where η_m denotes the maximal elevation of the event.

Another large wave event is obtained from computations of perturbed Stokes waves. The initial wave slope is $a_0k_0 = 0.1125$. The periodic wave train – five wavelengths long – develops into local breaking since a_0k_0 is above the threshold of 0.11. The Stokes wave, perturbed at wavenumbers, $0.8k_0, 1.2k_0$, with non-dimensional perturbation amplitude of 0.1 develops side-band instability. This in turn leads to a local nonlinear focussing of the wave, causing the formation of a large wave event. The resulting surface elevation is shown in figure 1b,c after 85 and $85\frac{1}{2}$ wave periods. The computation broke right after. The local values of ϵ and $k\eta_m$, estimated from (8.1) below, are $\epsilon = 0.32$ and $k\eta_m = 0.4$ in figure 1b, and $\epsilon = 0.38$ and $k\eta_m = 0.48$ in figure 1c. Computations of the wave induced velocities at the free surface are shown in figures 2a and 3a,b.

Particle Image Velocimetry (PIV) is an experimental technique that maps the velocity vectors in two-dimensional sections of the flow. The method is useful to extract the velocity and acceleration fields induced by the wave motion. In a laboratory campaign the wave induced velocities below the crests of 122 large wave events were measured. The wave events were realized in three different types of wave fields, including the leading, unsteady wave of a Stokes wave train, focussing waves, and irregular waves resulting from the JONSWAP spectrum. The experimental velocities were scaled by a reference velocity defined by $\epsilon\sqrt{g/k}$, where the local wavenumber, k, and an estimate of the wave slope, ϵ, were evaluated from the measurements as follows (Grue, Clamond, Huseby and Jensen, 2003): From the experimental wave record obtained at a fixed point, the local trough-to-trough period, T_{TT}, and the maximal elevation above mean water level, η_m, of an individual steep wave event were identified. Then k and ϵ were evaluated from

$$\frac{\omega^2}{gk} = 1 + \epsilon^2, \qquad k\eta_m = \epsilon + \frac{1}{2}\epsilon^2 + \frac{1}{2}\epsilon^3, \tag{8.1}$$

where $\omega = 2\pi/T_{TT}$ and g denotes the acceleration of gravity. In the numerical simulations shown in figure 1 the local wavenumber k was extracted directly by the trough-to-trough wave length of the event. The trough-to-trough period and trough-to-trough wave length are much used references in experimental investigations dealing with highly nonlinear wave phenomena, see e.g. Su (1982).

Wave induced velocity vectors. Data from the six largest focussing wave events of the campaign, with ϵ in the range 0.40–0.46, contain 36000 velocity vectors and were put on non-dimensional form using the scaling (8.1) and reference velocity $\epsilon\sqrt{g/k}$ (Grue et al., 2003; Grue and Jensen, 2006). Non-dimensional measurements of the horizontal velocities below the crests of the six waves (with $0.40 < \epsilon < 0.46$) are shown in figure 2b. The figure includes the measurement of the velocity below the crest of a non-breaking focussing wave with $\epsilon = 0.29$ from another laboratory campaign using Laser Doppler Anemometry (Baldock, Swan and Taylor, 1996). The measurements were performed in the perspective of supporting the theoretical development. The figure includes the computed velocity profile below the crest of the non-breaking wave in figure 1a (with $\epsilon = 0.29$) as well as the horizontal velocities at the crests of the waves in figure 1b,c (with $\epsilon = 0.32,\ 0.38$). The figure illustrates that the wave induced horizontal velocity below the crests of the various steep waves, with ϵ in the range 0.29–0.46, observed in the computations, and in different sets of experiments, fall along one curve, approximately, using the scaling (8.1). Figures 4e,f show the experimental waves and velocity vectors (there are 12000 vectors) from two of the events of the laboratory campaign with $\epsilon = 0.40$ and 0.41. Velocity vectors are plotted in the velocity vector plane (u, v) in figure 3, illustrating the following two features: Firstly, the maximal non-dimensional horizontal velocity is the same in the theoretical computation and the experiment, when the value of ϵ is about the same. The maximal horizontal velocity in figure 3b,c, with ϵ in the range 0.38–0.40, is $1.85\times\epsilon\sqrt{g/k}$, for example. Secondly, the horizontal velocities are increasing, while the vertical velocities are decreasing, during the prephase of a breaking event.

The wave propagation speed may be estimated by evaluating ω/k. In non-dimensional terms the wave speed becomes $[\omega/k]/[\epsilon\sqrt{g/k}] = \sqrt{1+\epsilon^2}/\epsilon$. For the steepest computed wave shown here we obtain $[\omega/k]/[\epsilon\sqrt{g/k}] \simeq 2.8$ for $\epsilon = 0.38$. For the steepest experimental wave shown in figure 4f with $\epsilon = 0.41$ we obtain $[\omega/k]/[\epsilon\sqrt{g/k}] \simeq 2.6$. The estimates of the wave speed significantly exceed the wave induced fluid velocities of the computed and measured waves (figure 3). Velocity vectors of larger experimental waves, with ϵ in the range 0.44–0.46, exhibit wave induced fluid velocities that are smaller than the estimated wave propagation speed, except for the strongest breaking case of the campaign, where the velocity vectors in the tip of the wave become equal to the wave speed. We note that computations of overturning waves in water of moderate depth by New, McIver and Peregrine (1985) showed velocities in the tip of the breaking wave that were 1.5 times larger than the wave speed.

Acceleration vectors. Sequences of PIV measurements of steep non-breaking waves, with ϵ in the range 0.40–0.45, exhibit horizontal accelerations up to $0.7g$ in the direction of

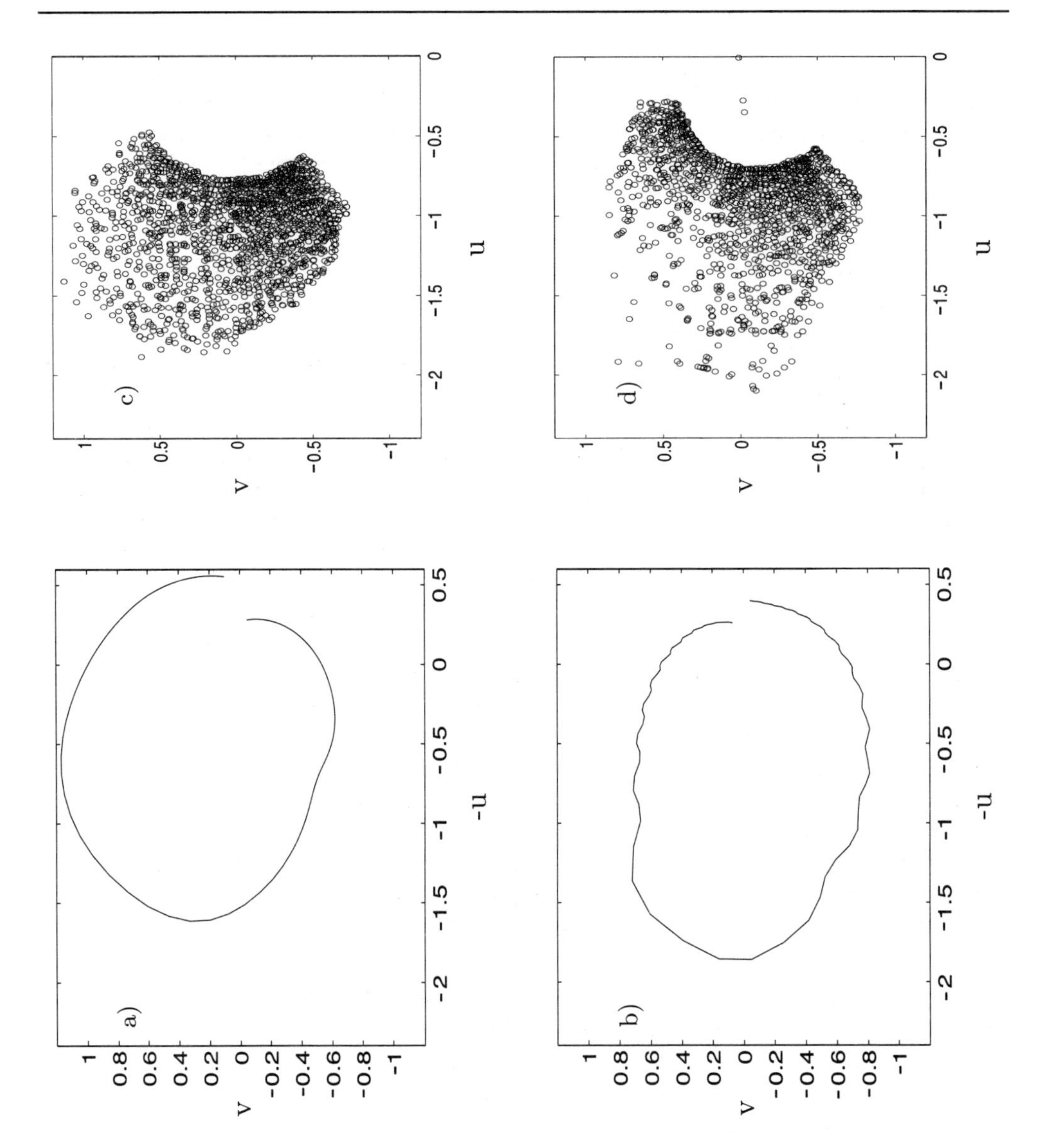

Figure 3. Non-dimensional velocity plane vectors $(\mathrm{u}, \mathrm{v}) = (u, v)/\epsilon\sqrt{g/k}$. a) Velocity vectors along the surface of the wave in figure 1, with $\epsilon = 0.32$, $[\omega/k]/\epsilon\sqrt{g/k} = 3.3$, and b) the wave in figure 1c-2a, with $\epsilon = 0.38$, $[\omega/k]/\epsilon\sqrt{g/k} = 2.8$. c) Velocity plane vectors of the experimental wave in figure 4e, with $\epsilon = 0.40$, $[\omega/k]/\epsilon\sqrt{g/k} = 2.7$, and d) velocity plane vectors of the experimental wave in figure 4f, with $\epsilon = 0.41$, $[\omega/k]/\epsilon\sqrt{g/k} = 2.6$. Experimental waves from Grue and Jensen (2006).

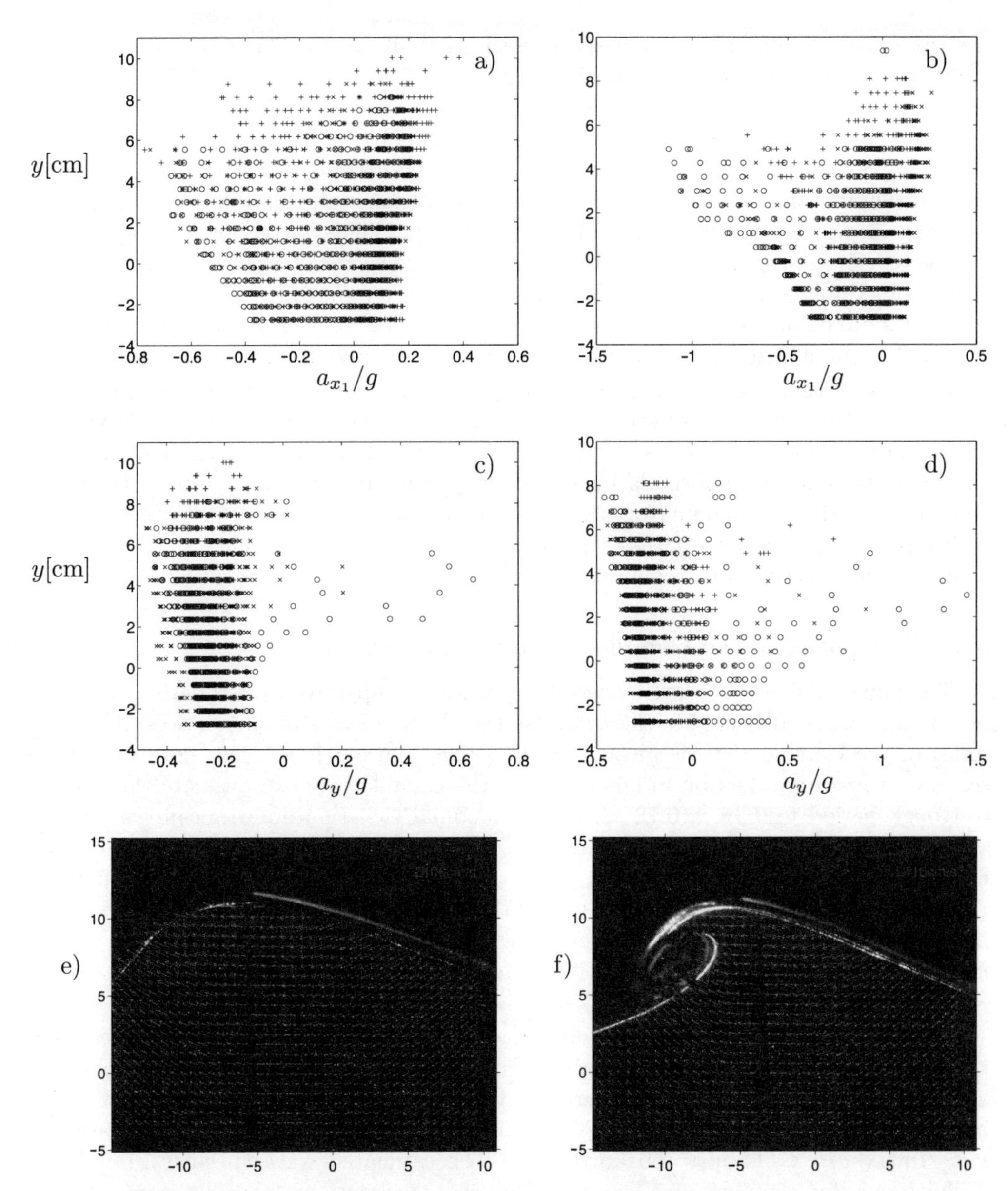

Figure 4. Accelerations in experimental waves a) Horizontal accelerations a_{x_1}/g, for non-breaking waves with $\epsilon = 0.4$ (+), $\epsilon = 0.44$ (x), $\epsilon = 0.45$ (o), b) a_{x_1}/g, for breaking waves with $\epsilon = 0.41$ (+), $\epsilon = 0.44$ (x), $\epsilon = 0.46$ (o), c) vertical accelerations a_y/g, for non-breaking waves with $\epsilon = 0.4$ (+), $\epsilon = 0.44$ (x), $\epsilon = 0.45$ (o), d) a_y/g, for breaking waves with $\epsilon = 0.41$ (+), $\epsilon = 0.44$ (x), $\epsilon = 0.46$ (o). e)–f) Experimental waves with velocity vectors. Non-breaking wave with $\epsilon = 0.40$, and breaking wave with $\epsilon = 0.41$. Velocity vectors are indicated in figure 3. Scales in e)-f) are in cm. From Grue and Jensen (2006).

the wave propagation. (These waves are breaking right after.) The acceleration vectors include both the temporal and convective terms. The horizontal acceleration vectors appear in a systematic way, with many occurrences in the acceleration vector plot, and seem to depend very little on the value of ϵ when this is in the range 0.40–0.45 and the events are non-breaking. A vertical acceleration of similar size occurs for relatively fewer acceleration vectors. The (few) high vertical accelerations take place in the front face of the wave at a vertical level about half way to the crest. This is true in the steepest case with $\epsilon = 0.45$. The group of waves with a developing overturning jet and ϵ in the range 0.41–0.46 exhibit accelerations in the base below the jet that are up to $1.1g$ along the horizontal direction and up to $1.5g$ along the vertical. Large horizontal accelerations are observed in significantly more points than the vertical accelerations (figure 4).

Computations using two-dimensional methods exhibit accelerations up to $6g$ in a thin region along the curved boundary of the wave right below the overturning jet (New et al. 1985). Computations of the wave induced velocities and accelerations in strongly overturning waves require a reformulation of the mathematical/numerical method outlined in this chapter. Finally, we note from the equation of motion that $\rho(\mathrm{D}\mathbf{v}/\mathrm{D}t + g\nabla y) = -\nabla p$, which means that $\mathrm{D}\mathbf{v}/\mathrm{D}t + g\nabla y$ is directed along the normal of the curved wave surface. This provides a check of the computations and the measurements.

8.2 Kinematics of the Camille and Draupner waves

The numerical and measured waves are put into perspective by estimating the value of ϵ of large waves observed in the field, like the Camille and Draupner waves. The time series of the Camille wave is given in Ochi (1998, figure 8.1, p. 218). From the time series that is a reproduction in laboratory of the Camille wave we estimate the trough-to-trough period to $T_{TT} = 0.72$ s and the maximal elevation above mean sea level to $\eta_m = 7.2$ cm. Putting these quantities into the set of equations (8.1) we obtain $\epsilon \simeq 0.38$ and $k\eta_m \simeq 0.49$ for the Camille wave. From the time series of the surface elevation of the Draupner wave, given in figure 1 in Ch. 2 of the volume, the trough-to-trough period and maximal elevation above mean sea level are estimated to $T_{TT} = 11.5$ s and $\eta_m = 18.5$ m, respectively. Putting these quantities into (8.1) we obtain $\epsilon \simeq 0.39$ and $k\eta_m \simeq 0.49$ for the Draupner wave.

The estimates of ϵ for the Camille and Draupner waves indicate that the large waves in the computations and experiments are useful to predict the magnitude of the velocities and accelerations induced by the large waves at sea. For the Draupner wave, with $\eta_m = 18.5$m, $\epsilon = 0.39$ and $k\eta_m = 0.49$, the maximal fluid velocity at crest becomes $u \simeq 1.85\epsilon\sqrt{g/k} \simeq 14$ m/s (50 km/hour). The estimated wave speed is 21 m/s, or 74 km/hour, for comparison. The wave induced accelerations have the same scale in laboratory and field and are extracted from the results in figure 4.

9 Computations of tsunami waves in three dimensions

It is a desire to numerically predict the nonlinear process of tsunami generation, propagation and shoaling, including in the modelling the combined effect of long and short waves. The waves may be generated by the motion of a slide or by a sudden vertical displacement of the sea floor. The equations presented in this chapter may be generalized to include also the motion of the sea bottom whereby the fully nonlinear, fully dispersive motion can be studied in a rapid fashion. Preliminary calculations of the waves due to an impulsive N-shaped bottom motion are presented in figure 5. The calculations are performed by the generalized three-dimensional method using the FFT-part only. This resolves the wave spectra accurately up to $kh = 3$ and are tested against the full equations which are valid for the wavenumbers that are resolved by the discretization.

10 Computations of three-dimensional wave patterns

Three-dimensional wave structures may occur at the sea surface in the form of crescent-shaped patterns. These play an important role in wave breaking and cause transfer of momentum and energy between the ocean and atmosphere. The three-dimensional structures may result from spontaneous instabilities of a two-dimensional water wave field. Particularly those emerging naturally from the instability of Stokes waves are considered here. The stability of Stokes waves has been studied for many decades. Two-dimensional instability was studied by Lighthill (1965) and Benjamin and Feir (1967), working with triads – or the side-band instability, while Zakharov (1968) also included three-dimensional instabilities in his analysis, working with quartets in the wavenumber space. Later, McLean et al. (1981) and McLean (1982) discovered a new kind of three-dimensional instability which, prior to this, had been suggested in a two-dimensional study by Longuet-Higgins (1978). The new instability analysis involved the interaction between quintets and higher resonances in the wavenumber space. While the side-band instability has been extensively studied in Ch. 2 and 3 of the volume, we study here 5-wave and higher interactions.

10.1 The stability analysis by McLean et al. (1981)

This analysis was motivated by experimental measurements of three-dimensional wave patterns by Su, later documented in Su (1982), and by preliminary calculations of the patterns by Saffman and Yuen, see reference 12 in McLean et al. (1981). The research group, including McLean, Ma, Martin, Saffman and Yuen at the Fluid Mechanics Department, TRW Defense and Space Systems Group in California, preformed analysis and computations demonstrating that there were two distinct types of instabilities for gravity waves of finite amplitude on deep water. They solved the fully nonlinear equations for water waves, finding that one type is predominantly two-dimensional, relating to all the previously known results for special cases. The other predominantly three-dimensional instability becomes dominant when the wave steepness is large.

The set of equations that McLean et al. solved, working with the velocity potential, φ, was: $\nabla^2\varphi = 0$ for $-\infty < y < \eta(x_1, x_2, t)$; and $\varphi_t + \frac{1}{2}|\nabla\varphi|^2 + gy = 0$, $\eta_t + \nabla\varphi\cdot\eta - \varphi_y = 0$ on $y = \eta$. They calculated steady Stokes waves up to $2a/\lambda = 0.131$ (corresponding to $ak = 0.412$) on the form $\eta_s = \sum_0^\infty A_n \cos[2n\pi(x_1 - Ct)]$, where the Fourier coefficients A_n and the wave speed C were functions of $2a/\lambda$. In the computations λ was put to 2π (giving $k_0 = 1$) and $g = 1$. $2a$ denoted the wave height.

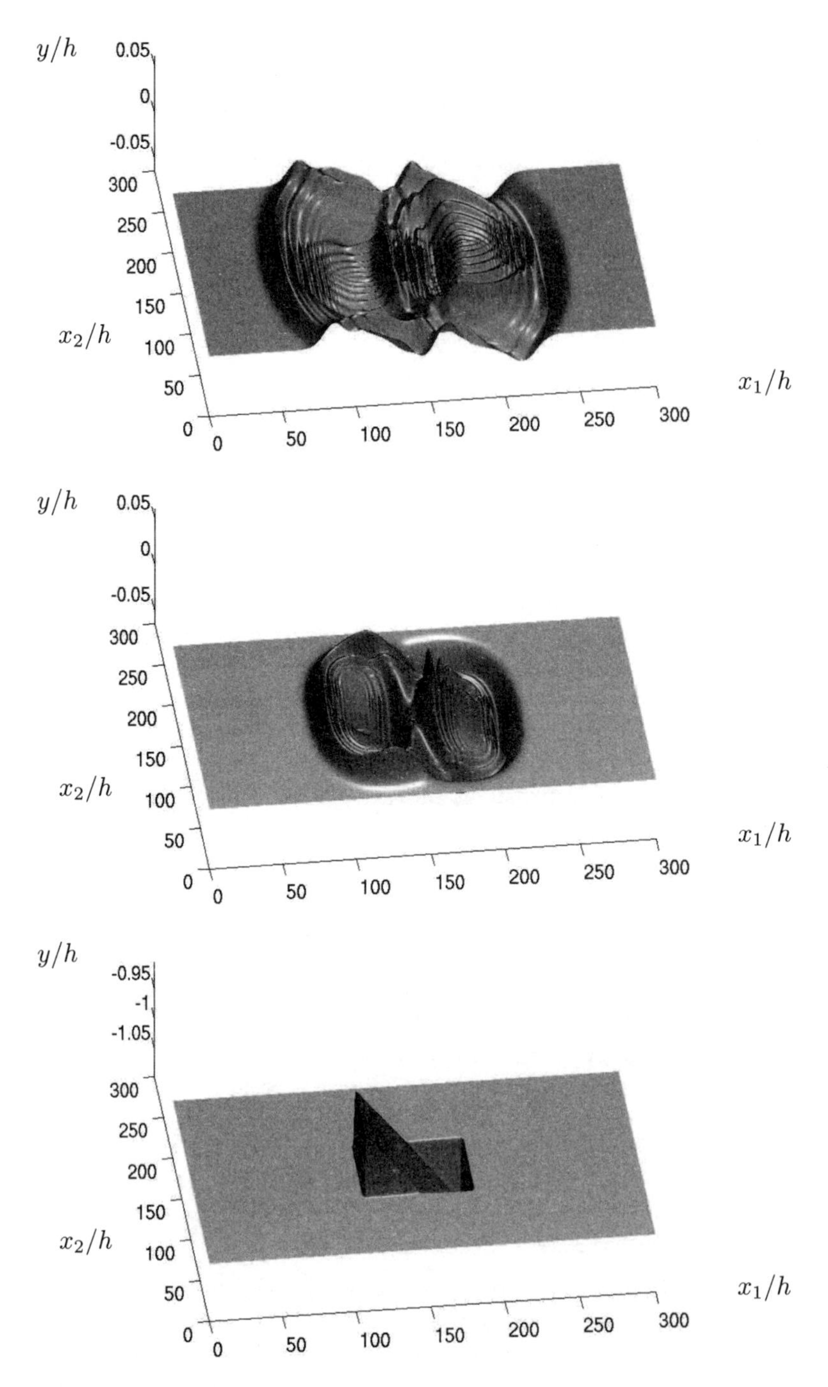

Figure 5. Motion of the sea floor (lower), snapshot of fully nonlinear free surface after $t\sqrt{g/h} = 25$ (mid) and $t\sqrt{g/h} = 50$ (upper). Vertical and horizontal scales: water depth (h). The characteristics for the moving bottom are: maximum amplitude: 0.1 h size of the displaced domain: [90h x 60h].

They perturbed the steady waves by infinitesimal perturbations η' of the form

$$\eta' = e^{\mathrm{i}[p(x_1 - Ct) + qx_2 - \sigma t]} \sum_{-\infty}^{\infty} a_n e^{\mathrm{i}n(x_1 - Ct)}, \tag{10.1}$$

where the perturbation wavenumbers p and q are real numbers. Eq. (10.1) is an eigenvector of the infinitesimal perturbations to the Stokes wave, with σ being the eigenvalue. Instability corresponds to $Im(\sigma) \neq 0$, since the perturbation also involves the complex conjugate. The task set by McLean et al. was to determine the eigenvalues, σ. This was done by inserting a truncated version of (10.1), and a corresponding perturbation of the velocity potential, φ', both including $2N+1$ modes, into the full equations. The kinematic and dynamic boundary conditions were satisfied at $2N+1$ nodes. The resulting homogeneous system of order $4N+2$ was solved as an eigenvalue problem for σ by means of standard methods. McLean et al. used $N = 20$ for $2a/\lambda < 0.1$ and $N = 50$ for the steepest case with $2a/\lambda = 0.131$.

In a frame of reference moving with the wave speed, C, the unperturbed wave, in the limit $2a/\lambda \to 0$, becomes (McLean, 1982): $\eta_s = 0, \varphi_s = -x_1$, where C has been put to unity. In this case, the eigenfunctions and eigenvalues are

$$\eta' = e^{-\mathrm{i}\sigma_n t + \mathrm{i}[(p+n)x_1 + qx_2]}, \tag{10.2}$$

$$\sigma_n = -k_x \pm |\mathbf{k}|^{\frac{1}{2}}, \tag{10.3}$$

where the latter represents the linear dispersion relation on a unit current (of negative speed), and where $\mathbf{k} = (k_{x_1}, k_{x_2}) = (p+n, q)$. Eq. (10.3) gives

$$\sigma_n = -(p+n) \pm [(p+n)^2 + q^2]^{\frac{1}{4}}. \tag{10.4}$$

McLean et al. noted that the σ's were degenerate in the sense that $\sigma_n(p,q) = \sigma_{n+1}(p-1, q)$. In the case of finite amplitude Stokes waves, the eigenvalues may become complex, meaning instability, if

$$\sigma^{\pm}_{n_1}(p,q) = \sigma^{\pm}_{n_2}(p,q). \tag{10.5}$$

The corresponding eigenvectors have dominant wave vectors $\mathbf{k}_1 = (p+n_1, q)$ and $\mathbf{k}_2 = (p+n_2, q)$. For deep water waves, the solution to (10.5) was divided into two classes:
Class I:

$$\mathbf{k}_1 = (p+m, q),\ \mathbf{k}_2 = (p-m, q),\ \sigma^{+}_{m}(p,q) = \sigma^{-}_{-m}(p,q), \tag{10.6}$$

$$[(p+m)^2 + q^2]^{\frac{1}{4}} + [(p-m)^2 + q^2]^{\frac{1}{4}} = 2m,\ (m \geq 1). \tag{10.7}$$

Class II:

$$\mathbf{k}_1 = (p+m, q),\ \mathbf{k}_2 = (p-m-1, q),\ \sigma^{+}_{m}(p,q) = \sigma^{-}_{-m-1}(p,q), \tag{10.8}$$

$$[(p+m)^2 + q^2]^{\frac{1}{4}} + [(p-m-1)^2 + q^2]^{\frac{1}{4}} = 2m+1,\ (m \geq 1). \tag{10.9}$$

Class I curves are symmetric about $(p,q) = (0,0)$. Class II curves are symmetric about $(p,q) = (\frac{1}{2}, 0)$.

The eigenvalues can alternatively be interpreted as resonance of two infinitesimal waves with the carrier wave, giving, in the fixed frame of reference:

$$\omega_1 = -\omega_2 + N\omega_0, \quad \mathbf{k}_1 = \mathbf{k}_2 + N\mathbf{k}_0. \tag{10.10}$$

Here, $\mathbf{k}_1 = (p' + N, q)$, $\mathbf{k}_2 = (p', q)$ (and $-\mathbf{k}_2 = (-p', -q)$), $\mathbf{k}_0 = (1, 0)$, and $\omega_i = |\mathbf{k}_i|^{\frac{1}{2}}$. Class I corresponds to N even, $m = \frac{1}{2}N$, $p' = p - m$. Class II corresponds to N odd, $m = \frac{1}{2}(N-1)$, $p' = p - m - 1$. $N = 2$ corresponds to 4-wave interaction, $N = 3$ to 5-wave interaction, $N = 4$ to 6-wave interaction, and so on. The resonance curves are visualized in figure 6.

For $p = \frac{1}{2}$ McLean et al. found that $Re(\sigma) = 0$, which means that the perturbation wave pattern co-propagates with the Stokes wave. They calculated the regions of nonlinear instability, and moreover, identified the growth rate of the most unstable perturbation (corresponding to maximum of $|Im(\sigma)|$). From McLean et al. (1981, figure 1), we obtain selected values of $(q, ak, |Im(\sigma)|\text{max})$ (and $p = \frac{1}{2}$):

q	ak	$\|Im(\sigma)\|\text{max}$
1.65	0.1	$6 \cdot 10^{-4}$
1.53	0.2	$5 \cdot 10^{-3}$
1.25	0.3	$2 \cdot 10^{-2}$
1.16	0.41	$1 \cdot 10^{-1}$

Table 4. Growth rate of the most unstable perturbation of class II instability; 5-wave interaction. $p = \frac{1}{2}$. Extracted from McLean et al. (1981, figure 1).

McLean (1982) improved the accuracy of the stability calculations. In his table 2 and figure 3, more accurate values of the lateral wavenumber and growth rate were presented. Particularly for $ak = 0.3$, a new value of $q = 1.33$ was obtained, and is used here in the calculations that are discussed below (we use $q = \frac{4}{3}$). It is important to note that improved computations of the stability analysis were required, still. This was particularly true for waves with ak exceeding 0.412, for which no values were given by McLean (1982). This was the motivation for the studies by Kharif and Ramamonjiarisoa (1988, 1990).

10.2 Computations of the classical horseshoe pattern

The stability analysis of McLean et al. (1981) was motivated by the three-dimensional patterns observed experimentally by Su, published in 1982, as mentioned. The waves were also investigated experimentally by Melville (1982), Kusuba and Mitsuyasu (1986) and Collard and Caulliez (1999). The experimental observations of the waves are here numerically reproduced in the sense that we compute the growing wave instabilities up to the point of breaking. In the original experiments the Stokes waves had an amplitude corresponding to $(ak)_0 = 0.33$. For such steep waves the class II instability is stronger than the class I instability. We here illustrate numerically the growth of the class II

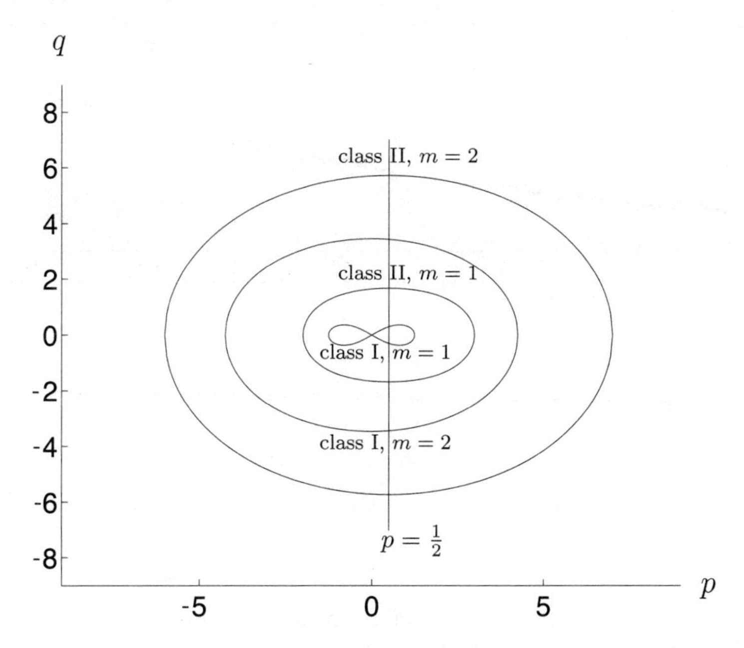

Figure 6. Plot of resonance curves for $ak = 0$. Class I instabilities, eq. (10.7) with $m = 1$ (inner curve – including points $(\pm\frac{5}{4}, 0)$) and $m = 2$ (curve including points $(\pm\frac{17}{4}, 0)$). Class II instabilities, eq. (10.9) with $m = 2$ (outer curve – including points $(-6, 0)$ and $(7, 0)$) and $m = 1$ (including points $(-2, 0)$ and $(3, 0)$). Class II instability curves are symmetric with respect to $p = \frac{1}{2}$.

instability, where in the computations the class I instability is suppressed. Description follows Fructus et al. (2005b). The waves were also studied numerically by Xue, Xü, Liu and Yue (2001) and Fuhrman, Madsen and Bingham (2004).

Stokes waves with frequency $\omega_0 = \omega(k_0)$ and wave slope $(ak)_0$ were computed using the procedure of Fenton (1988). To this wave train, a small perturbation, taking the form

$$\hat{\eta} = \epsilon a_0 \sin((1 + p)k_0 x_1) \cos(q k_0 x_2) \tag{10.11}$$

was superposed. Here, ϵ is a small number, making the amplitude of the initial perturbation field a fraction of the Stokes wave field, and $(1 + p, q)k_0$ denotes the directional wavenumber. Computations performed with $(ak)_0 = 0.2985$ and 0.33, have perturbation at $p = \frac{1}{2}$, $q = \frac{4}{3}$ and value of ϵ of 0.05.

Four periods of Stokes waves in the longitudinal direction and three periods of perturbation in the lateral direction were resolved by 128×64 nodes. The magnitude of the Fourier transform of the perturbation field, $|\mathcal{F}(\hat{\eta})|$, visualizes the growth of the modes with wavenumbers $(\frac{3}{2}, \pm\frac{4}{3})k_0$ up to the point of breaking, see figure 7a. The sum of the wave vectors of the satellites is $(3, 0)k_0$. The initial growth rate in the nonlinear calculations is 0.0194, slightly smaller than the analytical value of 0.021 computed by

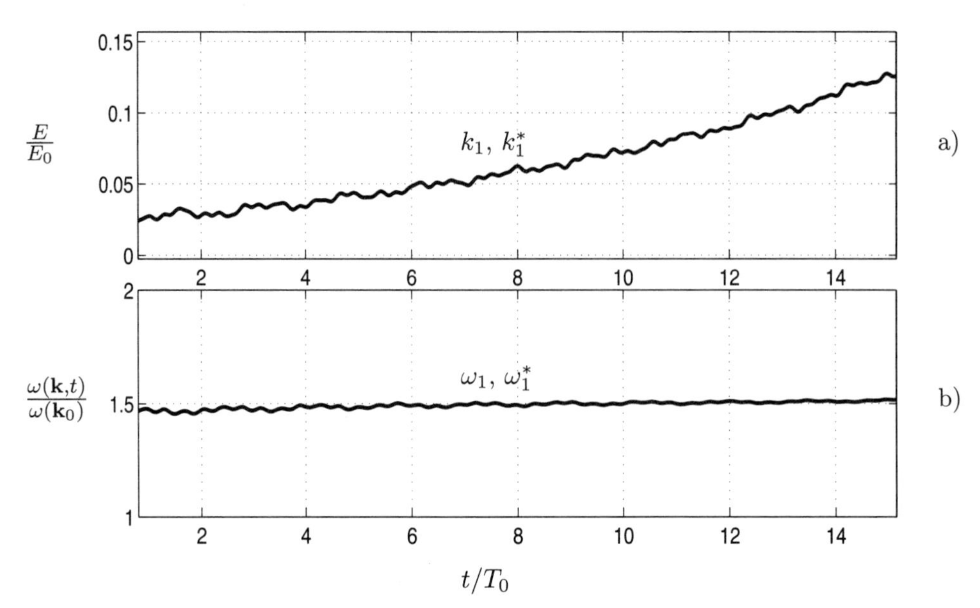

Figure 7. a) Energy evolution, E, relative to initial energy, $E_0 = \frac{1}{2}\rho g a_0^2$, of the dominant modes: $\mathbf{k}_1 = (\frac{3}{2}, \frac{4}{3})k_0$ and $\mathbf{k}_1^* = (\frac{3}{2}, -\frac{4}{3})k_0$. b) Evolution of frequency. Initial perturbation with $(p,q) = (\frac{1}{2}, \frac{4}{3})$. $(ak)_0 = 0.2985$, $\epsilon = 0.05$. Adapted from Fructus et al. (2005b). Reproduced by permission by J. Fluid Mech.

perturbation theory, when $(ak)_0 = 0.2985$. The frequency spectrum, $\omega(k_{x_1}, k_{x_2}, t)$, obtained from the Fourier transformed perturbation field by $\mathcal{F}(\hat{\eta}) = |\mathcal{F}(\hat{\eta})| \exp(i\chi)$ and $\omega(k_{x_1}, k_{x_2}, t) = \partial\chi/\partial t$, shows that $\omega(\frac{3}{2}k_0, \pm\frac{4}{3}k_0) = \frac{3}{2}\omega_0(1+\hat{\epsilon})$, where $\hat{\epsilon}$ represents a very small variation around zero and is visualized in figure 7b. Indeed, the figure illustrates that a quintet interaction is evident. The resulting wave field prior to breaking is visualized in figure 9a, and the wave frequency spectrum in figure 9c. The latter exhibits peaks at ω_0 – the fundamental frequency, and $n\omega_0$, $(n > 1)$ – the locked components of the Stokes wave. The dominant peaks corresponding to the satellites is observed at $\frac{3}{2}\omega_0$. Higher order motions of the satellites have small peaks at $\frac{1}{2}\omega_0$, $\frac{5}{2}\omega_0$, $\frac{7}{2}\omega_0$, ... The number of time steps in the computation of the waves is indicated in figure 7, 9a,c is visualized in figure 10.

The wave breaking appears in the computations by a rapid growth of the high wavenumbers of the spectrum. The computational wave breaking – when it is independent of the resolution – corresponds to breaking of the physical waves. The computed waves right before breaking resemble the steady state of the three-dimensional wave field in the experiments. Wave characteristics are defined in figure 8. The numerical waves compare very well to the waves in steady state, observed by Su (1982), see table 5. Su termed the pattern by L_2 pattern, since it repeated itself once per two wavelengths of the fundamental train. More generally, L_n patterns correspond to interactions between satellites $\mathbf{k}_1 = (1 + 1/n, q)k_0$ and $\mathbf{k}_2 = (2 - 1/n, -q)k_0$, meaning that $p = 1/n$, where

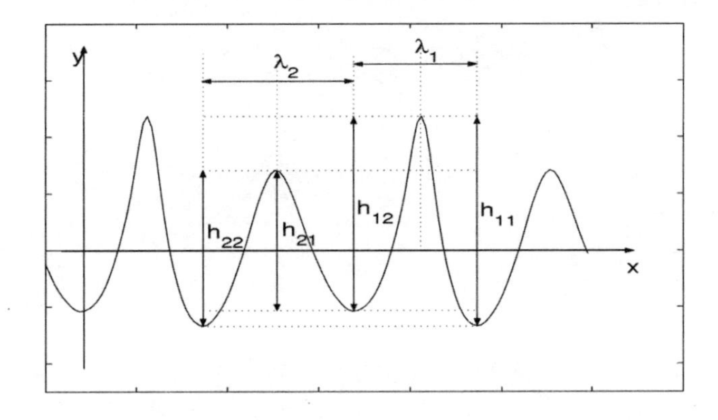

Figure 8. Definition of Su's geometric parameters.

$(ak)_0$	λ_2/λ_1	h_{11}/h_{12}	h_{21}/h_{22}	h_{11}/h_{21}	h_{22}/h_{11}	$\frac{h_{11}+h_{12}}{h_{22}+h_{21}}$	S_x,max
Su; 0.33	1.28	1.10	0.88	1.66	0.68	1.49	0.65
Simulations; 0.2985	1.28	1.11	0.88	1.56	0.73	1.38	0.66
Simulations; 0.33	1.28	1.12	0.85	1.53	0.76	1.33	0.67

Table 5. Characteristics of three-dimensional wave patterns. Numerical computations prior to breaking for $(ak)_0 = 0.2985$ and 0.33 compared to experiments of steady pattern by Su, $(ak)_0 = 0.33$. $S_x = \partial\eta/\partial x$. The numerical waves grow up to breaking during 18 wave periods ($(ak)_0 = 0.2985$) and 11 wave periods ($(ak)_0 = 0.33$).

the integer $n \geq 2$. The definition can also be used for $n = 1$ (L_1 pattern) corresponding to $p = 0$.

10.3 Oscillating horseshoe pattern. Computations of the experiments by Collard and Caulliez

A new surface wave pattern was experimentally observed by Collard and Caulliez (1999). They performed water wave experiments in a long, relatively wide indoor tank, finding an oscillatory, crescent formed, horse-shoe like pattern, riding on top of long-crested steep Stokes waves with wavenumber k_0. The crescents of the sideways oscillating pattern were always aligned, both in the longitudinal and transversal directions, see figure 9b for illustration. The wave patterns appeared due to an instability in the form of a quintet resonant interaction: wave satellites had longitudinal wavenumber k_0 and transversal wavenumber qk_0, and the other wavenumber $2k_0$ (longitudinal) and $-qk_0$ (transversal). These summed up to three times the wavenumber of the fundamental Stokes wave. The observed frequencies (in the experiments) were $1.36\omega_0$, $1.64\omega_0$, adding up to three times the fundamental frequency, ω_0 of the Stokes wave, satisfying also a quintet in the frequencies. Along the propagation direction, the pattern repeated itself once per fundamental wave length, and is so an L_1 pattern. The stability analysis above,

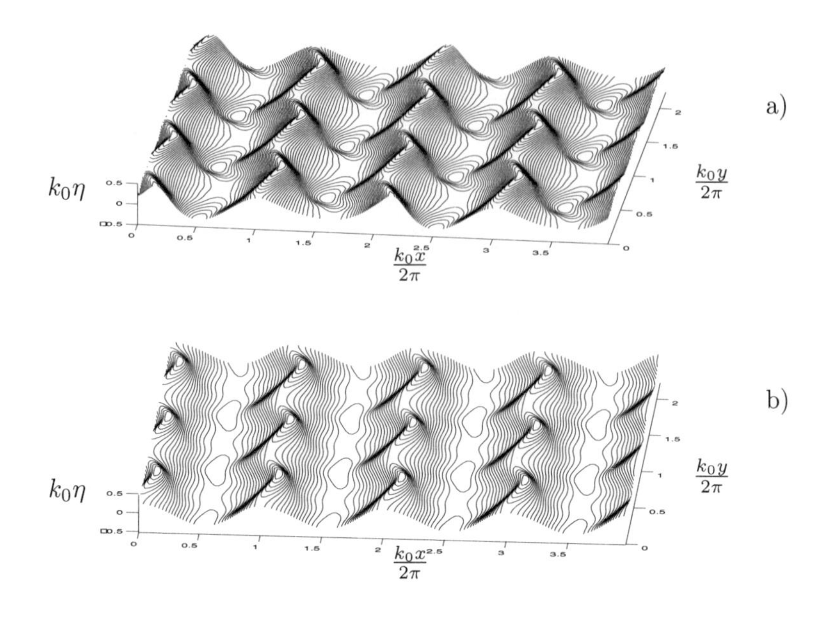

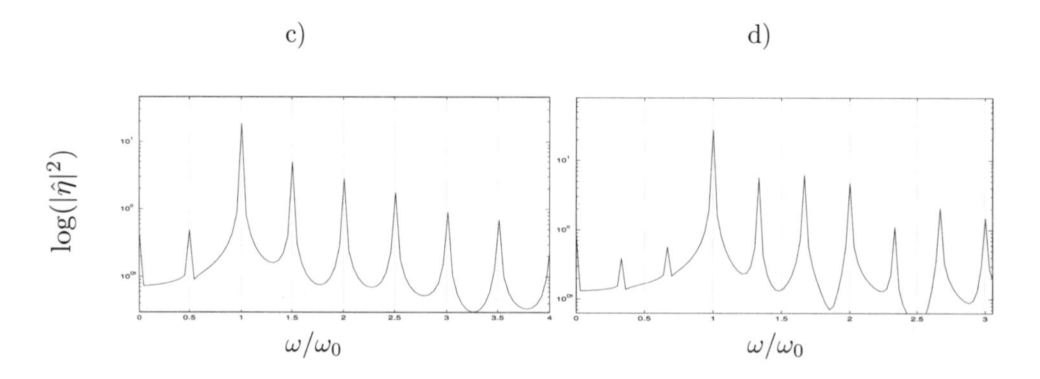

Figure 9. Three-dimensional wave patterns on Stokes waves with wave slope 0.30 of the fundamental mode. a) Steady (most unstable) horse shoe pattern, observed experimentally by Su in 1982. Snapshot at $t/T_0 = 16$ (T_0–wave period). b) Unsteady horse shoe pattern, observed experimentally by Collard and Caulliez in 1999 ($t/T_0 = 23$). c) Wave frequency spectrum corrsponding to a). d) Wave frequency spectrum corresponding to b). From Fructus et al. (2005a).

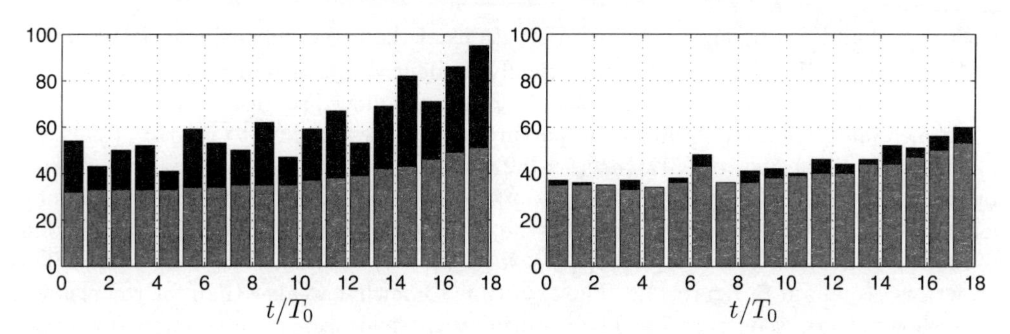

Figure 10. Number of time steps per wave period for the computation in figure 7. Time integration without (right) and with (left) stabilizing step size control. Number of accepted (grey) and rejected (black) steps per wave period. From Clamond, Fructus and Grue (2007).

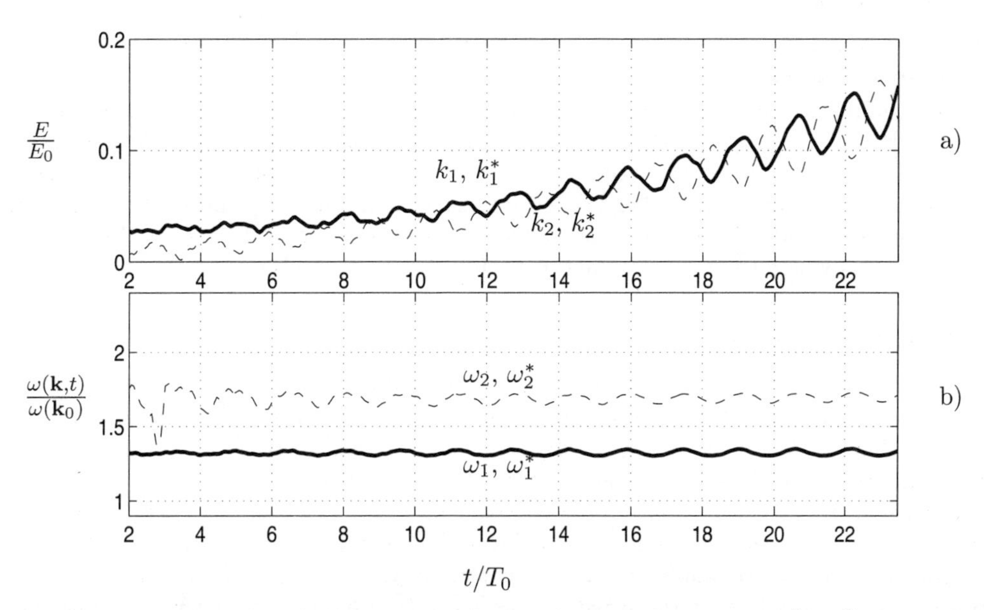

Figure 11. a) Energy evolution, E, relative to initial energy, $E_0 = \frac{1}{2}\rho g a_0^2$, of the dominant modes: $\mathbf{k}_1 = (1, \frac{4}{3})k_0$, $\mathbf{k}_1^* = (1, -\frac{4}{3})k_0$, and $\mathbf{k}_2 = (2, \frac{4}{3})k_0$, $\mathbf{k}_2^* = (2, -\frac{4}{3})k_0$. b) Evolution of frequency. Initial perturbation with $(p, q) = (0, \frac{4}{3})$. $(ak)_0 = 0.2985$, $\epsilon = 0.05$. Adapted from Fructus et al. (2005b). Reproduced by permission by J. Fluid Mech.

with $N = 3$ and $p = 0$, supports the observations of the three-dimensional pattern.

This aligned L_1 pattern was numerically evidenced by Skandrani (1996) and was also computed by Fuhrman et al. (2004). Here we follow Fructus et al. (2005b) who performed simulations with an initial perturbation corresponding to $(p, q) = (0, \frac{4}{3})$ and $\epsilon = 0.05$, for Stokes waves with $(ak)_0 = 0.2985$. The wave field was initially perturbed by $\mathbf{k}_1 = (1, \frac{4}{3})k_0$. This perturbation generated a motion with energy at wavenumber $\mathbf{k}_2^* = (2, -\frac{4}{3})k_0$. An interaction between the two modes then became evident. The pair of modes with wave vectors $\mathbf{k}_1 = (1, \frac{4}{3})k_0$, $\mathbf{k}_2^* = (2, -\frac{4}{3})k_0$ satisfied $\mathbf{k}_1 + \mathbf{k}_2^* = 3\mathbf{k}_0$. This growth is visualized in figure 11a. The growth is somewhat weaker than for the classical horseshoe pattern, see figure 7a. The resulting wave field prior to breaking is illustrated in figure 9.

The dominant frequencies of the satellites are visualized in figure 11b. The figure shows that $\omega(\mathbf{k}_1) + \omega(\mathbf{k}_2^*) = 3\omega_0(1 + \hat{\epsilon})$ where ϵ represents a small oscillation around zero. The oscillation has period $\frac{11}{7}$, approximately, of the fundamental period, T_0. The dominant frequencies are further illustrated in figure 9d, where peaks in the wave frequency spectrum is observed pair-wise, at $(n + a)\omega_0$, $(n + 1 - a)\omega_0$, for $n = 0, 1, 2, \ldots$, where $a = 0.33 \pm 0.02$. (Note that the value of $(ak)_0$ is different in the computation and the experiment by Collard and Caulliez, and thus makes a small difference in the value of the peak frequencies of the satellites.) The wave frequencies of the Stokes waves at $n\omega_0$, $n \geq 1$, are also present.

The computations by Fructus et al. (2005b) further showed that the modes ($\mathbf{k}_1 = (1, \frac{4}{3})k_0$, $\mathbf{k}_2^* = (2, -\frac{4}{3})k_0$) and ($\mathbf{k}_1^* = (1, -\frac{4}{3})k_0$, $\mathbf{k}_2 = (2, \frac{4}{3})k_0$) had the same development. A resonant interaction between the two pairs of modes was observed. The triggering mechanism of the waves was investigated, finding that the waves could be generated by parametric resonance due to the wave-maker. This L_1 pattern should be observed in coastal waters where the modulational instability becomes weaker.

10.4 Other features of class II instability

Class I instability may restabilize class II instability; $(ak)_0 = 0.10$. For moderate $(ak)_0$ the class II instability may be restabilized by class I instability. In an example, shown in figure 12a, the Stokes wave with $(ak)_0 = 0.10$ was perturbed by the most unstable modes of the class I and class II instabilities, corresponding to $p = \frac{1}{6}$ and $p = \frac{1}{2}$, with wave vectors $\mathbf{k}_1 = (\frac{5}{6}, 0)k_0$, $\mathbf{k}_2 = (\frac{7}{6}, 0)k_0$, $\mathbf{k}_3 = (\frac{3}{2}, 1.645)k_0$, $\mathbf{k}_3^* = (\frac{3}{2}, -1.645)k_0$. The explanation for the restabilization of the class II instability is that the modulational instability moves the very narrow class II instability region slightly (when $(ak)_0 = 0.1$), so that the pertubations at the wave vectors $\mathbf{k}_3 = (\frac{3}{2}, 1.645)k_0$, $\mathbf{k}_3^* = (\frac{3}{2}, -1.645)k_0$ fall outside the instability region. No breaking was observed in the simulation.

Class II instability may trigger class I instability, leading to breaking; $(ak)_0 \geq 0.10$. If the wave field is perturbed by class II instability only, the side-band instability becomes triggered. If $(ak)_0 \geq 0.10$, the interaction between the class II and class I instabilities and the fundamental wave train leads to breaking. An example with the initial unstable perturbation corresponding to the phase-locked crescent-shaped pattern, with wave satellites $\mathbf{k}_3 = (\frac{3}{2}, 1.645)k_0$, $\mathbf{k}_3^* = (\frac{3}{2}, -1.645)k_0$, is visualized in figure 12b.

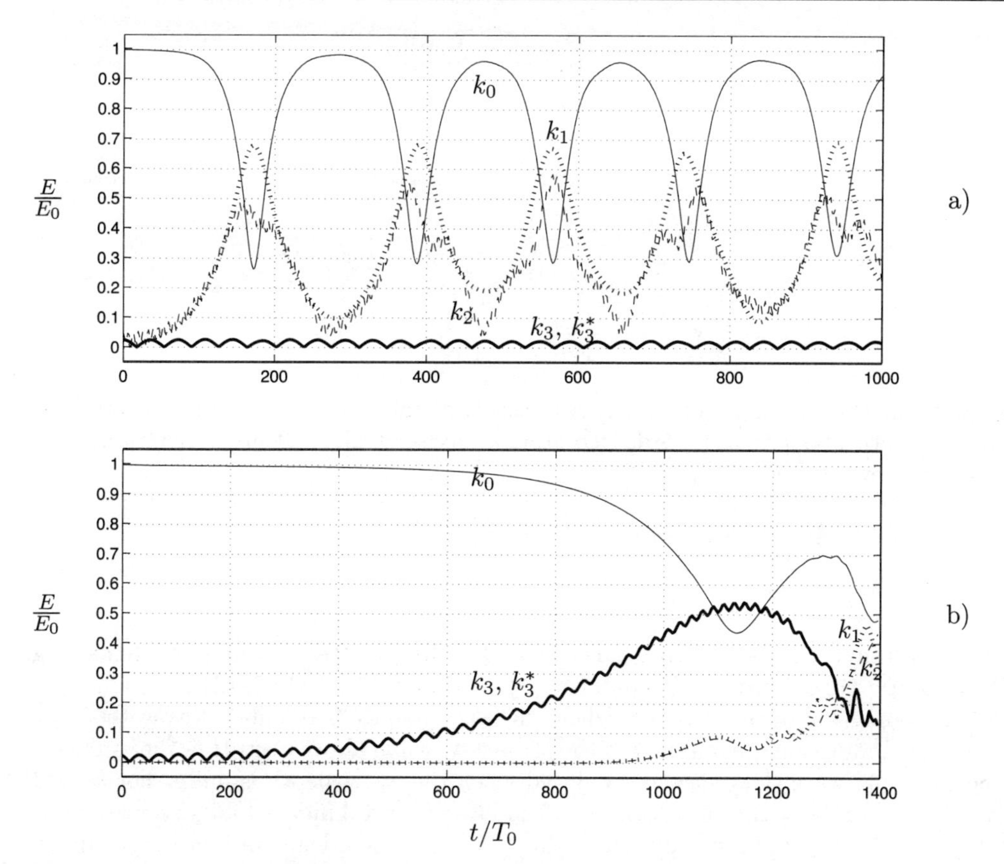

Figure 12. Energy evolution, E, relative to initial energy, $E_0 = \frac{1}{2}\rho g a_0^2$, of the main modes: $\mathbf{k}_0 = (1,0)k_0$, $\mathbf{k}_1 = (\frac{5}{6},0)k_0$, $\mathbf{k}_2 = (\frac{7}{6},0)k_0$, $\mathbf{k}_3 = (\frac{3}{2},1.645)k_0$, $\mathbf{k}_3^* = (\frac{3}{2},-1.645)k_0$. Stokes wave with $(ak)_0 = 0.1$ perturbed initually by a) $\mathbf{k}_1$, $\mathbf{k}_2$, $\mathbf{k}_3$, $\mathbf{k}_3^*$, and b) $\mathbf{k}_3$, $\mathbf{k}_3^*$ only. Adapted from Fructus et al. (2005b). Reproduced by permission by J. Fluid Mech.

Class I instability may trigger class II instability, leading to breaking; $(ak)_0 >$ **0.12.** Numerical simulations taking into account both class I and class II instabilities show that for moderately steep waves, namely $(ak)_0 > 0.12$, their nonlinear coupling (involving the fundamental of the Stokes wave) results in breaking of the wave when in the initial condition only the modulational instability was considered. This result is in agreement with the experiments conducted by Su and Green (1984).

Class II leading to breaking; $(ak)_0 > 0.17$**.** For steeper waves (in deep water), the strength of class II instability alone is found sufficient to trigger breaking of the wave. It is shown that the nonlinear dynamics of the most unstable class II perturbation lead to breaking when $(ak)_0 > 0.17$. For very steep waves (in deep water), with $(ak)_0 > 0.31$, class II instability dominates over class I instability, being the primary source of breaking

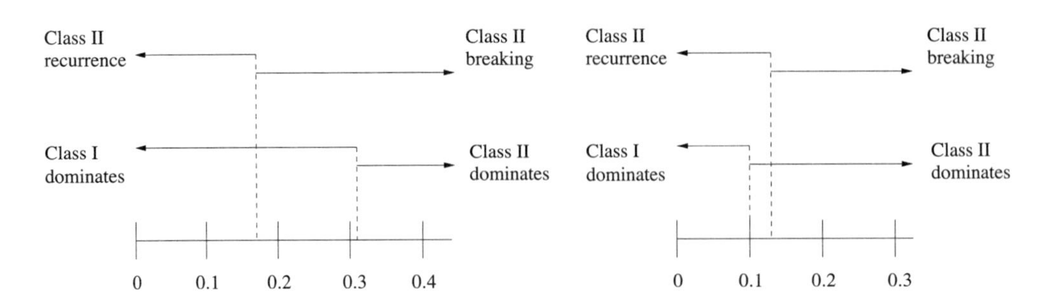

Figure 13. Predominance of class I and class II instabilities. Recurrence vs. breaking. Infinite depth and $kh = 1$. Adapted from Fructus et al. (2005b). Reproduced by permission by J. Fluid Mech.

of the waves.

Predominance of class I and class II instabilities. Recurrence vs. breaking. Wave slope thresholds. Computations show that class I and class II instabilities are equally strong for $(ak)_0 = 0.314$, when the water depth is infinite. Below this wave slope the modulational instability is the strongest, while above class II is the strongest. The threshold wave slope becomes reduced when the water depth is finite. For $kh = 1$, the value is $(ak)_0 = 0.1$, for example. The class II instability exhibits recurrence for $(ak)_0 < 0.17$ and breaking above this value (when $h = \infty$), while in finite water depth, for $kh = 1$, recurrence occurs when $(ak)_0$ is less than 0.13, approximately. The thresholds are summarized in figure 13. In general, higher order instabilities in the form of 5-, 6- and higher order wave interactions become more important for shallower water than in deep water, see also Francius and Kharif (2006), Kristansen, Fructus, Clamond and Grue (2005).

The author is grateful to Prof. C. Kharif (IRPHE, Marseille), Dr. H. Kalisch (Univ. of Bergen), Prof. K. Trulsen and Dr. Dorian Fructus for useful suggestions during the preparation of the manuscript. Drs. Dorian Fructus and Atle Jensen prepared several of the figures. This work was funded by the Research Council of Norway through the Strategic University Program: Modelling of currents and waves for sea structures 2002-6 at the University of Oslo.

Bibliography

T. E. Baldock, C. Swan and P. H. Taylor (1996). A laboratory study of nonlinear surface waves on water. Phil. Trans. R. Soc. A, Vol. 354, 649-76.

T. B. Benjamin and J. E. Feir (1967). The disintegration of wave trains on deep water. Part 1. Theory. J. Fluid Mech. 27, 417-30.

T. B. Benjamin and P. J. Olver (1982). Hamiltonian structure, symmetries and conservation laws for water waves. J. Fluid Mech. 1215, 137-85.

B. Büchmann, P. Ferrant and J. Skourup (2000). Run-up on a body in waves and current. Fully nonlinear and finite-order calculations. Appl. Ocean Res. Vol. 22, 349-60.

D. Clamond and J. Grue (2001). A fast method for fully nonlinear water wave computations. J. Fluid Mech. Vol. 447, 337-55.

D. Clamond and J. Grue (2002). Interaction between envelope solitons as a model for freak wave formation. C. R. Acad. Sci. Mech. Vol. 330, 575-80.

D. Clamond, D. Fructus, J. Grue and Ø. Kristiansen (2005). An efficient model for three-dimensional surface wave simulations. Part II Generation and absorption. J. Comp. Phys. Vol. 205, 686-705.

D. Clamond, D. Fructus and J. Grue (2007). A note on time integrators in water wave simulations. J. Engng. Math. (accepted).

F. Collard and G. Caulliez (1999). Oscillating crescent-shaped water wave patterns. Phys. Fluids Letters, Vol. 11, 3195-7.

J. D. Fenton (1988). The numerical solution of steady water wave problems. Computers Geosci. Vol. 14, 357-68.

M. Francius and C. Kharif (2006). Three-dimensional instability of nonlinear gravity waves in shallow water. J. Fluid Mech. (in press).

D. Fructus, D. Clamond, J. Grue and Ø. Kristiansen (2005a). An efficient model for three-dimensional surface wave simulations. Part I Free space problems. J. Comp. Phys. Vol. 205, 665-85.

D. Fructus, C. Kharif, M. Francius, Ø. Kristiansen, D. Clamond and J. Grue (2005b). Dynamics of crescent water wave patterns. J. Fluid Mech. Vol. 537, 155-86.

D. Fructus and J. Grue (2006). An explicit method for the nonlinear interaction between water waves and moving topography (submitted).

D. R. Fuhrman, P. A. Madsen and H. B. Bingham (2004). A numerical study of crescent waves. J. Fluid Mech. Vol. 513, 309-42.

D. Givoli (2004). High-order local non-reflecting boundary conditions: a review. Wave motion Vol. 39 (4), 319-24.

J. Grue (2002). On four highly nonlinear phenomena in wave theory and marine hydrodynamics. Applied Ocean Res. Vol. 24, 261-74.

J. Grue, D. Clamond, M. Huseby and A. Jensen (2003). Kinematics of extreme waves in deep water. Appl. Ocean Res. Vol. 25, 355-66.

J. Grue and A. Jensen (2006). Experimental velocities and accelerations in very steep wave events in deep water. Eur. J. Mech. B/Fluids, Vol. 5.

C. Kharif and A. Ramamonjiarisoa (1988). Deep-water gravity waves instabilities at large steepness. Phys. Fluids Vol. 31, 1286-8.

C. Kharif and A. Ramamonjiarisoa (1990). On the stability of gravity waves on deep water. J. Fluid Mech. Vol. 218, 163-70.

Ø. Kristiansen, D. Fructus, D. Clamond and J. Grue (2005). Simulations of of crescent water wave patterns on finite depth. Phys. Fluids. Vol. 17, 064101-1-15.

T. Kusuba and M. Mitsuyasu (1986). Nonlinear instability and evolution of steep water waves under wind action. Rep. Res. Inst. Appl. Mech. Kyushu Univ.Vol. 33 (101), 33-64.

M. J. Lighthill (1965). Contributions to the theory of of waves in nonlinear dispersive systems. J. Fluid Mech. Vol. 183, 451-65.

M. S. Longuet-Higgins (1978). The instabilities of gravity waves of finite amplitude in deep water. II. Subharmonics. Proc. R. Soc. Lond. A, Vol. 360, 489-505.

J. W. McLean, Y. C. Ma, D. U. Martin, P. G. Saffman and H. C. Yuen (1981). Three dimensional instability of finite amplitude water waves. Phys. Rev. Letters, Vol. 46, 817-20.

J. W. McLean (1982). Instabilities of finite-amplitude water waves. J. Fluid Mech. Vol. 114, 315-30.

W. K. Melville (1982). The instability and breaking of deep-water waves. J. Fluid Mech. Vol. 115, 165-85.

M. K. Ochi (1998). Ocean waves - the stochasitc approach. Camb. Univ. Press. 320 pp.

A. L. New, P. McIver and D. H. Peregrine (1985). Computations of overturning waves. J. Fluid Mech. Vol. 150, 233-51.

J. E. Romate (1992). Absorbing boundary conditions for free surface waves. J. Comp. Phys. Vol. 99 (1), 135-45.

C. Skandrani (1996). Contribution á l'étude de la dynamique non linéaire de champs de vagues tridimensionnels en profondeur infinie. PhD thesis, Univ. de la Méditerranée.

M. Y. Su (1982). Three-dimensional deep water waves. Part 1. Experimental measurements of of skew and symmetric wave patterns. J. Fluid Mech. Vol. 124, 73-108.

M. Y. Su and A. W. Green (1984). Coupled two- and three-dimensional instabilities of surface gravity waves. Phys. Fluids Vol. 27, 2595-7.

M. Xue, H. Xü, Y. Liu and D. K. P. Yue (2001). Computations of fully nonlinear three-dimensional wave-wave and wave-body interactions. Part 1. Dynamics of steep three-dimensional waves. J. Fluid Mech. Vol. 438, 11-39.

J. W. Wehausen and E. V. Laitone (1960). Surface waves. Handbuch der Physik, Vol. 9(3), 446-778.

V. E. Zakharov (1968). Stability of periodic wave of finite amplitude on the surface of a deep fluid. J. Appl. Mech. Phys. Engl. Transl. Vol. 9, 190-4.

Very large internal waves in the ocean – observations and nonlinear models

John Grue

Mechanics Division, Department of Mathematics, University of Oslo, Norway

Abstract This chapter describes the observations and modelling of very large internal waves in the ocean. We begin with a brief description of the dead-water phenomenon explained by Nansen, and the internal tides discovered by Pettersson. We continue by describing the very large oceanic solitary internal waves of depression observed in the field from 1965 and onwards. The main point of the mathematical analysis is to model the formation and propagation properties of the waves. Tidal generation of internal undular bores and solitary waves of amplitude comparable to the surface layer thickness at rest are exemplified by numerical simulations. The fully nonlinear and fully dispersive mathematical and numerical modelling is found to reproduce the wave motion taking place in the ocean, including the excursions of the mixed upper layer as large as 4-5 times the level at rest, as in the COPE experiment. The interface method – derived in two and three dimensions – is found to compare well with laboratory measurements of the waves using Particle Image Velocimetry. A fully nonlinear theory that accounts for the continuously stratified motion within the pycnocline is derived. The method is used to support experiments on internal wave breaking governed by shear. The latter model is also useful to compute internal wave motion when the upper part of the ocean is linearly stratified. The fully nonlinear modelling is put into perspective by deriving in parallel the weakly nonlinear Korteweg-de Vries, Benjamin-Ono and intermediate long wave equations. Recent observations in deep water reveal significant internal wave motion and corresponding strong bottom currents of magnitude about 0.5 m/s where the pycnocline meets the shelf slope. Future directions of internal wave research are indicated.

1 Introduction

Very large internal waves are a spectacular phenomenon taking place in the world's oceans. Their presence is well documented and are viewed from above through the wave induced signatures on the ocean surface. The internal wave motion may be divided into two main categories, including the dispersive wave trains that locally are characterized by a sinusoidal motion, and the solitary waves of depression that move with constant speed and shape. The latter are often termed internal solitons because of their dual wave and particle property. The wave pulses may also appear in trains. The amplitude of the individual waves are typically decaying with position in the rank-ordered wave train.

The internal waves have period ranging from a few minutes up to about half an hour, and wavelengths ranging from a few hundred meters up to some kilometers. The primary sources of the waves are wind and tide. Both effects uplift or depress the pycnocline in the ocean that separates the layer of light surface water from denser water below. More specifically, the tidal flow over topography generates undular bores and solitary waves of large amplitude, sometimes comparable to, or even exceeding, the depth of the mixed upper layer of the ocean. The effect of wind may uplift or depress the thermocline along the shelf-geometry, leading to subsequent strong internal run-up or run-down in the cross-wise topographical direction. The waves propagate along the pycnocline or in the form of beams in the stratified ocean. The internal waves we discuss in this chapter propagate horizontally. The waves have shorter period than the internal tides which are the subject of Ch. 6 of this volume.

In this chapter we begin with a short summary of some main observations of the very large internal waves in the ocean (section 1). The weakly nonlinear models are outlined in section 2, including the Korteweg-de Vries, Benjamin-Ono and finite-depth equations. The corresponding analytical solitary wave (soliton) solutions are given. In section 3 we describe a fully nonlinear and fully dispersive method that is useful to obtain highly accurate computations of interfacial solitary waves of large amplitude. The section contains a description of laboratory measurements of the waves that confirm the usefulness of the interfacial theory. Modelling of transient and fully nonlinear, fully dispersive interfacial waves is described in section 4, including numerical studies of tidally generated interfacial waves at geometry. The simulations are compared to observations of solitary waves at Knight Inlet in British Columbia and in the Sulu Sea. The waves generated in the dead-water problem are briefly commented on. The two-dimensional interfacial theory is complemented by a generalization to motion in three dimensions. In section 5 we outline a method to compute the wave motion in the case when the sea is continuously stratified. The particular objectives are to model the internal wave induced velocity field within the pycnocline, and the wave motion when the upper layer of the ocean is linearly stratified. We discuss how the simulations of the large waves in the continuously stratified sea can be used to study the onset of breaking. A short section 6 describes future directions of internal wave research.

1.1 The dead-water phenomenon

An old phenomenon relating to internal waves is dead-water. Dead-water was known by seamen to slow down sailing ships but was a mystery until the explanation came with Fridtjof Nansen's polar expedition with Fram 1893-6. Nansen was sailing in the waters north of Russia where he reported (see Ekman, 1904, pp. 9-10):

"On Tuesday, August 29th, 1893, the Fram got into open water in the sound between the Isle of Taimur and the Almvist Islands and steamed in calm water through the sound to the north-east ... We approached the ice to make fast to it, but the Fram had got into dead-water, and made hardly any way, in spite of the engine going at full pressure. It was such slow work that I thought I row ahead to shoot seal ... the speed must have reduced to 1–1.5 knots in the dead-water ... The water at the surface was almost fresh, whereas through the bottom-cock of the engine room we got perfectly salt water."

Upon return to Norway, Nansen discussed his observations with Vilhelm Bjerknes who attributed the wave resistance to the internal waves generated by the ship. The waves could propagate along the interface separating fresh water from saltier water below. Vilhelm Bjerknes put one of his Carnegie assistants – Vagn Walfrid Ekman – to perform experiments to confirm the connection between dead-water and internal waves made by the ship (Ekman, 1904). The reader is also referred to the illustrations in figures 29–30 in subsection 4.4. A description about Vilhelm Bjerknes and his students is found in Eliassen (1982).

1.2 The discovery of internal tides

Besides dead-water, Fridtjov Nansen recorded internal wave motion in the sea during the Fram expedition, a motion having a different cause than the ship. In Nansen (1902, pages 346-7) he described these observations (translated): *"So I understood, that there were oscillations of some kind on the interface between cold polar water and the water below from the Gulfstream."* What Nansen had encountered was the internal tide, but did not use this notion. Neither did he use the term internal waves.

The internal tides were discovered by the Swedish oceanographer Otto Pettersson in Gullmarfjord on the west coast of Sweden a few years later – in 1909 (Walter Munk, private communication). We include here – for clarification – a letter by V. W. Ekman, published in the June issue of Nature in 1945 (p. 669):

"I am just reading with great interest the important and excellent work by H. U. Sverdrup, M. W. Johnson and R. H. Fleming, "The Oceans: their Physics, Chemistry and General Biology". I should like to direct attention to a lapse which I am rather anxious to have pointed out. On page 592 the first observation of internal waves in the sea is attributed to Helland-Hansen and Nansen. The fact is that such waves were discovered in the Great Belt two years before by the nestor of Swedish oceanographers, the late Otto Pettersson – well known as the originator of the international organization for marine research – and he also proved their tidal periodicity. Furthermore, on page 600 reference is incidentally made to observations of internal waves in a Swedish fjord which Otto Pettersson is said to have carried out "during 2 months of 1909". As a matter of fact, these observations were continued over several years, and – although particular theoretical conclusions which Pettersson inferred are open to serious critisism – the observations themselves are in their method by far the most exhaustive ever made.

These comments may seem to be of minor importance, but since internal waves have, of late, attracted the intense and increasing interest of oceanographers, and since Otto Pettersson himself was from the first aware of the importance of his discovery and took particular interest in it, I think we owe him the justice of acknowledging his priority, at the same time as we recognize the important contributions of Nansen and Helland-Hansen."

It is worth noting that Otto Pettersson gave credit to Fridtjov Nansen to have first encountered internal waves in the field. See also the biography of Otto Pettersson by Svansson (2006, Ch. 15).

1.3 Internal waves in the ocean. Research up to 1960

A description giving the state of the art of the research on internal waves in the ocean up to 1960 may be found in Defant (1961, II, Ch. XVI). The description centres around linear theory (by Stokes from 1847), observations of internal motion driven by tide or wind, including Sandström's experiment from 1908, and stability. Contributions from scientists Helland-Hansen, Nansen, Ekman, Bjerknes, Pettersson, Fjeldstad, Solberg, Seiwell, Lek, Lisitzin, Defant, Zeilon and Haurwitz are among those referred.

1.4 Loss of submarines

The US Navy has been investigating the motion of internal pycnoclines in the ocean because of several losses of modern submarines. These may move along the pycnoclines where detection by surface vessels is avoided, since the pycnocline tends to reflect and refract high-energy sonar pulses. The submarines move vertically according to the motion of the pycnocline since the density is adjusted to that of the surrounding water. While the density of the water along the pycnocline is constant, the pressure varies according to the water column above the submarine. A well documented loss is the submarine USS Thresher that sank in 1963 with 129 crew members and civilians abroad. Thresher had been the most advanced submarine in the world of the time, capable of reaching depths and speeds unimaginable a decade before. According to Pinet (1992, p. 220) there had been no indication of equipment malfunction or of unusual storm weather. Navy scientists speculate that Thresher was probably cruising along a pycnocline when it encountered a large internal wave that moved the submarine down to a depth below the pressure capacity of the hull. The incident evidently occurred too rapidly for crew members to reduce the submarine's density to arrest their fall.

1.5 Very large internal waves

Descriptions of internal solitary wave motion in the ocean appeared in the literature from 1961 and onwards. Lee (1961) reported on a campaign in 1959 in the coastal water near San Diego, measuring solitary wave motion with vertical displacements – of size typically up to about 4 m – which were upward when the thermocline was near the bottom, and downward when the thermocline was higher. Gaul (1961) analyzed temperature records obtained near Hudson Canyon indicating that internal solitary waves were a common occurrence. The vertical excursions in those measurements were 10 m. Byshev, Ivanov and Morozov (1971) found occasional motion behaving like solitary waves of depression from data taken in the Black Sea. See also the book by Roberts (1975) summarizing internal wave research up to 1973, including 827 references.

Very large oceanic internal waves appear to first have been reported by Perry and Schimke (1965) who documented measurements of 80 m deep depression waves from a survey vessel in the Andaman Sea. Osborne, Burch and Scarlet (1978) recorded internal waves with surface currents up to 1 m/s and heights more than 50 m in the same region under contract to Esso Exploration Inc. and Exxon Production Research Co. They concluded that internal waves are an important factor in the design of drill operations and production in deep water. In an accompanying paper, Osborne and Burch (1980)

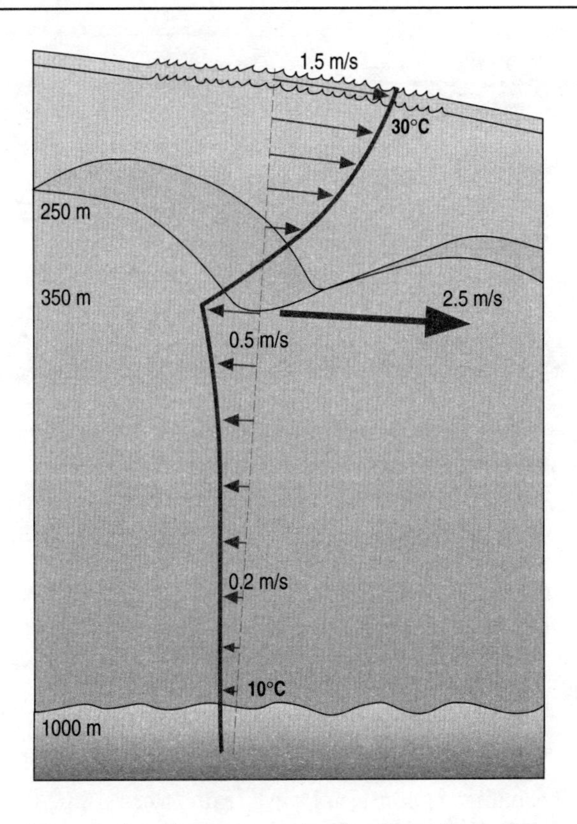

Figure 1. Sketch of internal solitary wave in the Sulu Sea. Waves were documented by Apel et al. (1985) to have an amplitude of about 90 m. Velocity profile is obtained by Froude number scaling of computations and laboratory measurements by Grue et al. (2000). The strong internal wave induced horizontal velocity below the surface effects the local wind-wave field, making signatures that are visible from above.

interpreted the waves of the previous measurement campaign in terms of solitary wave solutions of the Korteweg-de Vries equation, see also the description in section 2.4 below and in the Appendix. The very large internal waves were observed in the famous Sulu Sea experiment where the warm ocean surface layer typically is 250 m thick. Internal depression waves were characterized by an amplitude of 90 m, wave speed of 2.5 m/s and wavelength of 2-3 km (Apel et al., 1985). The wave induced velocities can be estimated from model studies of the waves and are felt through the entire water column, with a fluid velocity of 1.5 m/s at the ocean surface and about 0.1 m/s at the sea floor, see figure 1. Very large internal solitary waves were measured in the Celtic Sea by Pingree and Mardell (1985) and are characterized by an amplitude that corresponds to the upper layer thickness, see figure 2. The wave width that can be measured at the level corresponding to half-amplitude is 450 m. Measured wave speed of 0.7 m/s exceeded linear long wave speed by 22 %, as documented by subsequent computations by Ostrovsky and Grue

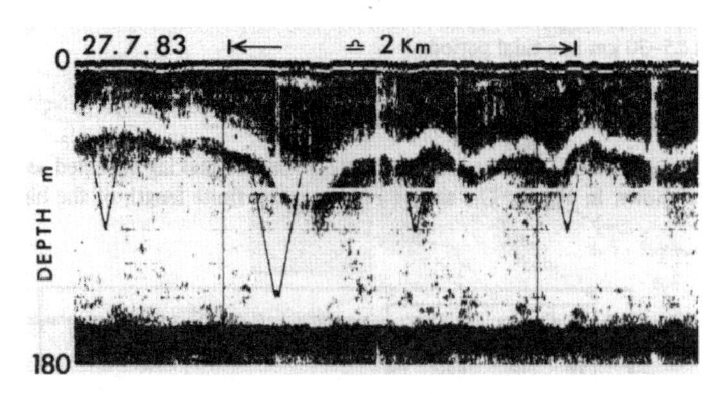

Figure 2. Internal solitary wave in the Celtic Sea. Reproduced from Pingree and Mardell (1985, fig 4.2) with permission.

(2003). The recent Coastal Ocean Probe Experiment (COPE) took place at the shelf break off Northern Oregon. The campaign exhibited solitary wave motion with excursions of the pycnocline as large as 4-5 times the level at rest (Stanton and Ostrovsky, 1998; Kropfli et al. 1999), see figure 3. The waves were about 90 m wide and wave speed exceeded linear long wave speed by about 100 %. Fluid velocities up to about 85 % of wave speed were measured.

The generation mechanisms of internal wave trains were studied by research groups in the US during the 1970s. Halpern (1971) measured the tidally generated internal waves at Stellwagen Bank north of Cape Cod. The investigation led Lee and Beardsley (1974) to suggest that the wave generation occurred due to partial blocking over the bank forming a relatively steep front of the pycnocline. Short periodic waves subsequently emerged, as suggested by Haury, Briscoe and Orr (1979). When the tidal current slowed down, or turned, the waves that were generated over, or on lee side of the topography, could propagate to the upstream side. Observation of upstream waves were also described by New, Dyer and Lewis (1987) where they credited Maxworthy (1979) – performing laboratory studies of the generation mechanism – with being the first to describe them. The prosess of the generation is illustrated numerically in figures 24-27 in section 4.3.

Internal bores formed by a transformation of internal tides may generate packets of short-periodic internal waves. This is observed, e.g. at the Strait of Gibraltar (Armi and Farmer, 1988; Morozov et al., 2002) and on the Australian North West Shelf (Holloway et al., 1997).

The short waves may be swept both downsteam and upstream, depending on the surrounding flow: if the flow speeds up, instead of slowing down, these same waves can be swept downstream. This has been observed in photos of the 2004 tsunami generating similar surface waves which were then swept onwards with the tsunami's flow as its strength increased (Peregrine, personal communication).

1.6 Mechanisms for internal wave – surface wave interaction

Internal waves are visible from above because they strongly modulate the surface wave spectrum, a mechanism described already by Ewing (1950) and Shand (1953), and further documented in LaFond (1966). Observations of internal waves from above received rather intense attention from the 1970s and onwards, including recordings from several satellites and space crafts, as reported by Apel et al. (1975). The astronauts on the joint US-USSR Apollo-Soyuz mission in 1975 were requested to make pictures of the ocean surface from above. In the Summary Science Report of the mission, Apel (1978) prepared a subchapter on the observations of the internal-wave surface signatures from photographs, reporting:

"At least three Apollo-Soyuz Test Project photographs showed clear indications of oceanic internal gravity waves; the features are indicated by periodic changes in the optical reflectivity of the ocean surface overlying the waves. Two packets (or groups) of waves seen off Cádiz, Spain, have characteristics similar to internal waves seen in satellite images taken off the U.S. eastern coast. In the Andaman Sea off the Malay Peninsula, several groups were observed having wavelengths of 5 to 10 km and interpacket separations on the order of 70 to 115 km. If these are indeed surface signatures of internal waves, the waves are among the longest and fastest observed to date. Earlier shipboard measurements have shown that large-amplitude internal waves exist in the area."

Later satellite images revealed that the Andaman Sea is rich in such excitations. The same is true in other places in the world's oceans, which through the US and Soviet space programs – and the growing number of satellites orbiting the earth – for the first time systematically was viewed from above. An increasing documentation of the internal waves by images from planes, space crafts and satellites is made available on the Internal Wave Atlas, see *http : //atlas.cms.udel.edu/search.html*.

The interaction between internal waves and surface waves was particularly emphasized in the Coastal Ocean Probe Experiment (COPE), see Kropfli et al. (1999). The relationships between isotherm vertical displacements, internal currents, radar backscatter cross-sections complemented by Doppler velocity signals at horizontal and vertical polarizations were analysed in the COPE experiment. Radar images showed bands of relatively high signal strength. This occurred in regions where the surface waves were enhanced by the internal waves. These bands were separated by low intensity bands. These reflected that the surface waves were suppressed. The measurements showed: the phase of the radar-signal modulation, with respect to the internal wave, was such that the minimum radar signal intensity, and lowest micro wave brightness temperature, occured close to the maximum of the internal wave thermocline depression. This is at the maximum horizontal current near the surface.

The research team did further note from their radar experiments that the direction of the measured near-surface current was opposite to that of the Doppler velocity excursions in the weakly reflecting region of the internal waves. This supported the interpretation that observed modulations in Doppler velocity excursions were not caused directly by the internal wave induced current. To choose between different candidate hydrodynamic prosesses they suggested that one should consider the main qualitative features of the signal modulation, in particular its depth and phasing between radar modulation and

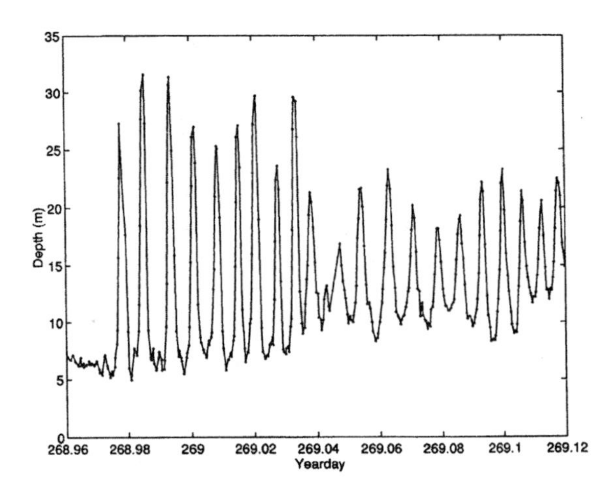

Figure 3. Internal solitary waves may have very large amplitudes that much exceed the thinner layer thickness. Depth of the 14° isoterm vs. time in days. The Coastal Ocean Probe Experiment (COPE) exhibited wave speed in the range about 0.7–0.9 m/s and wave induced surface current in the range 0.5–0.8 m/s. Surface waves with corresponding group velocities have wave lengths in the range of 0.8–2 m between solitons. Illustration reproduced from figure 3 in Kropfli et al. (1999), with permisson by Prof. L. A. Ostrovsky.

internal wave induced current variations.

Reduction of the surface wave amplitude caused by internal wave induced surface current. The interaction between surface and internal waves was studied by Gargett and Hughes (1972) on the basis of wave modulation due to a current. Hughes and Grant (1978) performed an extensive measurement campaign on the interaction between internal waves and surface waves. They provided experimental support for the idea that the surface slope statistics is closely related to the internal wave currents. Hughes (1978) outlined a linear theory for the interaction between the surface wave spectrum and the internal wave induced surface current. The theory shows that the modulation is in antiphase with internal wave induced current. This means that the surface waves are suppressed over internal solitary waves when surface and internal waves propagate in the same direction. The observations by Hughes and Gower (1983) led to the conclusion that strongest surface wave modulation occurs for those components propagating parallel to its crest. They suggested that internal waves act primarily on surface gravity waves of a few decimeters to a few meters in wavelength, causing modulation of the gravity-capillary waves. This again contributes to scattering at micro wave frequencies. Because more intense surface gravity waves tend to break and form capillary ripples or enhance the scattering, radar signal minima will coincide with those of the suppressed surface gravity waves (figure 4). In the case when the effect of wind is strong, the surface wave maximum will move toward the forward edge of the internal wave where the surface current gradient is maximum, a tendency that was been observed in parts of the COPE experiment.

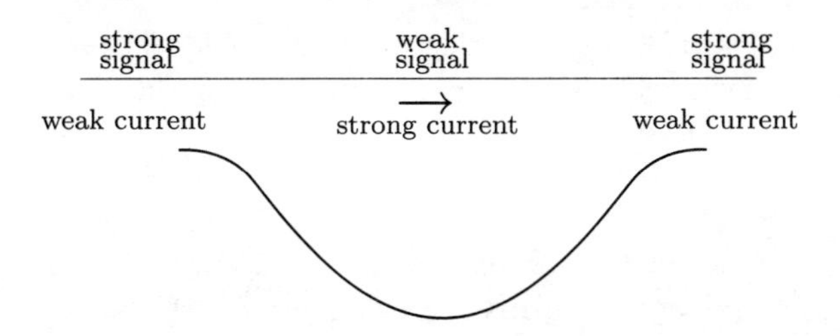

Figure 4. The surface waves – and the micro reflection – become reduced by a surface current in the direction of the waves.

The effect of a local current on the surface wave field may be illustrated in a simple way. Following an argument by Unna (1942), the wave frequency σ is constant when observed in the fixed frame of reference

$$\sigma = k(c+U) = k_0 c_0 = \text{const.} \tag{1.1}$$

In deep water, with $c^2 = g/k$ and $c_0^2 = g/k_0$, the relation becomes,

$$\frac{c}{c_0} + \frac{U}{c_0} = \frac{k_0}{k} = \frac{c^2}{c_0^2}, \tag{1.2}$$

giving

$$\frac{c}{c_0} = \frac{1}{2}\left[(1 + \left(1 + 4\frac{U}{c_0}\right)^{1/2}\right]. \tag{1.3}$$

Longuet-Higgins and Stewart (1961) showed that the work done by the rate of strain against the radiation stress gives the effect of the current variation on the wave energy, see also section 6.1 in Ch. 3 of the volume, i.e.

$$\frac{\partial}{\partial x}\Big[E(c_g + U)\Big] = -S_x \frac{\partial U}{\partial x}. \tag{1.4}$$

Here E denotes the wave energy, c_g group velocity, U current speed and $S_x = E(\frac{c}{c_g} - \frac{1}{2})$ radiation stress. In deep water $c_g = \frac{1}{2}c$ and $S_x = \frac{1}{2}E$. An exact integral of (1.1) and (1.4) reads, assuming deep water,

$$E(\tfrac{1}{2}c + U)c = \text{const.} \tag{1.5}$$

With $E = \frac{1}{2}\rho g a^2$ (ρ density, g acceleration of gravity, a wave amplitude), this gives

$$\frac{a}{a_0} = \left[\frac{c_0^2}{c(c+2U)}\right]^{1/2}, \tag{1.6}$$

where index zero refers to position where $U = 0$. The combination of (1.3) and (1.6) gives that $a/a_0 \sim 1 - 2U/c_0$ to leading order in U/c_0. This means that surface current moving in direction of surface waves weakens the wave field. The amplitude of the short waves become particularly reduced. In the case when the surface waves are opposing the internal waves the modulation is always such that the surface waves are amplified over internal solitary waves. The waves are blocked if the group velocity is less than the wave induced surface current.

The effect of surface active films. Surfactants on the sea surface have a wave reducing effect. Horizontal currents induced by internal waves may redistribute the surfactant – the film – causing surface wave modulations. A conservation law for the film concentration, Θ assumes the form

$$\Theta_t + (\Theta U)_x = 0, \tag{1.7}$$

where U denotes the internal wave induced current at the free surface. Assuming that $U = U(x - ct)$ the film concentration becomes $\Theta(x - ct)$ (with $U < c$)

$$\Theta = \frac{\Theta_0}{1 - \frac{U}{c}}. \tag{1.8}$$

This relation shows that the film concentration Θ has a maximum where U is maximum, i.e. over the depression maxima of internal solitary waves. The short wave damping rate is the greatest where film concentration is the greatest. The result of the film concentration should be observed as a minimum in the radar signal directly over the internal solitary waves, and agrees with the observations in the COPE experiment.

1.7 Transportation of biological and geological material

In-situ registrations from ship and buoy show internal wave motion with fish as markers. An example is shown in figure 5. In that observation it was indicated that the fish was eating plankton that was drifting with the motion in the ocean. The traces of the individual fish motion were obtained by data processing routines and are visible in the figure by small streaks. It can be concluded that the fish served as markers of an oscillatory vertical fluid motion in the sea with amplitude of 20 m at average depth of 100 m. Satellite images from the area reveal internal wave motion in the form of dispersive wave trains with wavelengths 300–500 m. Calculations show that the waves have a speed of 30 cm/s and period of 16 minutes – significantly longer than what appears on the log of the echo-sounder. The linear formulae predict a horizontal surface velocity that somewhat exceeds the wave speed. This indicates a highly nonlinear wave motion.

It is evidenced that organisms in the sea benefit from the vertical velocities induced by the waves. It is suggested that organisms that swim vertically – in opposition to the physical flow – will experience ephemereal changes in concentration throughout the water column as the wave is passing by (Lennert-Cody and Franks, 1999).

Internal solitary waves carry larvae and sediments. The strong unidirectional horizontal fluid velocities that are induced by the wave pulses mean that the waves represent a transport mechanism horizontally in the ocean. The waves act as a pump of nutrient water masses that are carried from the deep ocean to depleted water in shallow regions

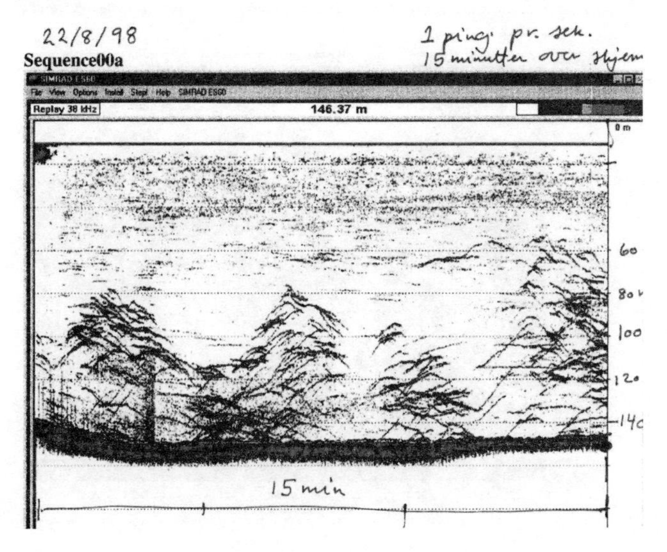

Figure 5. Registration of fish school by echo-sounders. Time variable along horizontal axis. Depth in meters. Total water depth at location: 146 m. Observation made in the Norwegian Sea at Sørøya in Finnmark, Norway. Position 71°N 22°10'E. Wavelength estimated from Syntetic Aperture Radar (SAR): 300-500 m, estimated wave speed: 30 cm/s. From Grue (2004).

and are important for the production taking place there (Stastna and Lamb, 2002; Carr and Davies, 2006).

1.8 Breaking internal waves and energy dissipation in the world's oceans

The large internal waves may become so strong that they break, see figure 6. Not only the wave pulses moving below the sea surface may become strong. Internal waves that are generated at pycnoclines in the deep ocean exhibit breaking during their interaction with submarine ridges or shelves. Reflectors placed on the moon on the Apollo 11 mission in 1969 determines accurately that moon is leaving earth by 4 cm per year (Dickey et al. 1994). This gives a precise estimate of the energy dissipation in the world's oceans. One third of this dissipation comes from tidal flow over topography, where internal wave breaking contributes with a significant portion (Munk and Wunsh, 1998). See also Ch. 6 of the volume.

1.9 Strong bottom currents due to internal waves

The offshore industry is facing new challenges when moving to deep water. An example is the gas field Ormen Lange which is a complete sub-sea development in deep water on the Norwegian shelf slope. The pipelines are laid in long spans along the uneven sea floor and may start to vibrate due to interaction with the relatively strong ocean currents. This has required the development of sub-sea forecasts of the currents and the

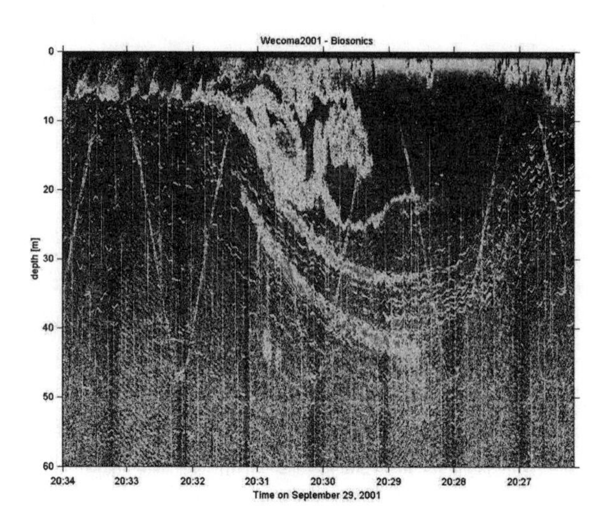

Figure 6. Very large internal wave motion may break. Observations by David Farmer et al., Wecoma 2001. Reproduced by permission by Prof. D. M. Farmer.

loads they impose on the pipelines.

At the location of Ormen Lange there is a horizontal thermocline at about 600 m depth. The thermocline separates an upper layer of warm, Atlantic water from a lower layer of denser water of sub-zero temperature. Very long internal waves of large amplitude can move along the thermocline. The strong currents at the sea floor are associated with internal wave motion and become magnified where the thermocline meets the shelf slope. Temperature measurements document that warm water may move down to the level of Ormen Lange. It has been measured that water with sub-zero temperature may move upward along the shelf slope, even up to the level of the shelf at 200 m depth, as illustrated in figure 7. The very long internal waves are running up the shelf slope, inducing an extreme fluid velocity of 0.5 m/s. The recording of a strong landward current event at the sea floor at Ormen Lange is illustrated in figure 8. Prior to the event the temperature at the sea floor exhibited a slow build-up phase reaching a maximum of 3 C°. This indicates that the thermocline had moved down the slope prior to the current event, beyond the measurement position. The sudden return of the temperature coincides with the rapid increase of the current at the sea floor. Atmospheric forcing is the most probable generation mechanism of the strong events (Alendal et al., 2005). Field measurements of near-bed currents at the shelf slope west of Norway are documented by e.g. Yttervik and Furnes (2004). Field measurements in the Faeroe-Shetland Channel reveal intense nonlinear wave trains propagating up the continental slope. Wave induced velocities near sea-bed are up to 0.5 m/s but have shorter duration than at Ormen Lange (Hosegood and van Haren, 2004; Helfrick, 1992).

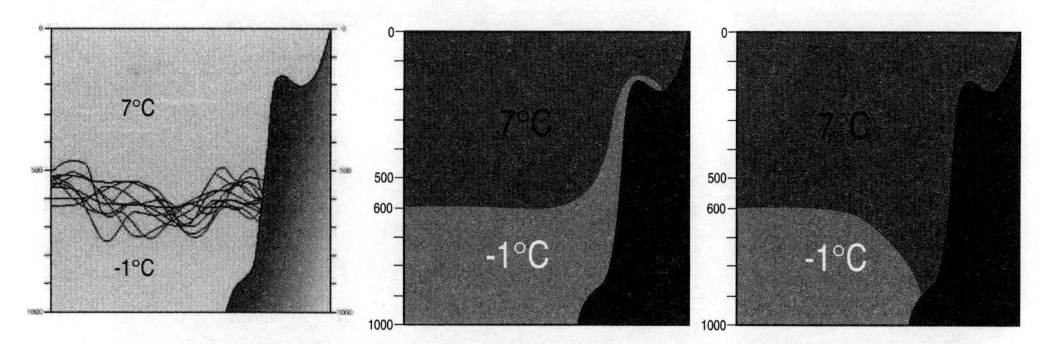

Figure 7. Motion of the 0°C isoterm at the Ormen Lange field (The Svinøy Section). Water depth in meters. Note that the horizontal distance in the figures is appr. 300 km. Left: Measurements summers 1977-1991 by Johan Blindheim. Mid and right: Strong run-up/down of pycnocline that separates warm Atlantic water of 7°C from cold polar water of -1°C.

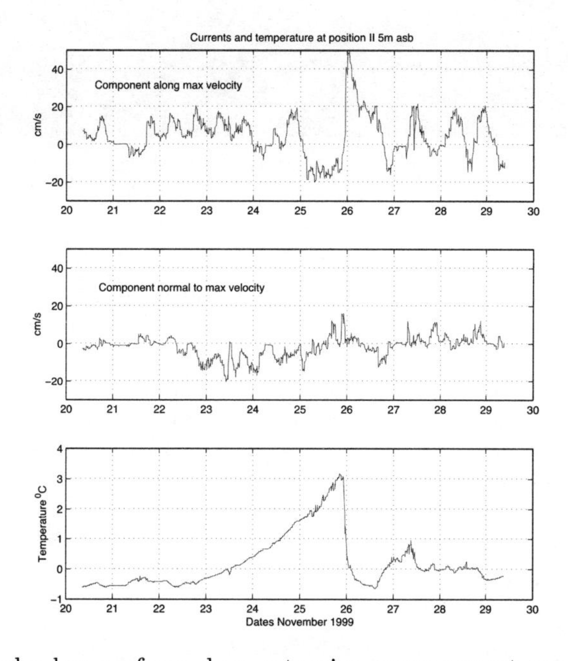

Figure 8. Norsk Hydro has performed an extensive measurement campaign of the currents along the trace of the pipelines. Strong current event at the position of Ormen Lange. Measurements 5 m above sea floor. Reproduced with permission by Norsk Hydro ASA.

Figure 9. The Ormen Lange gas field of Norsk Hydro. Ill: Norsk Hydro.

The gas-field Ormen Lange is located at 850 m depth, was discovered by Norsk Hydro in 1987, and opens summer 2007. Pipelines transport the gas up the slope and across the shelf to the landing terminal in Nyhamna close to the city of Molde on the coast of Norway (figure 9). The shelf slope is extremely uneven due to the submarine Storegga slide that occurred approximately 8000 years ago. The slide set up a big tsunami wave in the Norwegian Sea. The pipelines have a large number of free spans that are up to 500 m long (figure 10).

2 Long wave models

Linear theory of internal waves in the form of the motion of two fluids was first derived by Stokes in 1847 with nonlinear extensions derived by Boussinesq in the 1870s. In 1932 Dubreil-Jacotin derived the equations for nonlinear motion in a stratified fluid. Derivations of the Korteweg-de Vries equation for the motion in two-layer fluids were subsequently given by Keulegan (1953) and Long (1956). Weakly nonlinear and weakly dispersive theories are useful to model the internal wave motion when the lines of constant density – the isolines – have moderate excursions relative to the level at rest and the waves are long relative to depth.

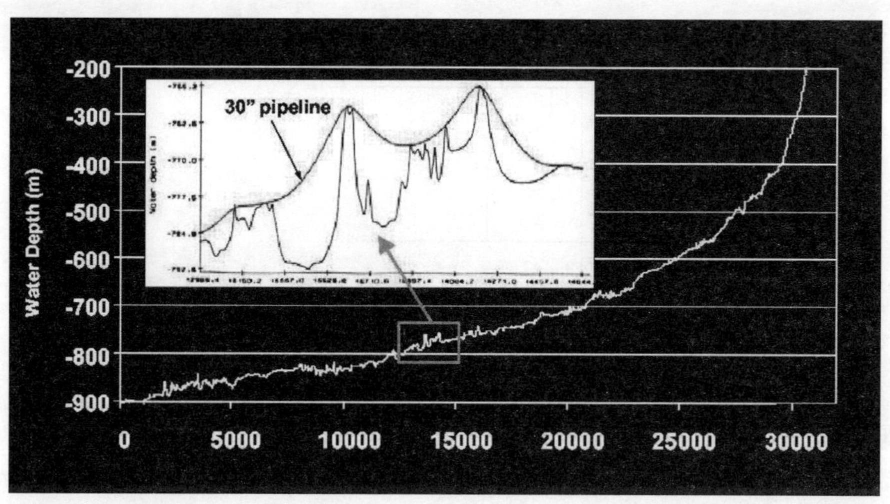

Figure 10. Pipelines of Ormen Lange. The sea-bed at Ormen Lange is extremely uneven. Ill: Norsk Hydro.

2.1 The Korteweg-de Vries equation

Continuous stratification. We revisit Benney (1966) who was the first to develop weakly nonlinear theory of internal waves in continuously stratified fluids leading to the Korteweg-de Vries equation and its higher-order extensions. A coordinate system Oxy is introduced where x is directed along the horizontal and y the vertical. The internal wave motion takes place in the fluid layer between two horizontal rigid walls at $y = 0$ and $y = -H$. The density field when there is no motion is determined by $\bar{\rho}(y)$. We consider two-dimensional motion with velocity field perturbation $\mathbf{v}(x, y, t) = (u(x, y, t), v(x, y, t))$ and density field perturbation $\rho(x, y, t)$ such that the total density is $\rho_s = \bar{\rho} + \rho$. The equations of continuity and motion read:

$$\rho_{st} + u\rho_{sx} + v\rho_{sy} = 0, \tag{2.1}$$

$$\rho_s(u_t + uu_x + vu_y) = -p_x, \tag{2.2}$$

$$\rho_s(v_t + uv_x + vv_y) = -p_y - g\rho_s, \tag{2.3}$$

$$u_x + v_y = 0, \tag{2.4}$$

where p denotes pressure and g acceleration of gravity. We let a denote the wave amplitude and λ wavelength. We introduce small parameters $\epsilon = a/H$ and $\mu = H^2/\lambda^2$ and look for a balance between nonlinear and dispersive effects, i.e. ϵ and μ are of the same order of magnitude. Dimensionless, primed variables are introduced by

$$u = Uu', \quad v = Vv', \quad p = \rho_0 gHp', \quad \rho = \rho_0\rho', \quad t = \omega^{-1}t', \tag{2.5}$$

where the balances $UH = V\lambda$ and $\omega\lambda = U$ are chosen. ρ_0 is a reference density. We further introduce the inverse Froude number $G = gh/U^2$ and a stream function $\psi'(x, y, t)$

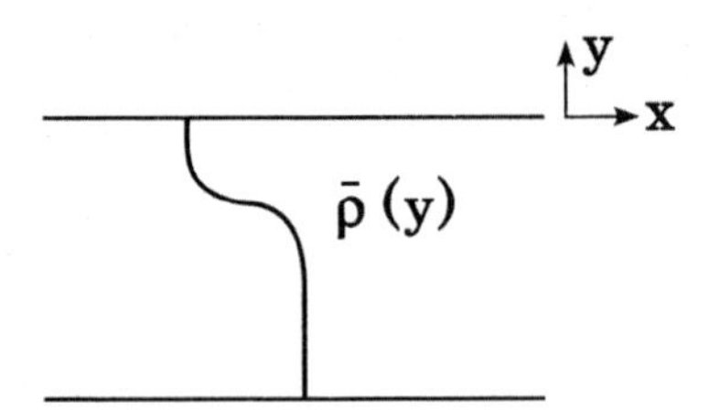

Figure 11. Density profile at rest.

such that $(u', v') = (\psi'_y, -\psi'_x)$. From now on we work with non-dimensional quantities and drop the primes. The equations (2.1)–(2.3) become

$$\rho_t - \bar{\rho}_y \psi_x + \epsilon(\rho_x \psi_y - \rho_y \psi_x) = 0, \tag{2.6}$$

$$\begin{aligned}&(\bar{\rho}\psi_{yt})_y - G\rho_x + \epsilon[\rho\psi_{yt} + \bar{\rho}(\psi_y\psi_{xy} - \psi_x\psi_{yy})]_y + \mu\bar{\rho}\psi_{xxt}\\ &\quad+\epsilon^2[\rho(\psi_y\psi_{xy} - \psi_x\psi_{yy})]_y + \mu\epsilon[\rho\psi_{xt} + \bar{\rho}(\psi_y\psi_{xx} - \psi_x\psi_{xy})]_x\\ &\quad+\mu\epsilon^2[\rho(\psi_y\psi_{xx} - \psi_x\psi_{xy})]_x = 0.\end{aligned} \tag{2.7}$$

The rigid lid condition, $\psi_x = 0$ is applied at $y = 0, -1$. We seek wave solutions by introducing the amplitude function $A(x,t)$, and expand $\rho(x,y,t)$ and $\psi(x,y,t)$ by

$$\rho(x,y,t) = A\varrho(y) + \epsilon A^2\tilde{\rho}(y) + \mu A_{xx}\hat{\rho}(y) + ..., \tag{2.8}$$

$$\psi(x,y,t) = A\phi(y) + \epsilon A^2\tilde{\psi}(y) + \mu A_{xx}\hat{\psi}(y) + ... \tag{2.9}$$

These expansions are introduced into (2.6)–(2.7), giving to leading order in nonlinearity and dispersion

$$A_t + cA_x + \epsilon\alpha_0 AA_x + \mu\beta_0 A_{xxx} = 0, \tag{2.10}$$

and

$$(\bar{\rho}\phi_y)_y - \frac{G\bar{\rho}_y}{c^2}\phi = 0, \qquad \phi(0) = \phi(-1) = 0. \tag{2.11}$$

The eigenvalue problem has an infinite set of eigenvalues $c_0 > c_1 > c_2 > ... > 0$ with corresponding eigenfunctions $\phi_0(y), \phi_1(y), \phi_2(y), ...$. The eigenvalues and eigenfunctions represent the wave speed and vertical structure of the n-th mode linear, hydrostatic internal wave motion. We shall here be interested only in the lowest mode corresponding to (c_0, ϕ_0), where these quantities are determined from (2.11).

The boundary value problems for $\tilde{\psi}(y)$ and $\hat{\psi}(y)$ are used to determine the constants α_0 and β_0 in (2.10), i.e.

$$(\bar{\rho}\tilde{\psi}_y)_y - \frac{G\bar{\rho}_y}{c^2}\tilde{\psi} = -\alpha_0\frac{G\bar{\rho}_y}{c^3}\phi + \frac{1}{2c}(\bar{\rho}_y\phi\phi_y)_y + \frac{1}{2c}[\bar{\rho}(\phi_y^2 - \phi\phi_{yy})]_y - \frac{1}{2c}\bar{\rho}_{yy}\phi^2, \tag{2.12}$$

$$(\bar{\rho}\hat{\psi}_y)_y - \frac{G\bar{\rho}_y}{c^2}\hat{\psi} = -\bar{\rho}\phi - \frac{2\beta_0 G\bar{\rho}_y\phi}{c^3}, \tag{2.13}$$

with $\tilde{\psi}(0) = \tilde{\psi}(-1) = 0$, and $\hat{\psi}(0) = \hat{\psi}(-1) = 0$ (and subscript zero is omitted for c and ϕ). The boundary value problems (2.12)–(2.13) have unique solutions provided that the coefficients α_0 and β_0 have values such that

$$\int_{-1}^{0} \phi[(\bar{\rho}\tilde{\psi}_y)_y - \frac{G\bar{\rho}_y}{c^2}\tilde{\psi}]dy = 0, \tag{2.14}$$

$$\int_{-1}^{0} \phi[(\bar{\rho}\hat{\psi}_y)_y - \frac{G\bar{\rho}_y}{c^2}\hat{\psi}]dy = 0. \tag{2.15}$$

Introducing the r.h.s. of (2.12) into (2.14) and the r.h.s. of (2.13) into (2.15) we obtain for the latter

$$\int_{-1}^{0} \phi[-(\bar{\rho}\phi - \frac{G\bar{\rho}_y}{c^3}\phi]dy = 0. \tag{2.16}$$

From (2.11) we have that $G\bar{\rho}_y\phi/c^2 = (\bar{\rho}\phi_y)_y$ which gives $2\beta_0 \int_{-1}^{0} \bar{\rho}\phi_y^2 dy = c_0 \int_{-1}^{0} \bar{\rho}\phi^2 dy$ where partial integration is used. A similar relation is derived from (2.14). Returning to dimensional variables (where ϵ and μ are put to unity in eq. 2.10), the coefficients are determined by

$$\alpha_0 = \frac{3}{2}\frac{\int_{-H}^{0} \bar{\rho}\phi_y^3 dy}{\int_{-H}^{0} \bar{\rho}\phi_y^2 dy}, \tag{2.17}$$

$$\beta_0 = \frac{c_0}{2}\frac{\int_{-H}^{0} \bar{\rho}\phi^2 dy}{\int_{-H}^{0} \bar{\rho}\phi_y^2 dy}. \tag{2.18}$$

Extended variants of the Korteweg-de Vries equation have been developed by Lee and Beardsley (1974), Miles (1979, 1981), Lamb and Yan (1996), Stanton and Ostrovsky (1998), Ostrovsky and Grue (2003).

Two-layer (interfacial) case. In the case when a sharp pycnocline is separating an upper layer of fluid with depth h_2 and density ρ_2 from a lower layer with depth h_1 and density ρ_1 the coefficients in equation (2.10) (with $\epsilon = 1$ and $\mu = 1$) read

$$\alpha_0 = -\frac{3}{2}\frac{c_0(\rho_2 h_1^2 - \rho_1 h_2^2)}{h_1 h_2(\rho_2 h_1 + \rho_1 h_2)} \simeq -\frac{3}{2}\frac{c_0(h_1 - h_2)}{h_1 h_2}, \tag{2.19}$$

$$\beta_0 = \frac{1}{6}\frac{c_0 h_1 h_2(\rho_1 h_1 + \rho_2 h_2)}{\rho_2 h_1 + \rho_1 h_2} \simeq \frac{1}{6} c_0 h_1 h_2, \tag{2.20}$$

$$c_0^2 = \frac{g h_1 h_2(\rho_1 - \rho_2)}{\rho_2 h_1 + \rho_1 h_2} \simeq \frac{g' h_1 h_2}{h_1 + h_2}, \tag{2.21}$$

where $g' = g(\rho_1 - \rho_2)/\rho_1$. The approximations to the right of the above equations are valid for $(\rho_1 - \rho_2)/\rho_1 << 1$.

An extension of the KdV equation has been explored by Ostrovsky and Grue (2003) in the following way:

$$A_t + c(A)A_x + [\beta(A)A_{xx}]_x = 0, \tag{2.22}$$

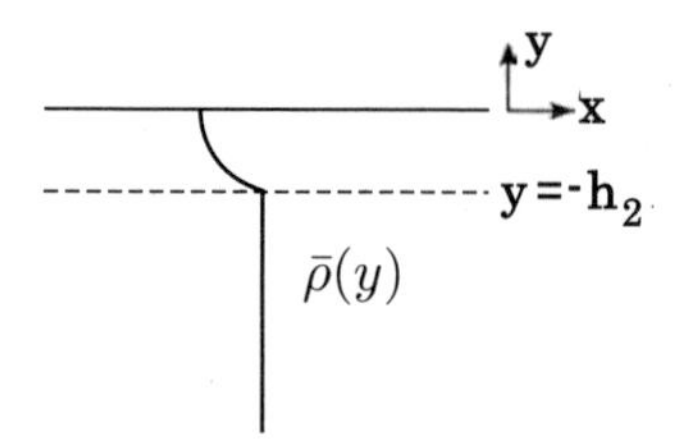

Figure 12. Density profile in the thin upper part of an infinitely deep fluid.

where

$$\beta(A) = \tfrac{1}{6}c(A)(h_2 + A)(h_1 - A), \tag{2.23}$$

and $c(A)$ is obtained by integrating

$$\frac{du_2}{dA} = \frac{c - u_2}{h_2 + A}, \tag{2.24}$$

$$c(A, u_2) = \frac{1}{H}\left[\frac{(h_1 - A)^2 - (h_2 + A)^2}{h_1 - A}u_2 + \sqrt{g'H(h_1 - A)(h_2 + A) - \frac{u_2^2 H^2 (h_2 + A)}{h_1 - A}}\right], \tag{2.25}$$

where $H = h_1 + h_2$. The nonlinear equation (2.22) quite successfully reproduces the highly nonlinear internal wave observations in the COPE experiment, see figure 21.

2.2 The Benjamin-Ono equation

In the case when the density has a variation in a thin region of depth h_2 of an infinitely deep fluid with constant density ρ_1 for $y < -h_2$, see figure 12, the evolution equation takes the form (Benjamin, 1967; Ono, 1975)

$$A_t + c_0 A_x + \alpha_0 A A_x - \beta_0 \frac{\partial^2}{\partial x^2}\frac{1}{\pi} PV \int_{-\infty}^{\infty} \frac{A(x', t)}{x - x'} dx', \tag{2.26}$$

where PV denotes principal value. The coefficient α_0 is determined by (2.17) and β_0 by

$$\beta_0 = \frac{\rho_1}{2\int_{-h_2}^{0} \bar{\rho}\phi_y^2 dy}. \tag{2.27}$$

The eigenfunction ϕ and wave speed c_0 are determined by

$$(\bar{\rho}\phi_y)_y - \frac{g\bar{\rho}_y}{c_0^2}\phi = 0, \qquad -h_2 < y < 0, \tag{2.28}$$

$$\phi(0) = 0, \qquad \phi(-h_2) = 1, \qquad \phi_y(-h_2) = 0. \tag{2.29}$$

2.3 The intermediate-depth equation

In the case of weakly nonlinear waves with length that is shorter than the depth of the fluid layer, Joseph (1977) and Kobuta, Ko and Dobbs (1978) developed an equation that in the case when $h_2/h_1 << 1$ and $\lambda/h_2 >> 1$ may be expressed in the following form:

$$A_t + c_0 A_x + \alpha_0 A A_x - \beta_0 \frac{\partial^2}{\partial x^2} PV \int_{-\infty}^{\infty} \frac{A(x'/h_1, t)}{\tanh\left(\frac{\pi}{2}\frac{x-x'}{h_1}\right)} dx' = 0, \tag{2.30}$$

where the coefficients c_0 and α_0 are determined as for the Benjamin-Ono equation, and β_0, in the case the pycnocline is sharp by (see Ostrovsky and Stepanyants, 2005)

$$\beta_0 = \frac{c_0 h_2}{4 h_1} \tag{2.31}$$

and $(\rho_1 - \rho_2)/\rho_1 << 1$.

2.4 Weakly nonlinear solitary waves

KdV soliton. Stratified case. In the case when there is balance between the effects of nonlinearity and dispersion, the equation (2.10) has solitary wave solutions of permanent shape. For the longest wave mode we obtain

$$A_0 = a_0 \operatorname{sech}^2[(x - ct)/\lambda], \tag{2.32}$$

where $c = c_0 + \Delta c$ and Δc denotes the excess speed. The coefficients are related by

$$\frac{1}{\lambda^2} = \frac{a_0 \alpha_0}{12 \beta_0}, \qquad \Delta c = \frac{a_0 \alpha_0}{3}, \tag{2.33}$$

where α_0 and β_0 are given in (2.17)–(2.18). While β_0 is always positive, α_0 may be positive or negative, depending on the stratification $\bar{\rho}(y)$. The product $a_0\alpha_0$ is always positive, however. This means that a positive α_0 gives positive a_0 and the wave is an elevation wave, while a negative α_0 gives a negative a_0 corresponding to a depression wave.

Interfacial KdV soliton. Various interfacial KdV models have been applied to investigate weakly nonlinear interfacial waves (Keulegan, 1953; Long, 1956; Koop and Butler, 1981; Segur and Hammack, 1982; Kao, Pan and Renouard, 1985). Consider the two-layer case with an upper layer of thickness h_2 and density ρ_2 and a lower layer of thickness h_1 and density $\rho_1 > \rho_2$. The coefficients of the solitary wave may be found in Long (1956, Part II):

$$c = c_0 + \Delta c,$$

$$\Delta c = -\frac{a_0 c_0 (\rho_2 h_1^2 - \rho_1 h_2^2)}{2 h_2 h_1 (\rho_2 h_1 + \rho_1 h_2)} \simeq -\frac{1}{2} a_0 c_0 \frac{h_1 - h_2}{h_1 h_2}, \tag{2.34}$$

$$\frac{1}{\lambda^2} = -\frac{3 a_0 [1 - \rho_1 h_2^2/(\rho_2 h_1^2)]}{4 h_2^2 h_1 (\rho_1/\rho_2 + h_2/h_1)} \simeq -\frac{3 a_0}{4} \frac{h_1 - h_2}{h_1^2 h_2^2}, \tag{2.35}$$

where the approximations to the right of the equations are valid for $(\rho_1 - \rho_2)/\rho_1 << 1$ and c_0 is given by (2.21). We note that $a_0 < 0$ when $\rho_2 h_1^2 - \rho_1 h_2^2 > 0$ and $a_0 > 0$ when $\rho_2 h_1^2 - \rho_1 h_2^2 < 0$. The wave induced fluid velocity at the maximum depression of the wave is obtained from KdV theory, giving for the velocity profile in the upper layer, i.e. for $-h_2 < y < 0$,

$$\frac{u_2(x=ct,y)}{c_0} \simeq \left[-\frac{\hat{\eta}}{h_2}\left[1+\frac{\Delta c}{c_0}\right] - \frac{\hat{\eta}^2}{h_2^2} - \frac{h_2\hat{\eta}''}{6} - \frac{h_2(\hat{\eta}')^2}{3} + \frac{\hat{\eta}'' y^2}{2h_2} \right]_{x=ct} \tag{2.36}$$

and for $-(h_1+h_2) = -H < y < -h_2$,

$$\frac{u_1(x=ct,y)}{c_0} \simeq \left[\frac{\hat{\eta}}{h_1}\left[1+\frac{\Delta c}{c_0}\right] - \frac{\hat{\eta}^2}{h_1^2} + \frac{h_1\hat{\eta}''}{6} - \frac{h_2(\hat{\eta}')^2}{3} - \frac{\hat{\eta}''(y+H)^2}{2h_1} \right]_{x=ct} \tag{2.37}$$

where Δc is specified above in (2.34), $\hat{\eta}(x,t) = a_0 \operatorname{sech}^2[(x-ct)/\lambda]$ ($a_0 < 0$ for the depression wave), $\hat{\eta}' = \partial\hat{\eta}/\partial x$ and terms $\mathcal{O}(a_0^3)$ are left out.

Algebraic soliton. Steady wave solution of the Benjamin-Ono equation (2.26) takes the form of an algebraic soliton:

$$\eta(x,t) = \frac{a_0}{1+(x-ct)^2/\lambda^2}, \tag{2.38}$$

where $c = c_0 + \frac{1}{4}a_0\alpha_0$ and $\lambda = 4\beta_0/(a_0\alpha_0)$. In the two-layer case the coefficients become (with $h_1 = \infty$ and $a_0 < 0$)

$$a_0\alpha_0 = -\frac{3a_0c_0}{2h_2}, \qquad \beta_0 = \frac{c_0h_2}{2}\frac{\rho_1}{\rho_2}, \tag{2.39}$$

giving that

$$c = c_0\left(1 - \frac{3a_0}{8h_2}\right), \qquad \lambda = -\frac{4h_2^2}{3a_0}\frac{\rho_1}{\rho_2}, \tag{2.40}$$

and c_0 is given by (2.21) in the limit $h_1 \to \infty$.

Intermediate depth soliton. The intermediate depth equation (2.30) has a solitary solution that may be expressed by (see Ostrovsky and Stepanyants, 2005)

$$\eta(x,t) = \frac{A_0\left(1+\cos\frac{2h_1}{\lambda}\right)}{1+\cos\frac{2h_1}{\lambda} + 2\sinh^2\frac{x-ct}{\lambda}}, \tag{2.41}$$

$$A_0 = -\frac{4}{3}\frac{h_2^2}{\lambda}\frac{\sin\frac{2h_1}{\lambda}}{1+\cos\frac{2h_1}{\lambda}}, \qquad c = c_0\left(1 - \frac{h_2}{\lambda\tan\frac{2h_1}{\lambda}}\right). \tag{2.42}$$

In the limit $h_1/\lambda \to 0$ (waves much longer than the fluid depth) the intermediate depth soliton tends to the KdV soliton. In the limit $h_1 \to \infty$ it tends to the BO-soliton. It has

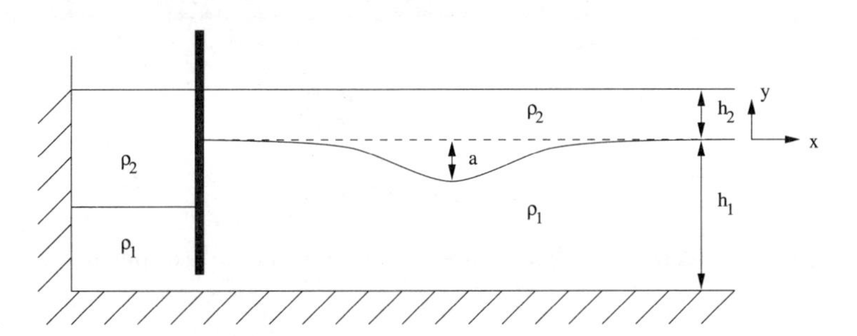

Figure 13. Sketch of wave tank experiments on internal solitary waves. Lower and upper layer depths: 62 cm and 15 cm. Densities: $\rho_1 \simeq 1.02$ g/cm^3, $\rho_2 \simeq 1$ g/cm^3. From Grue et al. (1999). Reproduced with permission by J. Fluid Mech. The first known experiment on internal waves was performed by Benjamin Franklin in 1762.

been documented that the intermediate depth soliton has very limited range of validity in terms of nonlinearity and is less useful as a model of the waves observed in the field as compared to the KdV soliton, which is a useful model for small to moderately large excursions, however.

3 Fully nonlinear interfacial solitary waves

Fully nonlinear and fully dispersive modelling is motivated by the very large excursions of the thermoclines in the ocean observed e.g. in the Celtic Sea and the COPE experiment (see figures 2 and 3). The motion is generally outside the range of validity of the integrable equations, and falls into the cathegory of non-integrable systems. We shall here discuss the motion of an interface separating two fluids of different density. The motion is two-dimensional. The upper fluid layer has thickness h_2 at rest and constant density ρ_2, and the lower layer has thickness h_1 at rest and constant density ρ_1, where ρ_1 is larger than ρ_2. Hereafter, index 1 refers to the lower fluid, and index 2 to the upper. From now on the coordinate system Oxy has x-axis along the interface at rest and y-axis pointing upward. Unit vectors $\mathbf{i}, \mathbf{j}$ are introduced accordingly. See figure 13 for illustration. We assume that the two fluids are homogeneous and incompressible and that the motion in each of the layers is irrotational such that the velocities may be obtained by potential theory. Description follows Grue et al. (1999) and Dold and Peregrine (1985). Other literature includes Turner and Vanden-Broeck (1988), Pullin and Grimshaw (1988), and Evans and Ford (1996).

We shall first consider wave solutions that propagate with permanent form and speed c. The motion is evaluated in the frame of reference that is moving with the wave. In this frame of reference there is a current with speed $-c$ in the far field directed along the negative x-axis. We assume that the wave induced velocity field is determined by

$$\mathbf{v}_1 = u_1\mathbf{i} + v_1\mathbf{j} = \nabla\phi_1, \qquad \mathbf{v}_2 = u_2\mathbf{i} + v_2\mathbf{j} = \nabla\phi_2, \tag{3.1}$$

where the potentials ϕ_1 and ϕ_2 vanish at $x = \pm\infty$. The potential functions satisfy the Laplace equation in their respective domains. The total velocity is given by $\mathbf{v}_j - c\mathbf{i}$.

Figure 14. Tangential velocities: $\gamma_1/|z_\xi|$ (lower layer) and $\gamma_2/|z_\xi|$ (upper layer).

3.1 Solution of the Laplace equation

It is convenient to employ complex function theory. Introducing the complex coordinate $z = x + \mathrm{i}y$ (i the imaginary unit), complex velocities are introduced by

$$q_j(z) = u_j(x, y) - \mathrm{i}v_j(x, y), \qquad j = 1, 2, \tag{3.2}$$

where $q_j(z)$ vanishes in the far field.

Let us apply Cauchy's theorem to the motion in the upper layer. Since q_2 tends to zero in the far field, we obtain for a point z' inside the fluid:

$$2\pi\mathrm{i}q_2(z') = \int_I \frac{q_2(z)\mathrm{d}z}{z - z'} + \int_I \frac{q_2^*(z^*)\mathrm{d}z^*}{z^* + 2\mathrm{i}h_2 - z'} \qquad \text{for } z' \text{ inside fluid 2,} \tag{3.3}$$

where in the first integral the integration is along the interface I (from minus to plus infinity) and the integration in the second term is an image term that accounts for an integration along the rigid lid boundary at $y = h_2$. A star means complex conjugate.

An integral equation to determine q_2 is obtained by letting z' approach the boundary I in (3.3). In the limit when z' is on I the factor 2π changes to π on the left hand side of (3.3), giving

$$\pi\mathrm{i}q_2(z') = PV\int_I \frac{q_2(z)\mathrm{d}z}{z - z'} + \int_I \frac{q_2^*(z^*)\mathrm{d}z^*}{z^* + 2\mathrm{i}h_2 - z'} \qquad \text{for } z' \text{ on } I, \tag{3.4}$$

where PV denotes the principal value. The equation is an integral equation for q_2.

A similar procedure is applied to the lower fluid. For a position z' in the interior of the fluid we obtain

$$-2\pi\mathrm{i}q_1(z') = \int_I \frac{q_1(z)\mathrm{d}z}{z - z'} + \int_I \frac{q_1^*(z^*)\mathrm{d}z^*}{z^* - 2\mathrm{i}h_1 - z'} \qquad \text{for } z' \text{ inside fluid 1,} \tag{3.5}$$

where the integration along I is from minus to plus infinity. By letting z' approach the boundary I from the side corresponding to fluid 1, we obtain

$$-\pi\mathrm{i}q_1(z') = PV\int_I \frac{q_1(z)\mathrm{d}z}{z - z'} + \int_I \frac{q_1^*(z^*)\mathrm{d}z^*}{z^* - 2\mathrm{i}h_1 - z'} \qquad \text{for } z' \text{ on } I, \tag{3.6}$$

which is an integral equation for q_1.

We parameterize the interface by $z(\xi)$ and introduce the tangential and normal derivatives along the interface by

$$\frac{\partial}{\partial s} - \mathrm{i}\frac{\partial}{\partial n} = \frac{1}{|z_\xi|}\left(\frac{\partial}{\partial \xi} - \mathrm{i}\frac{\partial}{\partial \nu}\right), \tag{3.7}$$

where s denotes the arclength and n the unit normal pointing out of fluid 1. The tangential velocities along I is obtained by $(q_1(z) - c)z_\xi/|z_\xi|$ for the lower fluid and $(q_2(z) - c)z_\xi/|z_\xi|$ for the upper, where $z_\xi = \mathrm{d}z/\mathrm{d}\xi$.

It is convenient to define the quantities $\gamma_1 = (q_1(z) - c)z_\xi$ and $\gamma_2 = (q_2(z) - c)z_\xi$ where $\gamma_1/|z_\xi|$ and $\gamma_2/|z_\xi|$ determine the tangential velocities on the lower and upper side of the interface I, respectively, see figure 14.

We multiply equation (3.4) by z'_ξ $(= [\mathrm{d}z/\mathrm{d}\xi]_{\xi=\xi'})$ and insert $q_2 z_\xi = \gamma_2 + cz_\xi$ into the equation. Further, we make use of the relation $\gamma_2 = c(\Gamma_2 + x_\xi)$ where $\Gamma_2 \to 0$ in the far field. The real part of the equation becomes

$$-\pi \mathrm{i}\Gamma_2' = PV\int_I Re\left(\frac{z'_\xi}{Z} - \frac{z'_\xi}{H_2}\right) y_\xi \mathrm{d}\xi + \int_I Im\left(\frac{z'_\xi}{Z} + \frac{z'_\xi}{H_2}\right)\Gamma_2 \mathrm{d}\xi, \tag{3.8}$$

where $\Gamma_2' = \Gamma_2(z')$, $Z = z' - z$ and $H_2 = z^* + 2\mathrm{i}h_2 - z'$. A similar relation for the motion in the lower fluid is obtained, i.e.

$$\pi \mathrm{i}\Gamma_1' = PV\int_I Re\left(\frac{z'_\xi}{Z} - \frac{z'_\xi}{H_1}\right) y_\xi \mathrm{d}\xi + \int_I Im\left(\frac{z'_\xi}{Z} + \frac{z'_\xi}{H_1}\right)\Gamma_1 \mathrm{d}\xi, \tag{3.9}$$

where $\Gamma_1' = \Gamma_1(z')$ and $H_1 = z^* - 2\mathrm{i}h_1 - z'$. Γ_1 is obtained from $\gamma_1 = c(\Gamma_1 + x_\xi)$ where $\Gamma_1 \to 0$ in the far field.

The third relation to determine γ_1, γ_2 and the shape $z(\xi)$ of I is derived from the dynamic condition expressing that the pressure is continuous at the interface. The pressure in each of the fluids is obtained at each side of the interface by

$$p_j = -\tfrac{1}{2}\rho_j\left(\gamma_j^2/|z_\xi|^2 - c^2\right) - \rho_j g y, \qquad j = 1, 2. \tag{3.10}$$

Putting $p_1 = p_2$ at I gives

$$\frac{1}{2|z_\xi|^2}(\rho_1\gamma_1^2 - \rho_2\gamma_2^2) - \tfrac{1}{2}(\rho_1 - \rho_2)c^2 + (\rho_1 - \rho_2)gy = 0. \tag{3.11}$$

Numerical procedure for the fully nonlinear two-layer model. Let I_+ denote the part of I where $x \geq 0$. Symmetric wave profiles with respect to $x = 0$ are obtained by replacing $z(-x) = -z^*(-x)$ for $x < 0$. Equation (3.8) becomes

$$\begin{aligned}-\pi\Gamma_2' = PV\int_{I_+} & Re\left(\frac{z'_\xi}{Z} - \frac{z'_\xi}{H_2} - \frac{z'_\xi}{z' + z^*} - \frac{z'_\xi}{z - 2\mathrm{i}h_2 + z'}\right) y_\xi \mathrm{d}\xi \\ + \int_{I_+} & Im\left(\frac{z'_\xi}{Z} + \frac{z'_\xi}{H_2} + \frac{z'_\xi}{z' + z^*} - \frac{z'_\xi}{z - 2\mathrm{i}h_2 + z'}\right)\Gamma_2 \mathrm{d}\xi \quad \text{for } z' \text{ at } I_+.\end{aligned} \tag{3.12}$$

The similar equation is derived for Γ_1 with z' at I_+.

We discretize the interface I_+ by N points (with $\xi = 1, 2, 3, ..., N$) and perform the integration using the trapezoidal rule. For the principal value integrals we expand the integrand in a series in the vicinity of the pole at $z' = z$. The discrete variant of integral equation (3.12) becomes

$$-\pi\Gamma_2(\xi') = \sum_{\xi=1}^{N} (\mathcal{A}(\xi',\xi) - \mathcal{A}_2(\xi',\xi))\, y_\xi(\xi) - y_{\xi\xi}(\xi') + \sum_{\xi=1}^{N} (\mathcal{B}(\xi',\xi) - \mathcal{B}_2(\xi',\xi))\, \Gamma_2(\xi) \tag{3.13}$$

where

$$\mathcal{A}(\xi',\xi) + \mathrm{i}\mathcal{B}(\xi',\xi) = \begin{cases} z_{\xi\xi}/2z_\xi, & \xi = \xi' = 1 \\ \frac{1}{2}\left[z'_\xi/Z - \left(z'_\xi/(z'+z^*)\right)^*\right], & \xi = 1,\ \xi' > 1 \\ z'_\xi/Z - \left(z'_\xi/(z'+z^*)\right)^*, & \xi > 1,\ \xi' \neq \xi \\ z_{\xi\xi}/2z_\xi - \left(z'_\xi/(z'+z^*)\right)^*, & \xi > 1,\ \xi' = \xi \end{cases} \tag{3.14}$$

and

$$\mathcal{A}_2(\xi',\xi) - \mathrm{i}\mathcal{B}_2(\xi',\xi) = \begin{cases} \frac{1}{2}\left(z'_\xi/H_2 + \left(z'_\xi/(z-2\mathrm{i}h_2+z')\right)^*\right), & \xi = 1 \\ z'_\xi/H_2 + \left(z'_\xi/(z-2\mathrm{i}h_2+z')\right)^*, & \xi > 1. \end{cases} \tag{3.15}$$

An equidistant distribution in x is used in computations of non-overturning waves, which means that $x(\xi)$ is specified. For computations of overhanging waves it is convenient with an equally spaced distribution along the arclength, where $x(\xi)$ becomes part of the solution (see section 3.5). Derivatives with respect to ξ are obtained by a five-point Lagrangian differentiation formula. The set of equations is closed by either giving the amplitude a or the volume of the solitary wave.

3.2 Fully nonlinear computations in the small amplitude limit

The numerical solution procedure of the fully nonlinear equations is initiated either by the KdV or Benjamin-Ono (BO) soliton. In the case when one layer is very much thicker than the other we expect the fully nonlinear solution to follow the BO soliton. This is confirmed by computations of the wave width, defined at the level corresponding to the half of the wave amplitude. The fully nonlinear solution approaches the BO-asymptote when a is very small and h_1/h_2 is very large but departs from the asymptote for a/h_2 growing beyond 0.1, see the computations in figure 15a for $h_1/h_2 = 500$.

The fully nonlinear solution tends to the KdV asymptote in the case when h_1/h_2 is moderate (finite) and $a \to 0$. Computations performed with $h_1/h_2 = 12$ are visualized in figure 15b. The fully nonlinear solution significantly departs from the KdV asymptote when a/h_2 exceeds a value about 0.4. In the log-log plot, the wave width of the BO-solution appears as a straight line with slope $-\log a$, while the corresponding line of the KdV solution has slope $-\frac{1}{2}\log a$. The asymptotes will always intersect.

3.3 Solitary waves of large amplitude

With the fully nonlinear method we obtain accurate computations of interfacial solitary waves of large amplitude. The computations have been supported by a measurement

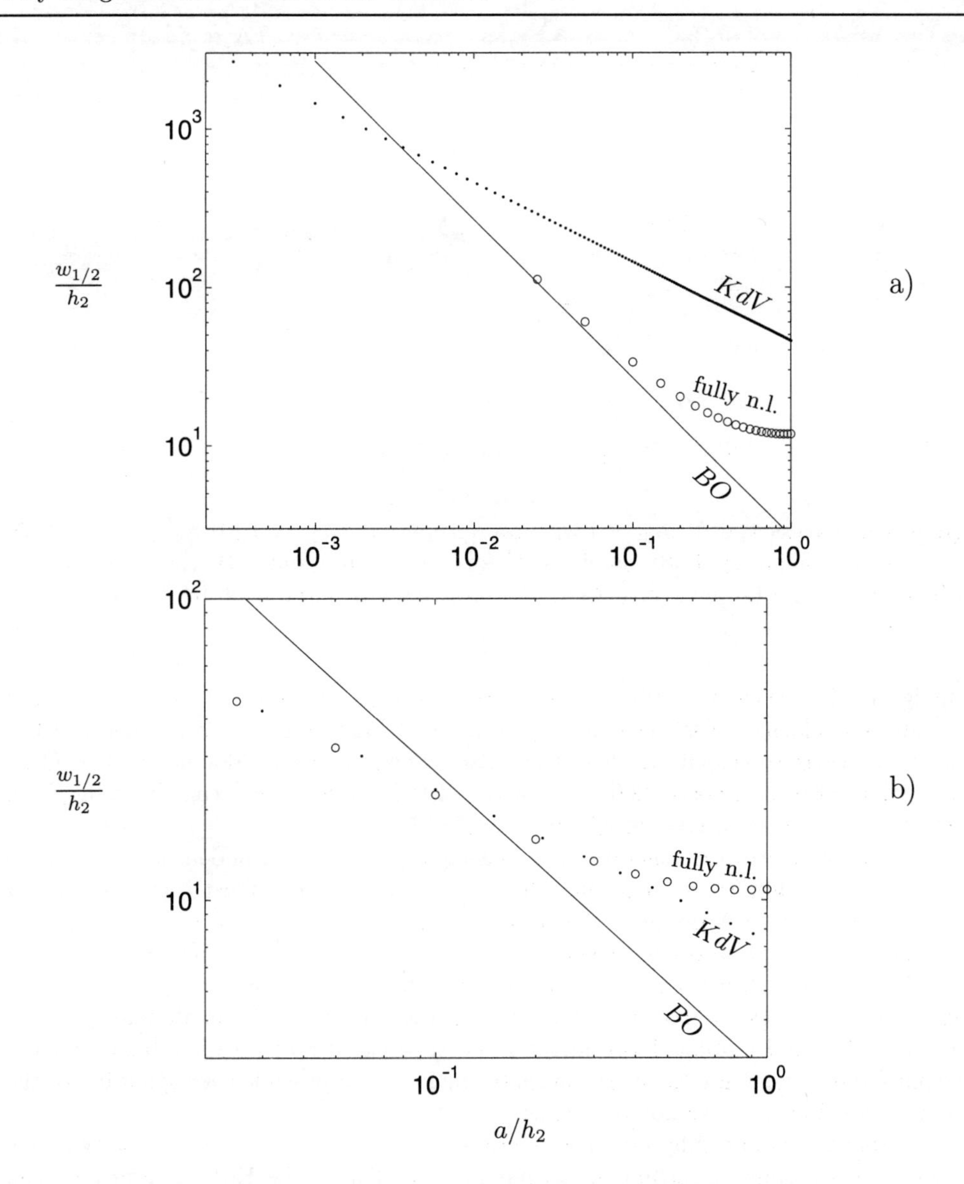

Figure 15. Wave width $w_{1/2}/h_2$ vs. a/h_2. KdV (dotted line), BO (solid line) and fully nonlinear ($\circ\ \circ\ \circ$). a) $h_1/h_2 = 500$ and b) $h_1/h_2 = 12$. From Grue (2005). Reproduced with permission by APS.

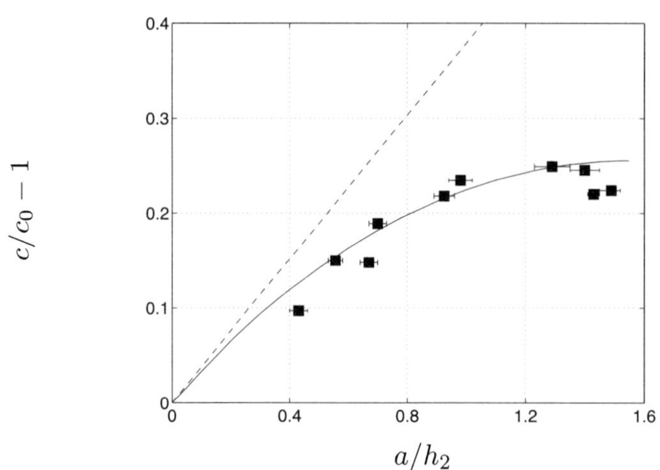

Figure 16. Excess speed $c/c_0 - 1$ vs. amplitude. $h_1/h_2 = 4.13$. $a_{max}/h_2 \simeq 1.55$. Measurements (squares). Fully nonlinear theory (solid line) and KdV theory, eq. (2.34) (dashed line). From Grue et al. (1999). Reproduced with permission by J. Fluid Mech.

campaign in the physical laboratory where properties like wave speed, wave shape and wave induced velocities were measured. The method Particle Image Velocimetry (PIV) gives highly accurate velocity fields of the waves. The campaign is documented by Grue et al. (1999, 2000) and Grue (2004). A sketch of the two-layer experiments using fresh water above a layer of brine is provided in figure 13.

The solitary waves were generated by releasing a pool of water of density corresponding to that of the thin layer. This generation mechanism is very robust since the solitary waves are self-focusing, a property that has been known since the first experiments were performed on solitary waves of elevation in a single layer fluid by John Scott Russell in the 1830s. An notable feature is that the excursions of the internal waves become much larger than those of free surface waves. This is because of the small density differences of the internal motion. Experimental and theoretical waves were obtained for the wave amplitudes ranging from small values to the maximal possible corresponding to the conjugate flow limit, see section 3.4 below.

The experiments and fully nonlinear computations of the wave phase velocity follow the same trend, while important deviations are noted from the KdV asymptote when a/h_2 exceeds the level of about 0.4 (figure 16). The excellent correspondence between experiments and fully nonlinear model is observed in the plots of the solitary wave shape (figure 17). Agreement with KdV theory is found when a/h_2 is small or moderate. The KdV wave length exhibits a monotoneous decay with amplitude, while the opposite is true in the fully nonlinear scenario.

For the largest possible wave (figure 17e) we note in particular the good correspondence between the theoretical prediction and the measurements in the front part of the wave. This confirms that the interfacial model provides useful prediction of the large internal waves moving along thin pycnoclines, up to the conjugate flow limit. Wave breaking develops from the location of maximum excursion and in the tail of the wave. The breaking is caused by shear instability. The estimated value of the Richardson number is 0.07 at the maximum excursion of the wave.

3.4 Solitary waves of maximum amplitude

The speed and excursion of interfacial solitary waves have theoretical upper bounds, c_{max} and δ_{max}, respectively, when h_1 and h_2 are finite (Amick and Turner, 1986). The volume of the wave may become infinitely large in this limit (Turner and Vanden-Broeck, 1988). The formulae for c_{max} and δ_{max} derived by Amick and Turner read

$$c_{max}^2 = \frac{g(h_1+h_2)(\rho_1-\rho_2)}{(\rho_1^{1/2}+\rho_2^{1/2})^2}, \quad \delta_{max} = \frac{h_2\rho_1^{1/2}-h_1\rho_2^{1/2}}{\rho_1^{1/2}+\rho_2^{1/2}}. \tag{3.16}$$

Alternatively, the expression for c_{max} may be written $c_{max}^2 = gh_1'h_2'(\rho_1-\rho_2)/(\rho_1 h_2' + \rho_2 h_1')$, where $h_2' = h_2 - \delta_{max}$ and $h_1' = h_1 + \delta_{max}$. The latter expression has precisely the same form as (2.21) for the linear long wave speed c_0^2, except that the local depths of the fluid layers are h_1' and h_2'.

The maximum wave induced fluid velocities in the layers are determined by conservation of mass, giving

$$h_2 c_{max} = h_2'(c_{max} - u_{2max}), \quad h_1 c_{max} = h_1'(c_{max} + u_{1max}). \tag{3.17}$$

The expressions for u_{2max} and u_{1max} become very simple in case $(\rho_1-\rho_2)/\rho_1 << 1$:

$$\frac{u_{2max}}{c_0} = -\frac{u_{1max}}{c_0} = \frac{h_1-h_2}{2(h_1h_2)^{1/2}}. \tag{3.18}$$

These expressions determine the maximum fluid velocities in figures 19–20. The broadening of the waves that is deduced from theory is observed in field measurements in the Celtic Sea and the COPE experiment (Pingree and Mardell, 1985; Stanton and Ostrovsky, 1998). The measured field data from COPE exhibit good correspondence with interfacial theory (figure 21). There is also good correspondence between observation and results of a modification of the KdV equation (Ostrovsky and Grue, 2003), taking into account the full, local layer thicknesses due to the propagation of the wave.

3.5 Overhanging waves

In the Boussinesq limit, and if breaking within the pycnocline can be avoided, the limiting form of solitary waves corresponds to the conjugate flow limit, i.e. broad wave which has an excursion extending to the mid-depth of the layer. This is true for interfacial waves of depression and of elevation. The conjugate flow limit is true also for interfacial depression waves with a finite (large) jump in the density (and rigid lids at the top and bottom of the layer). The limiting form of solitary waves of elevation exhibits an overhanging profile, however. This is true when the upper layer is much thicker than the lower one, and there is a finite density jump at the interface, see figure 22 for illustration. Experimental confirmation of the overhanging waves has not yet been given.

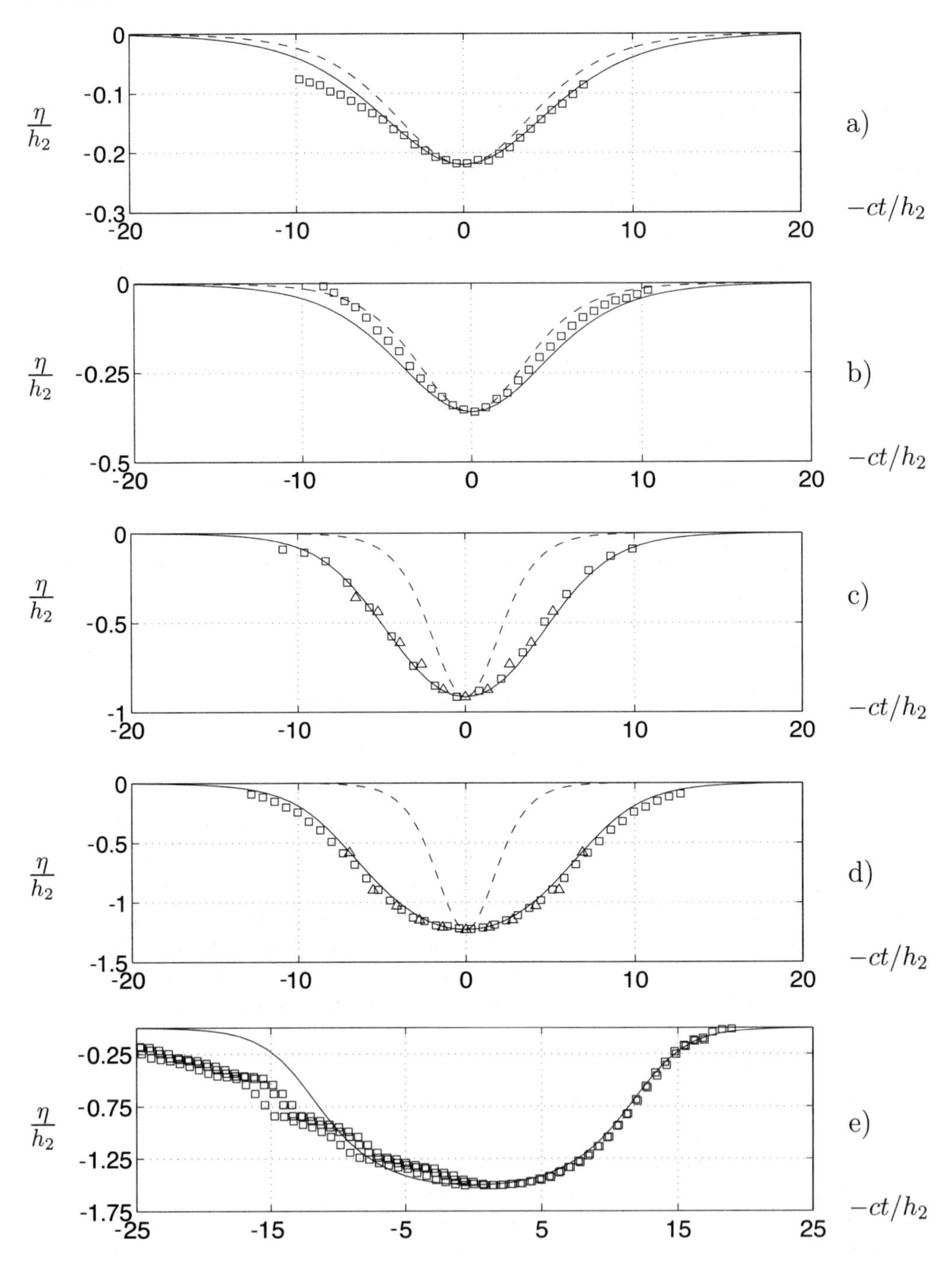

Figure 17. Wave profile $\eta(ct)$ at fixed position at camera 2 (10.5m). Squares: measured pycnocline. Triangles: measured jump in velocity. Solid line: fully nonlinear theory. Dashed line: KdV theory eq. (2.32). $h_1/h_2 = 4.13$. a) $a/h_2 = 0.22$. b) $a/h_2 = 0.36$. c) $a/h_2 = 0.91$. d) $a/h_2 = 1.23$. e) $a/h_2 = 1.51$. From Grue et al. (1999). Reproduced with permission by J. Fluid Mech.

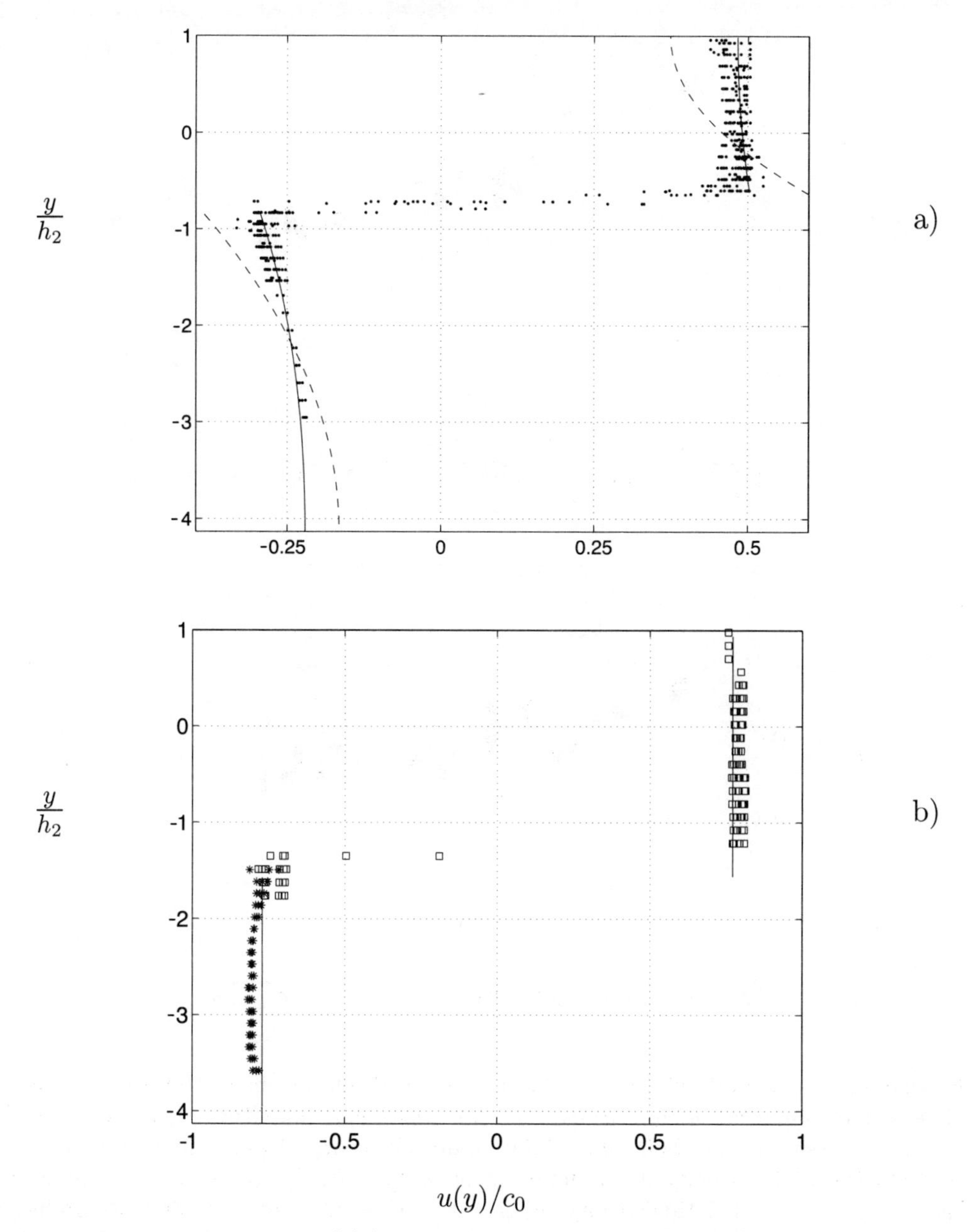

Figure 18. Velocity profile $u(y)/c_0$ at the maximum displacement of the wave vs. non-dimensional vertical coordinate. c_0 determined from eq. (2.21). Squares, stars or dots: measurements. Solid line: Fully nonlinear theory. $h_1/h_2 = 4.13$. Dashed line in a): KdV theory, eqs. (2.36)–(2.37). a) $a/h_2 = 0.78$. b) $a/h_2 = 1.51$. From Grue et al. (1999). Reproduced with permission by J. Fluid Mech.

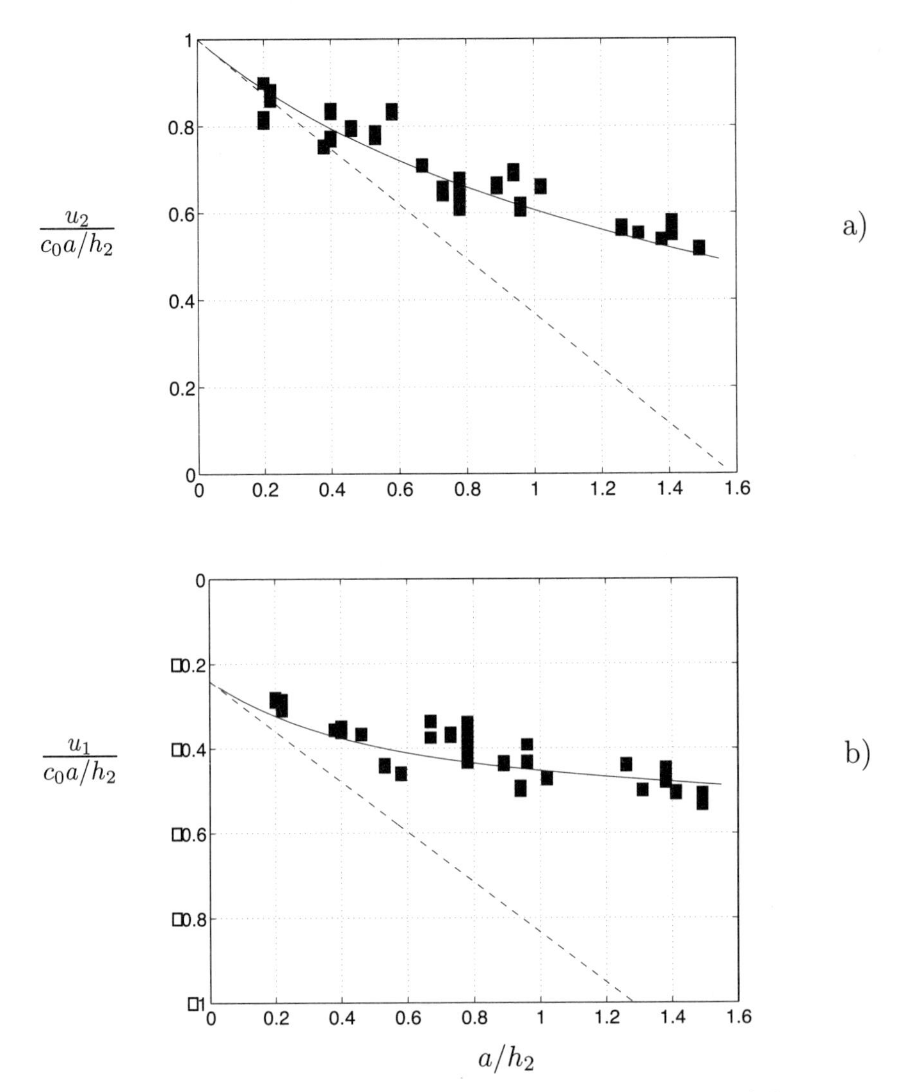

Figure 19. Horizontal velocities at the maximum displacement of the wave vs. non-dimensional amplitude a/h_2. $h_1/h_2 = 4.13$. Maximum amplitude $a_{max}/h_2 \simeq 1.55$. Squares: measurements. Solid line: fully nonlinear theory. Dashed line: KdV theory, eqs. (13)–(14). a) Velocity in the upper layer, $u_2(y = h_2/2)$. b) Velocity in the lower layer, $u_1(y = -a^-)$. For KdV theory: $u_1(y = 0^-)$. From Grue et al. (1999). Reproduced with permission by J. Fluid Mech.

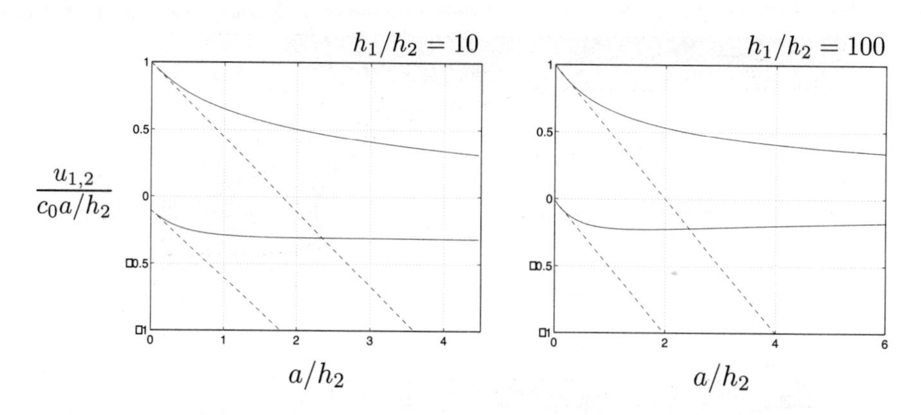

Figure 20. Horizontal velocities at the maximum displacement of the wave vs. non-dimensional amplitude a/h_2. Upper line in each figure: $u_2(y = h_2/2)$. Lower line in each figure: $u_1(y = -a^-)$. For KdV theory: $u_1(y = 0^-)$. Solid line: fully nonlinear theory. Dashed line: KdV theory eqs. (13)–(14). Left: $h_1/h_2 = 10$ ($a_{max}/h_2 \simeq 4.45$). Right: $h_1/h_2 = 100$ ($a_{max}/h_2 \simeq 49.2$). From Grue et al. (1999). Reproduced with permission by J. Fluid Mech.

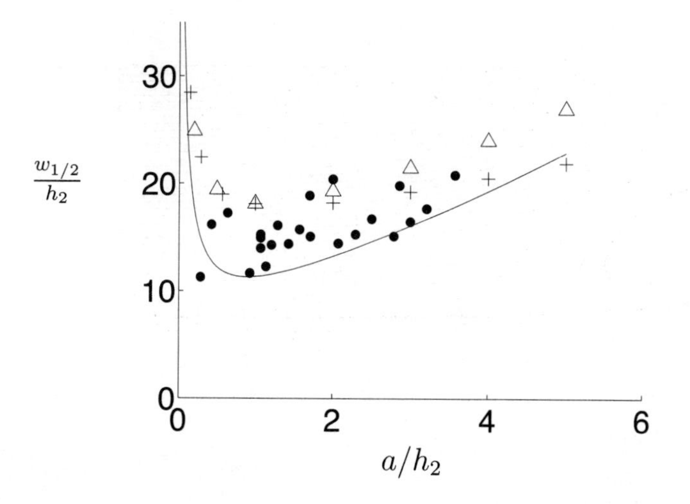

Figure 21. Wave width from the COPE experiment. From Ostrovsky and Grue (2003). Field observations (black small circles), fully nonlinear interfacial model, eqs. (3.8)–(3.9) (solid line), an extended KdV type model, eq. (2.22) (+), fully nonlinear long wave model by Choi-Camassa (1999) (Δ). Reproduced with permission by Phys. Fluids.

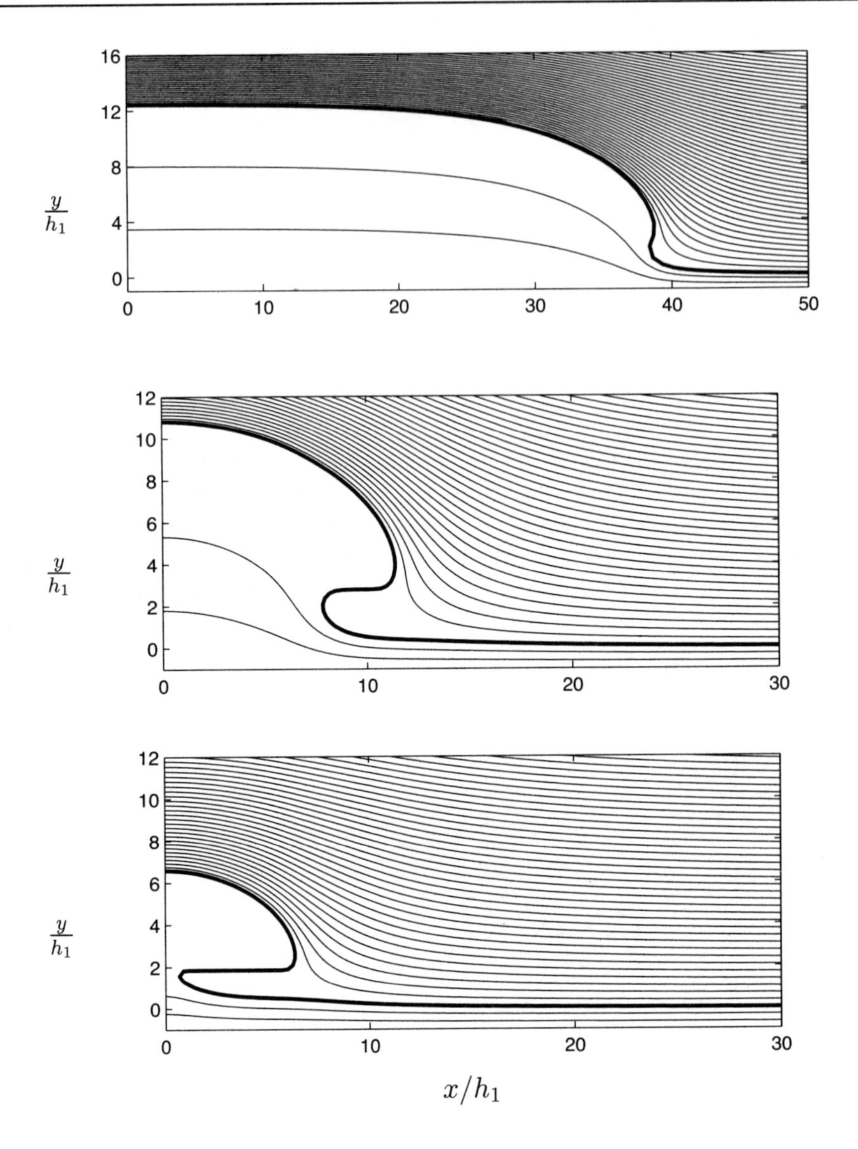

Figure 22. Extreme two-layer solutions with $\rho_2/\rho_1 = 0.4$. Upper plot: The tallest, broad solution found, for $h_2/h_1 = 21$. Mid: The maximum amplitude solution, with $h_2/h_1 = 30$. Lower: The most overhanging solution, with $h_2/h_1 = 30$. Propagation velocities: nonlinear, $c/c_0 = 2.2245, 3.0642, 2.4977$, and linear $c_0/\sqrt{gh_1} = 0.7673, 0.7695, 0.7695$ (upper, mid, lower). From Rusås and Grue, (2002). Reproduced with permission by Elsevier.

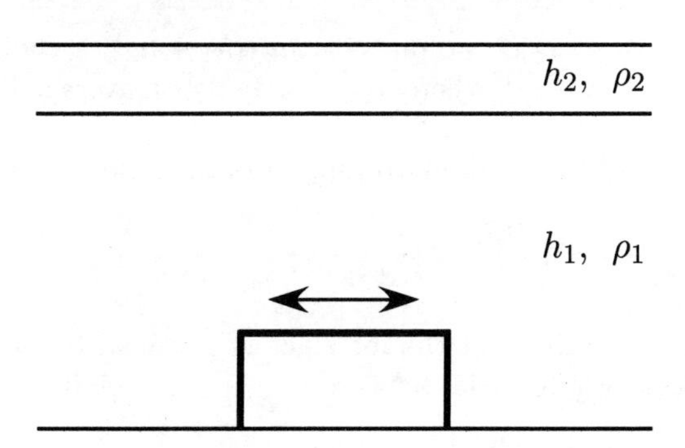

Figure 23. Sinusoidal motion of bottom geometry with horizontal velocity $U = U_0 \sin \Omega t$.

4 Transient computations of interfacial motion

Fully nonlinear and fully dispersive modelling of transient interfacial waves is outlined in this section. In the first part a two-dimensional motion of the fluid is assumed. Numerical examples illustrate the process of upstream formation of the solitary waves due to barotropic, tidal flow over submarine ridge. Description follows Grue et al. (1997). In the final subsection 4.5 a fully nonlinear and fully dispersive interfacial model in three dimensions is described.

4.1 Two-dimensional transient model

As in section 4, the two-layer fluid has an upper layer of density ρ_2 and thickness h_2 and a lower layer of density ρ_1 and thickness h_1 at rest. In the lower layer there is a geometry that may move horizontally, see figure 23. Coordinates (x, y) are chosen as before where the x-axis coincides with the interface at rest and the y-axis is vertical. The velocities in each of the layers are determined by potential functions that are Laplacian. We shall first develop the equations to integrate forward the motion of the interface. Let $\mathbf{R} = (X, Y)$ be a position on the interface. The motion of this point is governed by the equation

$$\frac{D_\times \mathbf{R}}{Dt} = \mathbf{v}_\times \tag{4.1}$$

where

$$\frac{D_\times}{Dt} = \frac{\partial}{\partial t} + \mathbf{v}_\times \times \nabla \tag{4.2}$$

and

$$\mathbf{v}_\times = (1 - \alpha)\mathbf{v}_1 + \alpha \mathbf{v}_2. \tag{4.3}$$

α is a real number between 0 and 1. From the computational point of view it is important that the marker particles – pseudo particles – with position $\mathbf{R}$ have a motion that as close as possible is vertical. In practice we put $\alpha = h_2/(h_1 + h_2)$, a choice which is justified from the balance $\bar{u}_1 h_1 + \bar{u}_2 h_2 \simeq 0$, where $\bar{u}_{1,2}$ denote depth averaged horizontal velocities induced by the wave motion.

The pressure in each of the fluids is obtained from the Bernoulli equation

$$p_j = -\rho_j \left(\frac{\partial \phi_j}{\partial t} + \tfrac{1}{2} \mathbf{v}_j{}^2 + gy \right), \quad j = 1, 2. \tag{4.4}$$

The dynamic boundary condition at the interface is obtained by balancing the pressure difference there, neglecting interfacial tension, i.e. $p_1 = p_2$, giving

$$\frac{D_\times}{Dt}(\phi_1 - \mu\phi_2) = \mathbf{v}_\times \cdot (\mathbf{v}_1 - \mu\mathbf{v}_2) - \tfrac{1}{2}(\mathbf{v}_1{}^2 - \mu\mathbf{v}_2{}^2) - (1-\mu)gY \quad \text{at } I, \tag{4.5}$$

where $\mu = \rho_2/\rho_1$. As in the previous subsection 3.1 we represent the interface by $z(\xi)$, see eq. (3.7) and work with the following quantities at I

$$\psi = \phi_1 - \mu\phi_2, \tag{4.6}$$

$$\Upsilon = \phi_1 + \phi_2, \tag{4.7}$$

$$\phi_{1\nu} = \phi_{2\nu}, \tag{4.8}$$

where $\phi_{1\nu} = \partial\phi_1/\partial\nu$ and $\phi_{2\nu} = \partial\phi_2/\partial\nu$. The quantities ψ and $\mathbf{R} = (X, Y)$ are integrated forward in time by the prognostic equations. The other quantities – the Υ and $\phi_{j\nu} = (\partial\phi_j/\partial n)|z_\xi|$, where n denotes the normal vector pointing out of the lower fluid layer – are obtained by solving the Laplace equation in each of the fluids, see below.

In the computations of interfacial motion we have applied higher order derivatives of the prognostic equations in a time stepping procedure for ψ and $\mathbf{R}$:

$$\psi(t + \Delta t) = \psi(t) + [D_\times\psi/Dt](t)\Delta t + \tfrac{1}{2}[D_\times^2\psi/Dt^2](t)(\Delta t)^2 + ..., \tag{4.9}$$

and similar for $\mathbf{R}$, taking into account terms up to the 7th derivative.

4.2 Solution of the Laplace equation

The quantities Υ and $\phi_{1\nu}$ are obtained by solving the Laplace equation in each of the fluids, given the updated $\psi = \phi_1 - \mu\phi_2$ and position of the interface $\mathbf{R}$. The mathematical formulation and the numerical implementation accounts for a geometry moving with a translatory velocity $\mathbf{V}$ in the lower layer. The latter imposes the boundary condition

$$\frac{\partial\phi_1}{\partial n} = \mathbf{V} \cdot \mathbf{n} \tag{4.10}$$

at the boundary B of the geometry, where $\mathbf{n}$ denotes the normal to B pointing out of the lower fluid. The upper and lower boundaries of the fluid are represented by rigid lids. The fluid flow is assumed to vanish in the far field.

The solution of the Laplace equation in each of the fluids follows the description in subsection 3.1. By multiplying equation (3.4) by z'_ξ $(= [\mathrm{d}z/\mathrm{d}\xi]_{\xi=\xi'})$ we obtain

$$\pi \mathrm{i} q_2(z')z'_\xi = \pi\mathrm{i}\phi_{2\xi} + \pi\phi_{2\nu} \tag{4.11}$$

on the left hand side of the equation, where $\phi_{2\xi} = \partial\phi_2/\partial\xi$. Similarly, by multiplying equation (3.6) by z'_ξ we obtain

$$-\pi \mathrm{i} q_1(z')z'_\xi = -\pi \mathrm{i}\phi_{1\xi} - \pi\phi_{1\nu} \tag{4.12}$$

on the left hand side of the equation, where $\phi_{1\xi} = \partial\phi_1/\partial\xi$. An integral equation for $\Upsilon_\xi = \phi_{1\xi} + \phi_{2\xi}$ is obtained from the imaginary part of the difference between the full equations corresponding to (4.11) and (4.12), giving for z' on I,

$$\begin{aligned}\pi\Upsilon'_\xi = &\frac{1}{\mu+1}\int_I Im\left(\frac{z'_\xi}{Z}\right)(2\psi_\xi + (\mu-1)\Upsilon_\xi)\mathrm{d}\xi + \int_I Re\left(\frac{z'_\xi}{H_1} - \frac{z'_\xi}{H_2}\right)\phi_{1\nu}\mathrm{d}\xi \\ &+\frac{1}{1+\mu}\int_I Im\left(\frac{z'_\xi}{H_1} + \frac{z'_\xi}{H_2}\right)\psi_\xi\mathrm{d}\xi + \frac{1}{1+\mu}\int_I Im\left(\mu\frac{z'_\xi}{H_1} - \frac{z'_\xi}{H_2}\right)\Upsilon_\xi\mathrm{d}\xi \\ &+\int_B Re\left(-\frac{z'_\xi}{Z} + \frac{z'_\xi}{H_1}\right)\phi_{1\nu}\mathrm{d}\xi + \int_B Im\left(\frac{z'_\xi}{Z} + \frac{z'_\xi}{H_1}\right)\phi_{1\xi}\mathrm{d}\xi,\end{aligned} \tag{4.13}$$

where $Z = z' - z$, $H_1 = z^* - 2\mathrm{i}h_1 - z'$ and $H_2 = z^* + 2\mathrm{i}h_2 - z'$.

Similarly, an integral equation for $\phi_{1\nu}$ is obtained by taking the real part of the equation corresponding to (4.11) multiplied by μ minus the real part of the equation corresponding to (4.12), giving, for z' on I,

$$\begin{aligned}\pi(1+\mu)\phi_{1\nu}(\xi') = &PV\int_I Re\left(\frac{z'_\xi}{Z}\right)\psi_\xi\mathrm{d}\xi + (1-\mu)\int_I Im\left(\frac{z'_\xi}{Z}\right)\phi_{1\nu}\mathrm{d}\xi \\ &+\frac{1}{1+\mu}\int_I Re\left(\frac{z'_\xi}{H_1} + \mu\frac{z'_\xi}{H_2}\right)\psi_\xi\mathrm{d}\xi + \frac{\mu}{1+\mu}\int_I Re\left(\frac{z'_\xi}{H_1} - \frac{z'_\xi}{H_2}\right)\Upsilon_\xi\mathrm{d}\xi \\ &-\int_I Im\left(\frac{z'_\xi}{H_1} - \mu\frac{z'_\xi}{H_2}\right)\phi_{1\nu}\mathrm{d}\xi + \int_B Re\left(\frac{z'_\xi}{Z} + \frac{z'_\xi}{H_1}\right)\phi_{1\xi}\mathrm{d}\xi + \int_B Im\left(\frac{z'_\xi}{Z} - \frac{z'_\xi}{H_1}\right)\phi_{1\nu}\mathrm{d}\xi,\end{aligned} \tag{4.14}$$

where PV denotes principal value and Z, H_1, H_2 are defined above.

Finally, an integral equation for the tangential derivative of the velocity potential at the moving geometry in the lower fluid layer is obtained, for z' on B:

$$\begin{aligned}\pi\phi_{1\xi}(\xi') = &\int_B Im\left(\frac{z'_\xi}{Z} + \frac{z'_\xi}{H_1}\right)\phi_{1\xi}\mathrm{d}\xi + \int_I Re\left(-\frac{z'_\xi}{Z} + \frac{z'_\xi}{H_1}\right)\phi_{1\nu}\mathrm{d}\xi \\ &+\frac{1}{1+\mu}\int_I Im\left(\frac{z'_\xi}{Z} + \frac{z'_\xi}{H_1}\right)(\psi_\xi + \mu\Upsilon_\xi)\mathrm{d}\xi + PV\int_B Re\left(-\frac{z'_\xi}{Z} + \frac{z'_\xi}{H_1}\right)\phi_{1\nu}\mathrm{d}\xi.\end{aligned} \tag{4.15}$$

4.3 Solitary wave generation

The formulation of the interfacial motion is fully nonlinear and fully dispersive and is useful to investigate the transient formation of the relatively short interfacial waves at a submarine ridge, a process that is driven by the barotropic tide. We assume here that the ridge has very long crest, that the tidal flow is orthogonal to the ridge crest, and

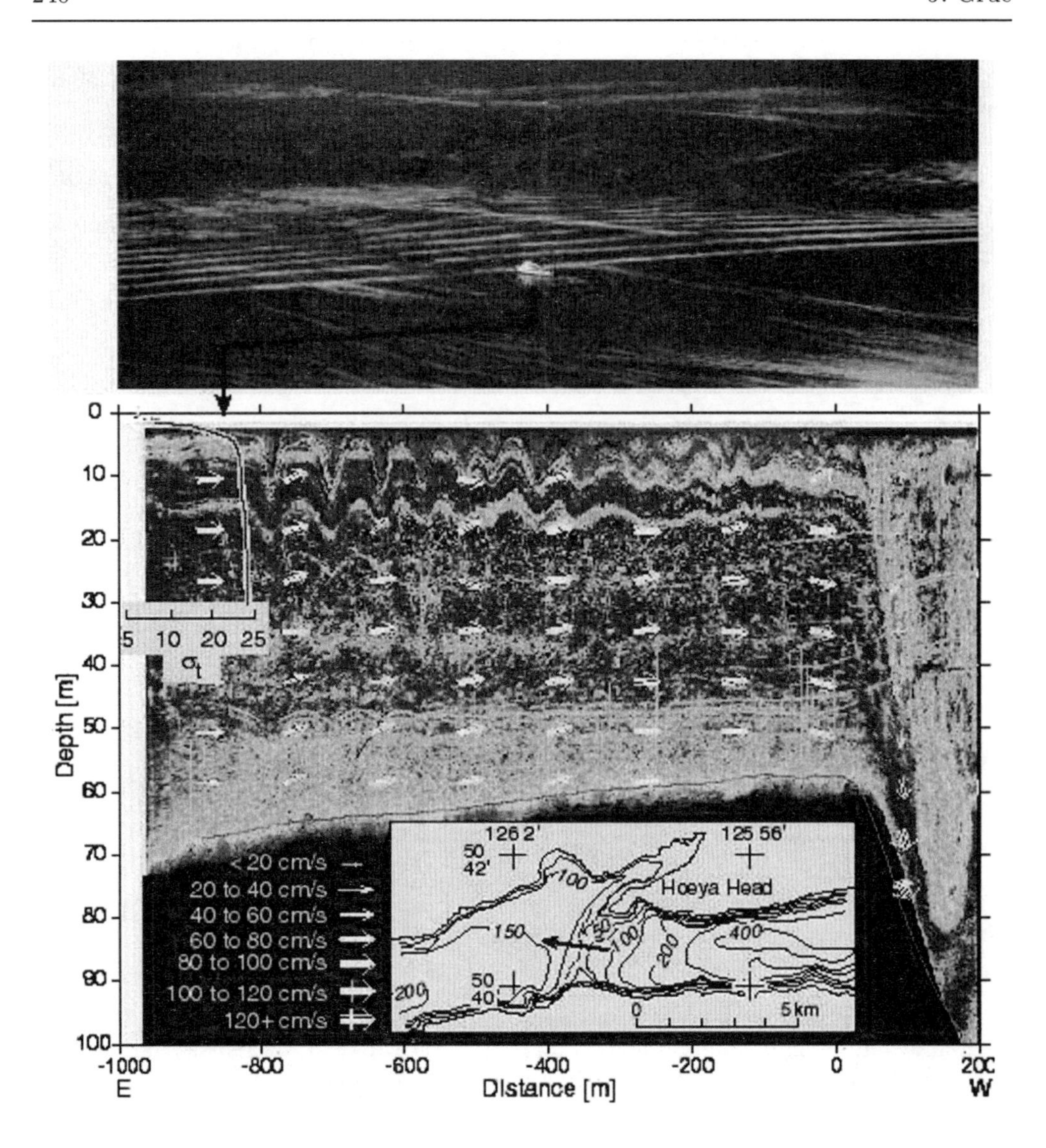

Figure 24. Observations of strong internal waves generated at Knight Inlet in British Columbia. From Farmer and Armi (1999a). Reproduced by permission by Prof. D. M. Farmer.

that the fluid motion can be approximated by a two-dimensional motion, see also Grue (2005). Further, we disregard the effect of the rotation of the earth.

The model simulations presented in figures 25–27 are characterized by that the interface becomes lifted on the upstream side of the topography and depressed above the topography during an initial phase when the tidal current is increasing in strength. The barotropic tidal current has a maximum strength that is somewhat less than the internal linear long wave speed of the fluid. The interface develops a relatively steep jump with height comparable to the upper layer thickness at rest, between the upstream elevation and the depression above the topography, depending on the interaction between the barotropic flow and the topography. The interfacial jump is almost topographically trapped during a rather long phase of the motion.

Lee waves are formed behind the jump. The individual waves emerge at the jump and propagate downstream along the depression, according to the group velocity of the waves which is smaller than the wave phase velocity which equals the local current speed. The waves satisfy the nonlinear dispersion relation for internal waves. Within the linear approximation this is given by

$$\tilde{\omega}^2 = g'k \Big/ \left[\frac{\rho_2}{\rho_1} \coth k\tilde{h}_2 + \coth k\tilde{h}_1\right], \tag{4.16}$$

where $g' = g(\rho_1 - \rho_2)/\rho_1$, $\tilde{\omega} = U(t)k$, $U(t)$ denotes the local barotropic current, k the wavenumber and $\tilde{h}_2$, $\tilde{h}_1$ local depths of the layers.

The waves propagate upstream when the tide turns. The waves develop into a train of depression waves. The simulations here show that the leading depression wave eventually develops into a solitary wave of depression which matches with the steady solitary wave computations (figure 27d). The simulations are exemplified in figure 25 for the waves that are observed upstream at Knight Inlet in British Columbia and in figures 26–27 for the solitary waves in the Sulu Sea. The main trend of the simulations fit much with the observations by Halpern (1971) at the Stellwagen Bank at Cape Cod. His field observations were later analyzed by Lee and Beardsley (1974) using extended KdV theory. They suggested that the wave generation is due to partial blocking over the sill or bank, forming a relatively steep front of the pycnocline. Waves of short period were generated subsequently, see Haury et al. (1979). Ch. 6 of the volume contains descriptions of internal waves of short period that are generated by the transformation of the internal tide at the Kara Gates and in the Strait of Gibraltar. These waves are similar to the ones simulated in figures 25–27. The field observations and the numerical simulations illustrate that the generation process is both highly nonlinear and highly dispersive.

Simulations of the waves observed upstream at Knight Inlet. Knight Inlet has a prominent stratification and a strong tidal flow which generates internal waves on both sides of the sill, see Farmer and Armi (1999b). The topographically trapped wedge of mixed fluid behind the sill may be associated to the splitting of stream lines in large amplitude topographic flow (figure 24). Model simulations by Lamb (2004) of the lee wave formation exhibit a stable flow over the sill until the lee wave overturns. The breaking process develops further into a high-drag state with a downsloping jet beneath the overturning wave. The model simulations by Lamb were inconclusive whether the small scale instabilities observed in the sea have an active role in forming a wedge of stagnant flow behind the sill. Cummins et al. (2003) have reported a detailed set of large scale observations and numerical simulations of the upstream waves at the sill. The continuously stratified nonlinear model simulations showed that the flow is controlled over the crest during an initial phase, but becomes uncontrolled for strong forcing. Cummins et al. (2003) have also reported measurements of the barotropically forced current over the sill that was almost sufficient to keep the bore from moving until it had developed into a wave train, all with height corresponding to the thickness of the ocean surface layer.

To produce the waves in figure 25 the transient two-layer model outlined above was run with an upper layer of thickness $h_2 = 8$ m and depth ratio $h_1/h_2 = 18$ outside the sill. The contour of the real geometry at Knight Inlet with truncation at $h_2 + h_1 = 152$ m depth was defined. At this level the width was extending the interval $-109h_2 < x < 91.8h_2$. The flow at the sill was characterized by an oscillatory fluid velocity $U(t) = 0.673c_0 \sin \Omega t$, where U_0 denotes the maximum fluid velocity during the oscillation. This is less than the linear long wave speed c_0 of the layered fluid outside the sill, where $c_0 = [g'h_2/(\rho_2/\rho_1 + h_2/h_1)]^{1/2}$, $g' = g(\rho_1 - \rho_2)/\rho_1$. $\Omega = 2\pi/T$ is determined by the tidal period of $T = 12\frac{1}{2}$ hours. The non-dimensional frequency of the oscillatory flow was $\Omega h_2/c_0 = 10^{-3}$.

Simulations of the waves in the Sulu Sea. The Sulu Sea is characterized by an upper layer with a linear stratification extending down to about 250 m. A lower layer has almost constant density and depth in the range 600 – 2000 m. The actual sill is 7000 – 8000 m wide with a smallest depth of 340 m. To produce the results in figures 26–27 the transient two-layer model was run with $h_1/h_2 = 4$. An elongated half-ellipse of total width $2b = 30h_2$ and vertical half-width that equals the upper layer thickness h_2 represented the sill geometry. By putting $h_2 = 250$ m the length of the sill became 7500 m comparing with the sill in the Sulu Sea. The non-dimensional frequency of the oscillatory flow was $\Omega h_2/c_0 = 10^{-2}$. The given motion of the sill geometry was horizontal with speed $U(t) = 0.8c_0 \sin \Omega t$ for $0 < t < \frac{1}{2}T$ (i.e. half the tidal period).

4.4 Upstream waves: geometry in the thin layer

Consider an elongated geometry moving in the thin layer with speed U along the horizontal boundary (figure 28). The flow will be steady in a frame of reference moving with the geometry in the subcritical ($U/c_0 << 1$) and supercritical ($U/c_0 >> 1$) regimes, where c_0 is the linear long wave speed of the two-fluid system determined by (2.21). In the case when U is comparable to c_0 the flow becomes transcritical. The geometry then blocks the flow in the thin layer. This widens to a thicker level that expands horizontally on the upstream side of the geometry. The layer above the geometry becomes thinner. The most interesting part of the flow occurs on the upstream side of the geometry. In the case when U/c_0 slightly exceeds unity the thicker layer upstream develops into an expanding train of solitary waves. Fully nonlinear computations illustrate the phenomenon in figure 29. The waves have an amplitude that is comparable to the thin layer. The solitary

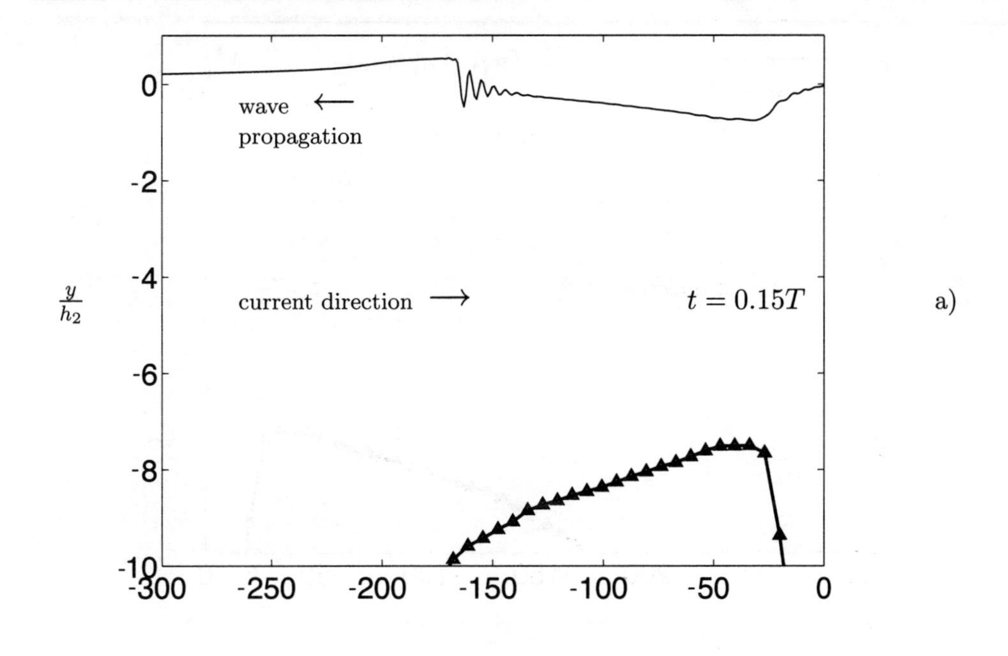
wave ⟵
propagation
current direction ⟶
$t = 0.15T$
$\frac{y}{h_2}$
0
-2
-4
-6
-8
-10
-300
-250
-200
-150
-100
-50
0
a)

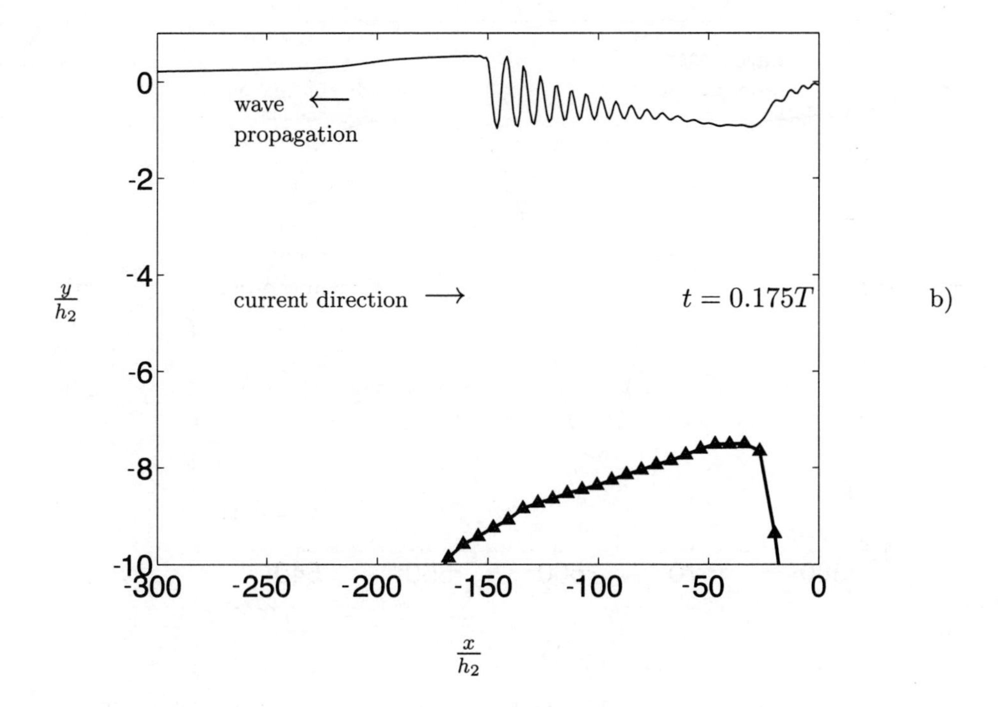
wave ⟵
propagation
current direction ⟶
$t = 0.175T$
$\frac{y}{h_2}$
0
-2
-4
-6
-8
-10
-300
-250
-200
-150
-100
-50
0
$\frac{x}{h_2}$
b)

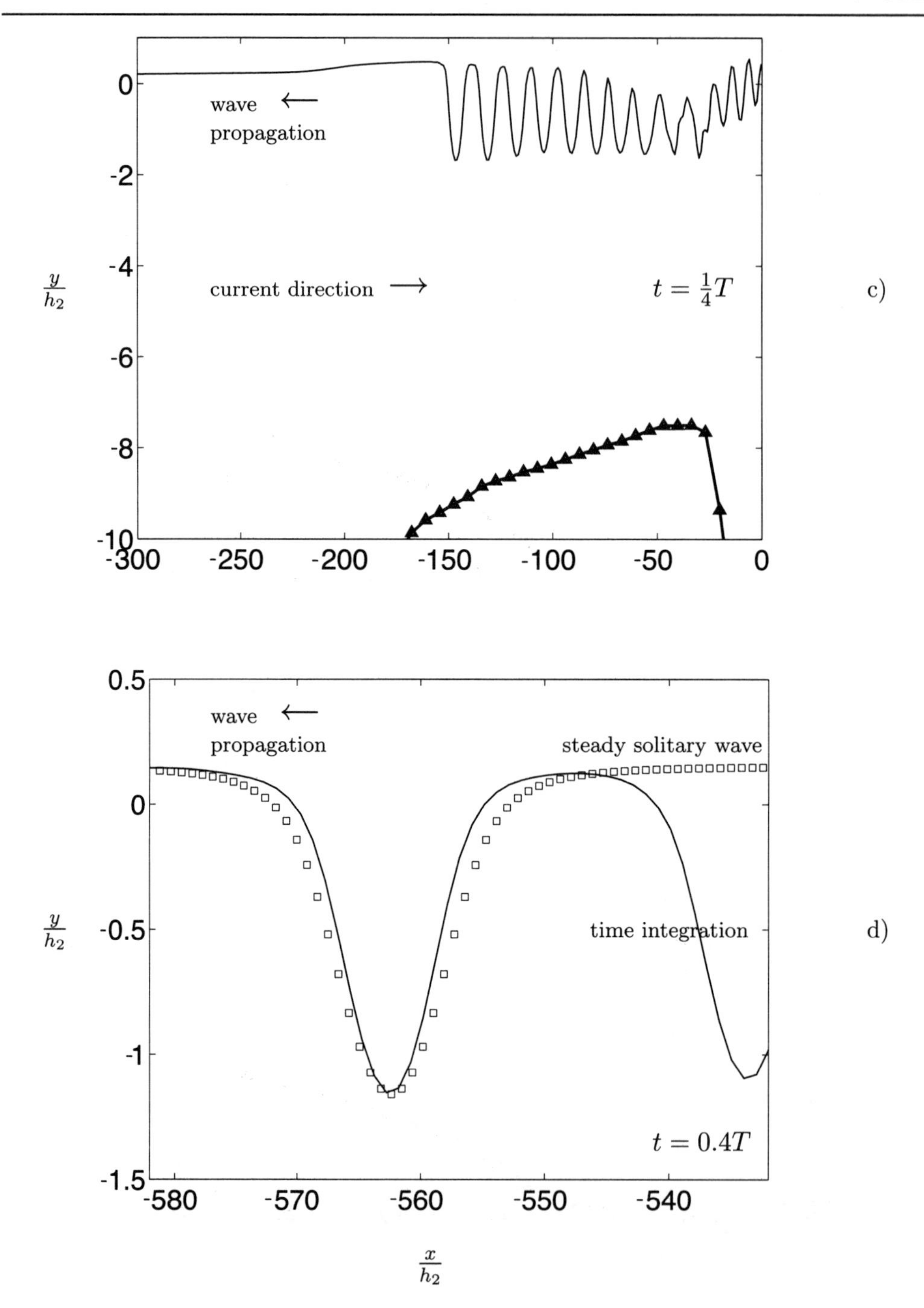

Figure 25. Formation of waves at Knight Inlet. Tidal flow from left ro right. Top part of geometry is visible in plots a)–c). Interface discretized by 1401 nodes with $\Delta x/h_2 = 1$ and the contour of the sill by 66 nodes. $\Delta t\sqrt{g/h_2} = 1$. $\rho_2/\rho_1 = 0.986$. Times: a) $t = 0.15T$, b) $t = 0.175T$, c) $t = \frac{1}{4}T$, d) $t = 0.4T$. In d): Steady solitary wave (small squares). From Grue (2005).

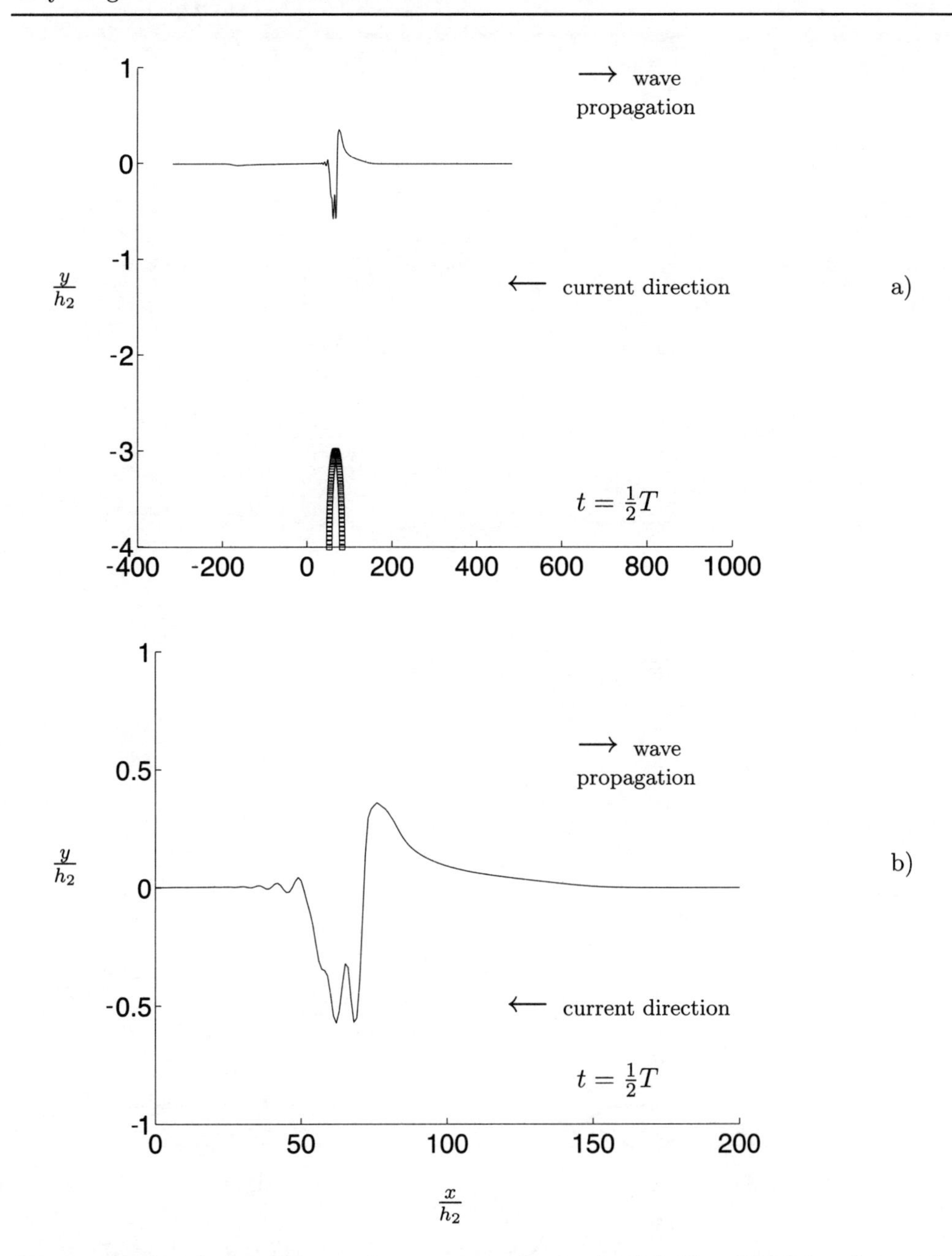

Figure 26. Simulation of internal wave generation in the Sulu Sea. Tidal flow from right to left. Interface at rest is at level $y = 0$. Upper layer thickness h_2. Lower layer thickness h_1 with $h_1/h_2 = 4$. Bottom topography is an ellipse with horizontal axis of $30h_2$ and vertical half-axis of h_2. Interface after time $t = \frac{1}{4}T$. Bottom topography is indicated in plot a). Plot b) is blow-up of plot a). 67 points are used on the half-ellipse, 801 points on the interface, horizontal resolution $\Delta x/h_2 = 1$, time resolution $\Delta t\sqrt{g/h_2} = 0.25$, and $\rho_2/\rho_1 = 0.96$. From Grue (2005).

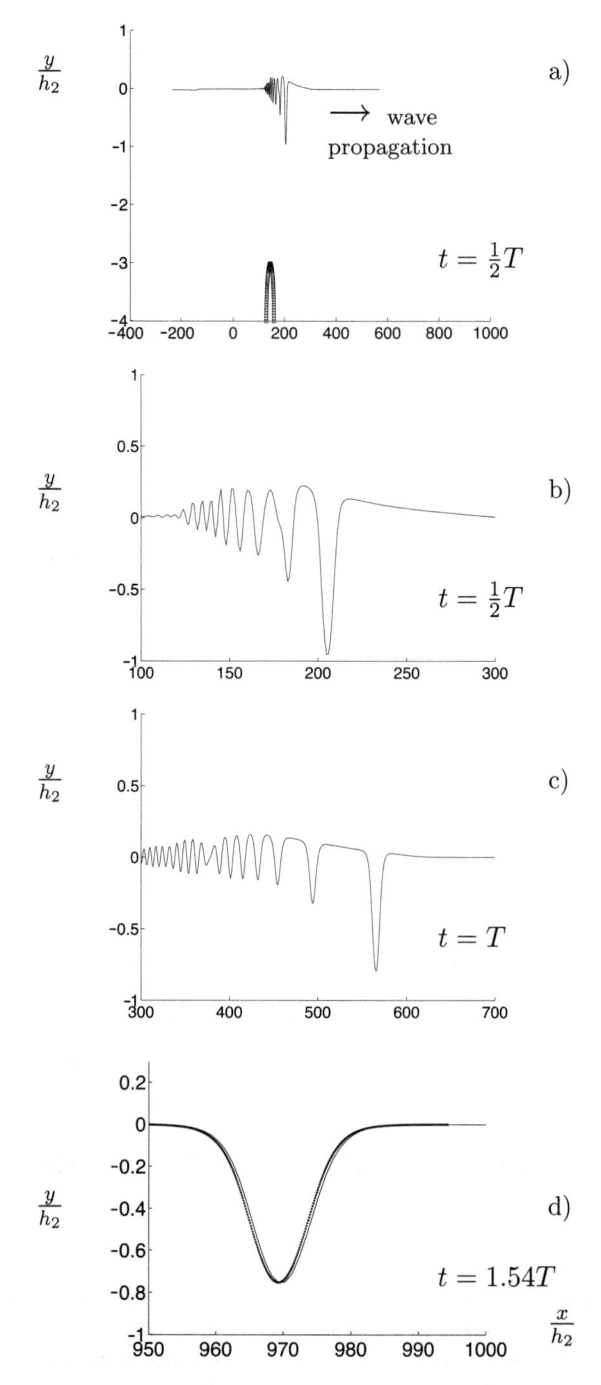

Figure 27. Same as previous figure but interface after a) and b) $t = \frac{1}{2}T$, c) $t = T$, d) $t = 1.54T$. Thick solid line in d) is computation of steady solitary wave with amplitude $a/h_2 = 0.752$. From Grue (2005).

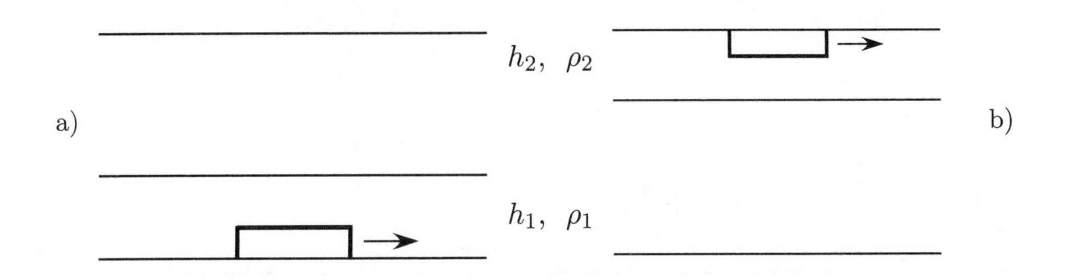

Figure 28. a) Geometry moving with constant speed U in the thin lower layer and b) thin upper layer.

waves are generated for a rather limited range of speeds where $1+\epsilon_1 < U/c_0 < 1+\epsilon_2$ and $0 < \epsilon_1 < \epsilon_2$. In range $1-\epsilon_0 < U/c_0 < 1+\epsilon_1$ an undular bore is generated on the upstream side of the geometry (figure 30). The value of $1-\epsilon_0$ may be a rather small positive number. In all cases a train of lee waves is formed behind the region of reduced thickness of the thin layer. This region expands horizontally above and behind the geometry. The values of ϵ_0, ϵ_1 and ϵ_2 depend on the shape of the geometry, depth ratio h_1/h_2 and density ratio ρ_2/ρ_1.

The case when the geometry is moving in a thin upper layer (figure 28b) was investigated experimentally by Melville and Helfrich (1987). Their experiments were reproduced numerically by integrating the equations in subsections 4.1–4.2 for a moving geometry with the shape corresponding to the body in the experiments, given by $y = h_2 - H_0 \text{sech}^2 Kx$ with $KH_0 = 0.1989$, see Grue et al. (1997). The regime of unsteady flow occurred for $0.38 < U/c_0 < 1.2$. Calculations exhibited upstream solitary waves for $U/c_0 = 1.05, 1.09, 1.14$, while an upstream undular bore was obtained for U/c_0 less than 1.02. Supercritical flow occurred for $U/c_0 > 1.2$. The transient waves due to the upstart of the geometry were then swept downstream and steady state was reached.

The simulations with different geometries have shown that weakly nonlinear theories have quite limited application in modelling unsteady transcritical two-layer flows when $h_1/h_2 >> 1$ (or $h_2/h_1 >> 1$), and that a fully nonlinear method in general is required for this purpose. There may be exceptions for very long geometries with very small height. For more results on upstream generation, see Helfrich and Melville (2006).

The results in figure 29 represent the two-dimensional counterpart of the dead-water phenomenon, see subsection 1.1. It is evident that the waves are generated on the interface both downsteam and upstream for forward speeds U that are comparable to c_0, the internal linear long wave speed of the two-layer fluid. The moving geometry becomes an excellent wave maker in such cases. Strongly nonlinear simulations of the dead-water phenomenon for a real ship in three dimensions have not yet been given. An investigation of the weakly nonlinear internal ship waves at high densimetric Froude numbers was performed by Tulin, Yao and Wang (2000).

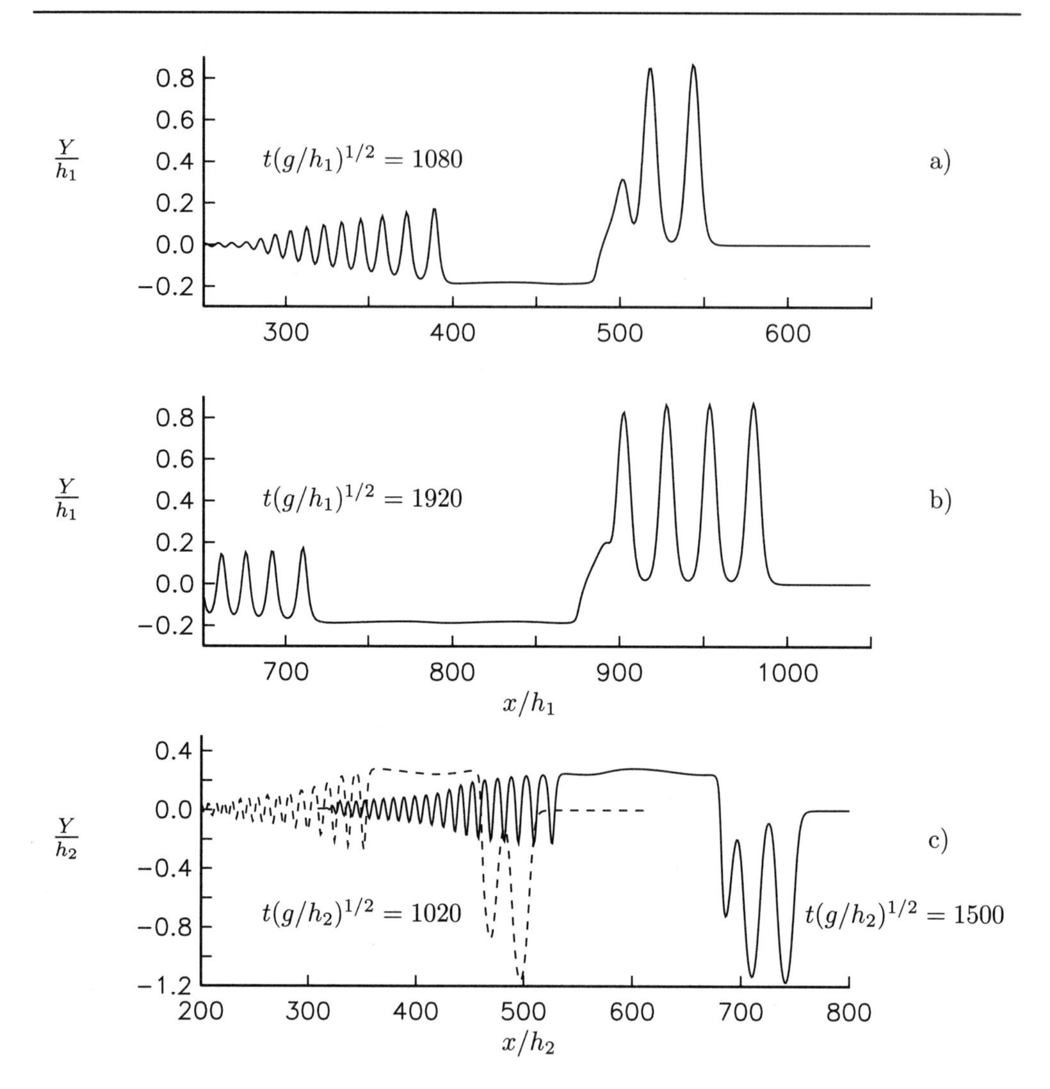

Figure 29. Generation of solitary waves. Moving bottom topography in the lower (thin) layer. The topography is a half ellipse, horizontal half-axis $10h_1$, vertical half-axis $0.1h_1$. $U/c_0 = 1.1$, $\rho_2/\rho_1 = 0.7873$, $h_2/h_1 = 4$. a) $t(g/h_1)^{1/2} = 1080$. b) $t(g/h_1)^{1/2} = 1920$. $|Y|_{max}/h_1 = 0.869$. c) Elevation of interface. $U/c_0 = 1.09$. Moving geometry with profile $H_0 \text{sech}^2 Kx$ and $KH_0 = 0.1989$. $t(g/h_2)^{1/2} = 1020$ (dashed line) $t(g/h_2)^{1/2} = 1500$ (solid line). $\rho_2/\rho_1 = 0.8114$, $h_1/h_2 = 4$. From Grue et al. (1997). Reproduced by permission of J. Fluid Mech.

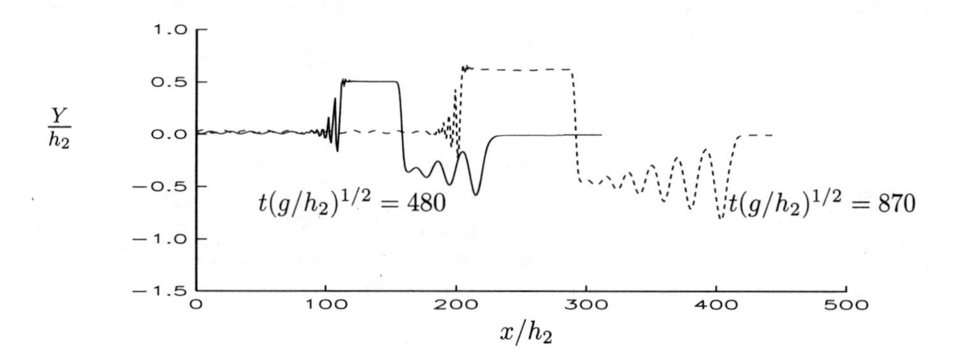

Figure 30. Same as figure 29c, but upstream undular bore and $U/c_0 = 0.81$. From Grue et al. (1997). Reproduced by permission of J. Fluid Mech.

4.5 Fully nonlinear interfacial motion in three dimensions

We here outline the transient interfacial formulation in three dimensions derived by Grue (2002, section 6). We consider a two-layer model with an upper layer of thickness h_2 and density ρ_2 and lower layer of thickness h_1 and density ρ_1. Coordinates are introduced with $\mathbf{x} = (x_1, x_2)$ being horizontal and y vertical. The level $y = 0$ separates the layers at rest. As above we assume that the flow may be modelled by potential theory, with ϕ_j $(j = 1, 2)$ as velocity potentials in the layers. The rigid wall condition applies at the bottom at $y = -h_1$ of the lower fluid. We shall assume that the rigid wall condition also applies at the free surface, i.e. at $y = h_2$.

The motion of a non-overturning interface I at $y = \eta$ is considered. The potentials evaluated at the interface are introduced by $\widetilde{\phi}_1(\mathbf{x}, t) = \phi_1(\mathbf{x}, y = \eta, t)$ and $\widetilde{\phi}_2(\mathbf{x}, t) = \phi_2(\mathbf{x}, y = \eta, t)$, i.e. a tilde means the value of the quantity at the actual position of the interface. The kinematic and dynamic boundary conditions at the interface gives

$$\eta_t - V = 0, \tag{4.17}$$

$$(\widetilde{\phi}_1 - \mu\widetilde{\phi}_2)_t + (1-\mu)g\eta + \frac{|\nabla_1\widetilde{\phi}_1|^2 - \mu|\nabla_1\widetilde{\phi}_2|^2 - (1-\mu)V^2}{2 + 2|\nabla_1\eta|^2}$$
$$+ \frac{-2V\nabla_1\eta \cdot \nabla_1(\widetilde{\phi}_1 - \mu\widetilde{\phi}_2) + |\nabla_1\eta \times \nabla_1\widetilde{\phi}_1|^2 - \mu|\nabla_1\eta \times \nabla_1\widetilde{\phi}_2|^2}{2 + 2|\nabla_1\eta|^2} = 0, \tag{4.18}$$

where $\mu = \rho_2/\rho_1$ and $V = \phi_{1n}\sqrt{1 + |\nabla_1\eta|^2}$ denotes a scaled normal velocity of the interface. Further, $\nabla_1 = (\partial_{x_1}, \partial_{x_2})$ denotes the horizontal gradient. The prognostic equations (4.17)–(4.18) may be integrated once relations between $(\widetilde{\phi}_1 + \widetilde{\phi}_2,\ V)$ and $(\widetilde{\phi}_1 - \mu\widetilde{\phi}_2,\ \eta)$ are obtained. The latter may be derived using Green's theorem, giving for the upper and lower layer, respectively,

$$\int\!\!\int_I \left(\frac{1}{r} + \frac{1}{r_2}\right)\frac{\partial\phi_2'}{\partial n'}\,\mathrm{d}S = -2\pi\widetilde{\phi}_2 + \int\!\!\int_I \widetilde{\phi}_2'\frac{\partial}{\partial n'}\left(\frac{1}{r} + \frac{1}{r_2}\right)\mathrm{d}S, \tag{4.19}$$

$$\int\!\!\int_I \left(\frac{1}{r} + \frac{1}{r_1}\right)\frac{\partial\phi_1'}{\partial n'}\,\mathrm{d}S = 2\pi\widetilde{\phi}_1 + \int\!\!\int_I \widetilde{\phi}_1'\frac{\partial}{\partial n'}\left(\frac{1}{r} + \frac{1}{r_1}\right)\mathrm{d}S. \tag{4.20}$$

In (4.19)–(4.20) the distance r is given by $r=\sqrt{R^2+(y-y')^2}$, where $R=|\mathbf{R}|$ denotes the scalar value of $\mathbf{R}=[(x_1',x_2')-(x_1,x_2)]$, and the latter denotes the vector distance between two points in the horizontal plane. Further, $r_2^2=R^2+(y'+y-2h_2)^2$ and $r_1^2=R^2+(y'+y+2h_1)^2$. The normal vector $\mathbf{n}$ is pointing out of the lower fluid and into the upper.

A function $D=(\eta'-\eta)/R$ is introduced. A suitable reorganization and inversion procedure of the integral equations (4.19)–(4.20) follows the same procedure as in Ch. 4, section 2 of the volume. The inversion by Fourier transform of (4.20) is expressed by

$$\begin{aligned}\mathcal{F}(V) &= kE_1\mathcal{F}(\widetilde{\phi}_1-\eta V^{(1)})-\mathrm{i}\mathbf{k}\cdot\mathcal{F}(\eta\nabla_1\widetilde{\phi}_1)\\ &+kC_1\mathcal{F}\left[T(\widetilde{\phi}_1)+T_1(\widetilde{\phi}_1)+N(V)+N_1(V)\right]\\ &+kC_1\left[e_1\mathcal{F}(\eta(V-V^{(1)}))+\mathcal{F}(\eta\mathcal{F}^{-1}(e_1\mathcal{F}(V-V^{(1)})))\right], \quad (4.21)\end{aligned}$$

where $\mathcal{F}$ denotes Fourier transform, $\mathcal{F}^{-1}$ inverse transform, $\mathbf{k}$ wavenumber (in Fourier space), $k=|\mathbf{k}|$, $e_1=e^{-2kh_1}$, $E_1=(1-e_1)/(1+e_1)=\tanh kh_1$, $C_1=1/(1+e_1)$ and

$$\mathcal{F}(V^{(1)})=kE_1\mathcal{F}(\widetilde{\phi}_1)=k\tanh(kh_1)\mathcal{F}(\widetilde{\phi}_1). \quad (4.22)$$

The functions $N(V)$ and $T(\phi_1)$ are obtained by (see eqs. (2.22)–(2.23) in Ch. 4)

$$N(V)=\frac{1}{2\pi}\int\int_{-\infty}^{\infty}\frac{V'}{R}[1-(1+D^2)^{-1/2}]\,\mathrm{d}\mathbf{x}', \quad (4.23)$$

$$T(\widetilde{\phi}_1)=\frac{1}{2\pi}\int\int_{-\infty}^{\infty}\widetilde{\phi}_1'[1-(1+D^2)^{-3/2}]\nabla_1'\cdot\left[(\eta'-\eta)\nabla_1'\frac{1}{R}\right]\mathrm{d}\mathbf{x}', \quad (4.24)$$

and the integration is over the horizontal plane. The cubic contribution to $N(V)$ may be obtained by a sum of convolutions which are computed via Fourier transform, plus an integral with a kernel that decays quickly, i.e.

$$N(V)=\frac{1}{2\pi}\int\int_{-\infty}^{\infty}\frac{V'D^2}{2R}\mathrm{d}\mathbf{x}'+\frac{1}{2\pi}\int\int_{-\infty}^{\infty}\frac{V'}{R}\left[1-\tfrac{1}{2}D^2-(1+D^2)^{-\frac{1}{2}}\right]\mathrm{d}\mathbf{x}' \quad (4.25)$$

where the first term on the right of (4.25) is expressed by

$$\begin{aligned}&\frac{1}{2\pi}\int\int_{-\infty}^{\infty}\frac{V'D^2}{2R}\mathrm{d}\mathbf{x}'=\\ &-\tfrac{1}{2}\left[\eta^2\mathcal{F}^{-1}\{k\mathcal{F}\{V\}\}-2\eta\mathcal{F}^{-1}\{k\mathcal{F}\{\eta V\}\}+\mathcal{F}^{-1}\{k\mathcal{F}\{\eta^2V\}\}\right]. \quad (4.26)\end{aligned}$$

The functions N_1 and T_1 are expressed by integrals by

$$N_1(V)=\frac{1}{2\pi}\int\int_{-\infty}^{\infty}V'\left(\frac{1}{R_1}-\frac{2(\eta'+\eta)h_1}{R_1^3}-\frac{1}{r_1}\right)\mathrm{d}\mathbf{x}', \quad (4.27)$$

$$\begin{aligned}T_1(\widetilde{\phi}_1) &= -\frac{1}{2\pi}\int\int_{-\infty}^{\infty}\widetilde{\phi}_1'\,\frac{12h_1^2(\eta'+\eta)}{R_1^5}\mathrm{d}\mathbf{x}'\\ &+\frac{1}{2\pi}\int\int_{-\infty}^{\infty}\widetilde{\phi}_1'\,(\mathbf{R}\cdot\nabla_1\eta-(\eta+\eta')-2h_1)\left(\frac{1}{r_1^3}-\frac{1}{R_1^3}\right)\mathrm{d}\mathbf{x}'. \quad (4.28)\end{aligned}$$

where $R_1^2 = R^2 + (2h_1)^2$.

Correspondingly, Fourier transform of (4.19) gives

$$\mathcal{F}(V) = -kE_2\mathcal{F}(\widetilde{\phi}_2 - \eta W^{(1)}) - \mathbf{ik}\cdot\mathcal{F}(\eta\nabla_1\widetilde{\phi}_2) + kC_2\mathcal{F}\left[T(\widetilde{\phi}_2) + T_2(\widetilde{\phi}_2) + N(V) + N_2(V)\right] + kC_2\left[e_2\mathcal{F}(\eta(V - W^{(1)})) + \mathcal{F}(\eta\mathcal{F}^{-1}(e_2\mathcal{F}(V - W^{(1)})))\right], \quad (4.29)$$

where $e_2 = e^{-2kh_2}$, $E_2 = (1-e_2)/(1+e_2) = \tanh kh_2$, $C_2 = 1/(1+e_2)$,

$$\mathcal{F}(W^{(1)}) = -kE_2\mathcal{F}(\widetilde{\phi}_2) = -k\tanh(kh_2)\mathcal{F}(\widetilde{\phi}_2), \quad (4.30)$$

and we have used that $\phi_{1n} = \phi_{2n}$ at the interface. In (4.29) the integrals N and T are given by (4.23) and (4.24), respectively, and N_2 and T_2 by

$$N_2(V) = \frac{1}{2\pi}\iint_{-\infty}^{\infty} V'\left(\frac{1}{R_2} + \frac{2(\eta'+\eta)h_2}{R_2^3} - \frac{1}{r_2}\right)\mathrm{d}\mathbf{x}', \quad (4.31)$$

$$T_2(\widetilde{\phi}_2) = -\frac{1}{2\pi}\iint_{-\infty}^{\infty}\widetilde{\phi}_2'\,\frac{12h_2^2(\eta'+\eta)}{R_2^5}\mathrm{d}\mathbf{x}' + \frac{1}{2\pi}\iint_{-\infty}^{\infty}\widetilde{\phi}_2'\,(\mathbf{R}\cdot\nabla_1\eta - (\eta+\eta') + 2h_2)\left(\frac{1}{r_2^3} - \frac{1}{R_2^3}\right)\mathrm{d}\mathbf{x}'. \quad (4.32)$$

Final set of equations. In order to obtain a set of equations expressing $\mathcal{F}(\widetilde{\phi}_1 + \widetilde{\phi}_2)$ and $\mathcal{F}(V)$ in terms of $\mathcal{F}(\widetilde{\phi}_1 - \mu\widetilde{\phi}_2)$, (4.29) is first subtracted from (4.21). By next adding (4.29) multiplied by μ to (4.21) we obtain another equation for the unknowns. This organization is the same as in subsection 4.2. The resulting equations are suitable for a procedure of successive approximations and iterations. The numerical implementation of the method is currently being tested out. Results in figures 26–27 are reproduced by the 3D method.

Global evaluation using FFT. An important feature of the method is that the leading contributions to the equations for $\mathcal{F}(\widetilde{\phi}_1 + \widetilde{\phi}_2)$ and $\mathcal{F}(V)$ are obtained by global convolution terms evaluated by FFT. This gives a very rapid procedure. Where to divide between global or local evaluation is determined by the stability of the time-stepping scheme: the inclusion of the first term in (4.25) in the global FFT part has been successfully tested out for the motion of a single layer fluid (Fructus et al., 2005). Extracting more terms from N and T was found to lead to numerical instabilities. The integrals N_1, T_1, N_2, T_2 may be brought on convolution form. Energy at high wavenumbers are damped out in these integrals because of the factors $e_1 = e^{-2kh_1}$ and $e_2 = e^{-2kh_2}$.

Local, truncated integration. The highly nonlinear terms represent nonlinear processes that also are very local. The kernels of the inner integrals of (4.24) and of the second term on the right hand side of (4.25) decay like R^{-4} and R^{-5}, respectively. The integrals (4.27)–(4.28) and (4.31)–(4.32) have rapidly decaying kernels. To leading order we have: $N_j(V) \sim \int\int \mathcal{O}(V\eta^2 R_j^{-5})\mathrm{d}\mathbf{x}'$, while $T_j(\widetilde{\phi}_j) \sim \int\int \mathcal{O}(\widetilde{\phi}_j \eta^2 R_j^{-5})\mathrm{d}\mathbf{x}'$, $j = 1, 2$. This means that the contributions to the integrals are very local. These integrals may be evaluated over a very limited region of the $\mathbf{x}$-plane, still keeping high accuracy. Even though this evaluation is highly local, it is much slower (about 100 times) than the global evaluations using FFT.

5 Fully nonlinear wave motion in a continuously stratified fluid

5.1 Basic equations

The purpose of this section is to describe a mathematical approach that is useful to model a) motion within the pycnocline of finite thinkness and b) wave motion in cases of a linearly stratifed upper layer above a homogeneous layer, or the combination of the two. A typical density profile $\rho(y)$ at rest is sketched in figure 11. The equations are derived along the lines of works by Dubreil-Jacotin (1932), Long (1958), Yih (1960), Tung, Chan and Kubota (1982), Turkington, Eydeland and Wang (1991), Grue et al. (2000), and Fructus and Grue (2004). The equations assuming a continuous Brunt-Väisälä frequency are much explored in recent developments (Lamb, 1994, 2004; Stastna and Lamb, 2002; Cummins et al. 2003).

In this subsection the fluid layer extends between the the rigid lids at $y = 0, -H$. The motion is two-dimensional, and waves of permanent form are propagating with speed c horizontally in the fluid. The problem is viewed in a frame of reference which follows the waves where the motion becomes steady, with a horizontal current with speed c along the negative x-axis in the far field. We assume that the fluid is incompressible and inviscid. The former means that $\nabla \cdot \mathbf{v} = 0$ where $\mathbf{v} = (-c + u, v)$ denotes the fluid velocity. Conservation of mass, $\nabla \cdot (\rho \mathbf{v}) = 0$ then gives that $\mathbf{v} \cdot \nabla \rho = 0$. Following the procedure of Yih (1960) we introduce a pseudo velocity by

$$\mathbf{v}' = \left[\frac{\rho}{\rho_0}\right]^{1/2} \mathbf{v}, \tag{5.1}$$

where ρ_0 is a (constant) reference density. Furthermore we introduce a pseudo stream function Ψ' such that

$$\mathbf{v}' = \nabla \Psi' \times \mathbf{k} \tag{5.2}$$

where $\mathbf{k} = \mathbf{i} \times \mathbf{j}$ and $\mathbf{i} = \nabla x$, $\mathbf{j} = \nabla y$. It follows that $\rho = \rho(\Psi')$. The equation of motion becomes

$$\rho_0 \mathbf{v}' \cdot \nabla \mathbf{v}' = -\nabla p - \rho g \nabla y. \tag{5.3}$$

We employ that

$$\mathbf{v}' \cdot \nabla \mathbf{v}' = \nabla \tfrac{1}{2} |\mathbf{v}'|^2 + \boldsymbol{\omega}' \times \mathbf{v}' \tag{5.4}$$

where

$$\boldsymbol{\omega}' = \nabla \times \mathbf{v}' = -\nabla^2 \Psi' \mathbf{k} \tag{5.5}$$

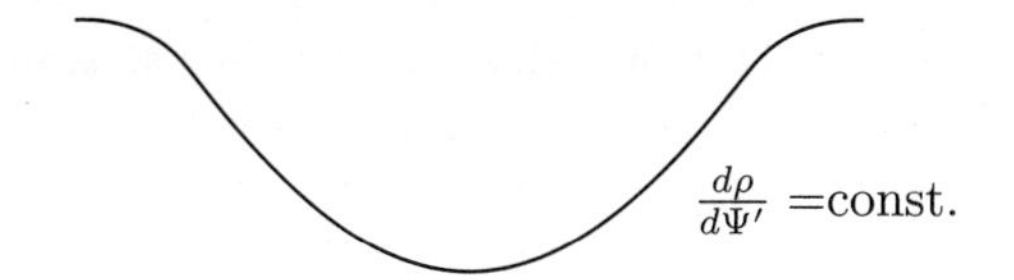

Figure 31. $d\rho/d\Psi'$ =const. along each pseudo stream line. So are $H(\Psi')$ and $dH(\Psi')/d\Psi'$.

denotes pseudo vorticity. The equation of motion then takes the form

$$\rho_0 \boldsymbol{\omega}' \times \mathbf{v}' = -\nabla H + yg\nabla\rho, \tag{5.6}$$

where

$$H = p + \tfrac{1}{2}\rho(u^2 + v^2) + \rho g y \tag{5.7}$$

denotes the total head (and p pressure and g acceleration due to gravity). We make use of

$$\boldsymbol{\omega}' \times \mathbf{v}' = -\nabla^2\Psi'\nabla\Psi' \tag{5.8}$$

and

$$\nabla\rho = \frac{d\rho}{d\Psi'}\nabla\Psi'. \tag{5.9}$$

The equation of motion then becomes

$$\left[\rho_0\nabla^2\Psi' + yg\frac{d\rho}{d\Psi'}\right]\nabla\Psi' - \nabla H = 0. \tag{5.10}$$

This means that H is constant along a pseudo streamline where Ψ' =const. and ρ =const. It follows that $H = H(\Psi')$. Further, this means that

$$\left[\rho_0\nabla^2\Psi' + yg\frac{d\rho}{d\Psi'} - \frac{dH}{d\Psi'}\right]\nabla\Psi' = 0, \tag{5.11}$$

giving

$$\rho_0\nabla^2\Psi' + gy\frac{d\rho}{d\Psi'} = \frac{dH(\Psi')}{d\Psi'}. \tag{5.12}$$

Thus, the l.h.s. and r.h.s. are functions of Ψ' and are constant along each pseudo streamline. The value of $dH/d\Psi'$ may be determined at any position of the pseudo stream line. Particularly, we have in the far field:

$$\frac{dH}{d\Psi'} = \left(\frac{dp}{dy} + \rho g\right)\frac{dy}{d\Psi'} + \frac{c^2}{2}\frac{d\rho}{d\Psi'} + gy\frac{d\rho}{d\Psi'}. \tag{5.13}$$

The vertical component of the equation of motion becomes in the far field $p_y + \rho g = 0$, which means that the first term on the right of (5.13) is zero. The pseudo stream function is then decomposed by $\Psi' = \Psi'_\infty + \psi'$, where Ψ'_∞ satisfies

$$\frac{d\Psi'_\infty}{dy} = -c\left(\frac{\rho}{\rho_0}\right)^{1/2} = -c' \tag{5.14}$$

giving

$$\nabla^2 \Psi'_\infty = \frac{c^2}{2\rho_0}\frac{d\rho}{d\Psi'}. \tag{5.15}$$

Since $d\rho/d\Psi'$ is constant along each streamline, (5.12) becomes

$$\rho_0 \nabla^2 \psi' + (y - y_\infty) g \frac{d\rho}{d\Psi'} = 0, \tag{5.16}$$

where y and y_∞ are vertical coordinates on the same streamline, with y_∞ in the far field.
We evaluate

$$\frac{\partial \rho}{\partial y} = \frac{d\rho}{d\Psi'}\frac{\partial \Psi'}{\partial y}. \tag{5.17}$$

Now, $\partial\Psi'/\partial y = -c' + u'$, giving

$$\frac{d\rho}{d\Psi'} = \frac{\partial \rho}{\partial y}\left[\frac{1}{-c' + u'}\right]. \tag{5.18}$$

At infinity we have

$$\frac{d\rho}{d\Psi'} = \left[\frac{\partial \rho}{\partial y}\right]_\infty \left[\frac{1}{-c'}\right]. \tag{5.19}$$

5.2 The vorticity

The equation (5.16) is useful to interpret the amount of wave induced pseudo vorticity at any position of the fluid:

$$-\omega' = \frac{\delta\, g}{\rho_0}\frac{\partial\rho/\partial y}{\partial \Psi'/\partial y} = \frac{\delta\, \hat{N}^2}{c - u}\left[\frac{\rho}{\rho_0}\right]^{\frac{1}{2}} = \frac{\delta\, N_\infty^2}{c}\left[\frac{\rho}{\rho_0}\right]^{\frac{1}{2}}. \tag{5.20}$$

In this equation we have introduced $\omega' = -\nabla^2\psi'$, the wave induced pseudo vorticity, and $\delta = -(y - y_\infty) > 0$, the local excursion of the pseudo stream line of the depression wave relative to its value in the far field. Further,

$$\hat{N}^2 = -\frac{g}{\rho}\frac{\partial \rho}{\partial y} \tag{5.21}$$

determines the local Brunt-Väisälä frequency and N_∞^2 its value in the undisturbed fluid in the far field.

From now on we apply that $\Delta\rho/\rho$ is small. Pseudo velocities and pseudo stream functions are then replaced by the corresponding unprimed quantities. Equation (5.20) may be used to express the local Richardson number in terms of c, $c-u$, N_∞ and δ, giving

$$\frac{1}{\omega} = \frac{c-u}{\delta \hat{N}^2}, \tag{5.22}$$

and, alternatively,

$$\frac{1}{\omega} = \frac{c}{\delta N_\infty^2}. \tag{5.23}$$

Equation (5.22) shows that a large, local Brunt-Väisälä frequency, a large excursion δ of the stream line (the wave), and a large, local wave induced velocity, u – such that $c-u$ becomes small – contribute to a large vorticity. Likewise, equation (5.23) shows that a large Brunt-Väisälä frequency in the far field (or in the fluid at rest) and a large excursion δ of the stream line (the wave) contribute to a large vorticity.

5.3 The local Richardson number

The motion within a stratified fluid may break if the wave amplitude exceeds a certain level. There are two dominating breaking mechanisms of stratified flows: convective instability and shear instability. The former takes place when the wave induced velocity increases beyond the wave speed. We shall discuss this below. The shear instability is a competition between the stabilizing effect of a stable density profile and the destabilizing effect of an unstable velocity profile. The inverse ratio between these effects is expressed in terms of the Richardson number, defined by

$$Ri = \frac{\hat{N}^2}{\omega^2}. \tag{5.24}$$

According to the theorem by Miles and Howard from 1961 a sufficient condition for the stability of a stationary, stratified flow is that the Richardson number everywhere should exceed $\frac{1}{4}$. Assuming that the wave induced motion is slowly varying in time and space, it is tempting to use the Richardson number as an indication of the stability of internal wave induced flow. We may use (5.22)-(5.23) to express the local Richardson number in terms of the global quantities, i.e.

$$Ri = \frac{\hat{N}^2}{\omega^2} = \frac{c(c-u)}{\delta^2 N_\infty^2}. \tag{5.25}$$

Equation (5.25) shows that a small Richardson number is obtained when $c-u$ is small, δ is large and the Brunt-Väisälä frequency of the undisturbed (far field) fluid is large.

While $N_\infty^2 = -(g/\rho)/(\partial\rho/\partial y)$ is determined by the density profile at rest, the other variables c, $c-u$ and δ may be quantified by experimental measurements, computations of the waves, or by a combination of the two. The formula (5.25) has the advantage that the local $\partial\rho/\partial y$ is obtained analytically by (5.17), and need not be measured directly in the experiment, for example.

5.4 The field equation

Equation (5.23) provides the field equation of the flow. Integrating (5.14) in the limit $\Delta\rho/\rho << 1$ we find

$$\Psi_\infty = -cy. \tag{5.26}$$

Using that

$$\Psi = \Psi_\infty + \psi = \text{const.} \tag{5.27}$$

along a stream line, we obtain that $-cy + \psi = -cy_\infty$, giving that $\delta = -(y - y_\infty) = \psi/c$. Equation (5.23) gives that the motion is governed by

$$\nabla^2\psi + \frac{N_\infty^2}{c^2}\psi = 0. \tag{5.28}$$

The task is to solve (5.28) for ψ, given the boundary conditions for ψ.

We shall here show results for the case of a three-layer model where the Brunt-Väisälä frequency in the far field is constant in each of the three layers. There are several reasons for this choice. Some of the experimental and computational results become relatively easier to interpret. Moreover, it is easier to calibrate experiments in the physical laboratory with a constant Brunt-Väisälä frequency in the layer (at rest). Finally, there are several examples in the field with constant Brunt-Väisälä frequency in the layers. In the case of constant N_∞^2 in the layers, the method of integral equations becomes a natural choice. In cases where N_∞^2 has a continuous variation throughout the water column other methods are employed, as investigated by, e.g., Tung et al. (1982), Turkington et al. (1991), Lamb (1994, 2004), Cummins et al. (2003).

5.5 The linear long wave speed

Three-layer case. We shall determine the linear long wave speed of the internal motion. In case the Brunt-Väisälä frequency in the far field is constant in each of three layers, i.e.

$$N_\infty = N_3 \quad \text{for} \quad -h_3 < y < 0, \tag{5.29}$$

$$N_\infty = N_2 \quad \text{for} \quad -h_2 - h_3 < y < -h_3, \tag{5.30}$$

$$N_\infty = N_1 \quad \text{for} \quad -H < y < -h_2 - h_3, \tag{5.31}$$

solution of the field equation – assuming linear motion – takes the form $A_j\cos(K_jy) + B_j\sin(K_jy)$ in each layer, where $K_j^2 = N_j^2/c_0^2$, $j = 1, 2, 3$, and A_j and B_j are constants. See figure 32. In addition the following boundary conditions apply: $\psi_3 = 0$ at $y = 0$ and $\psi_1 = 0$ at $y = -H$. The stream function and its vertical derivative are continuous at the two interfaces separating the layers, giving

$$K_2^2 - T_1T_2 - T_1T_3 - T_2T_3 = 0, \qquad T_j = K_j\cot(K_jh_j). \tag{5.32}$$

In the case when $N_1 = N_3 = 0$ (5.32) simplifies to

$$\cot Y + \frac{h_2/(h_1Y) - Yh_3/h_2}{1 + h_3/h_1} = 0 \tag{5.33}$$

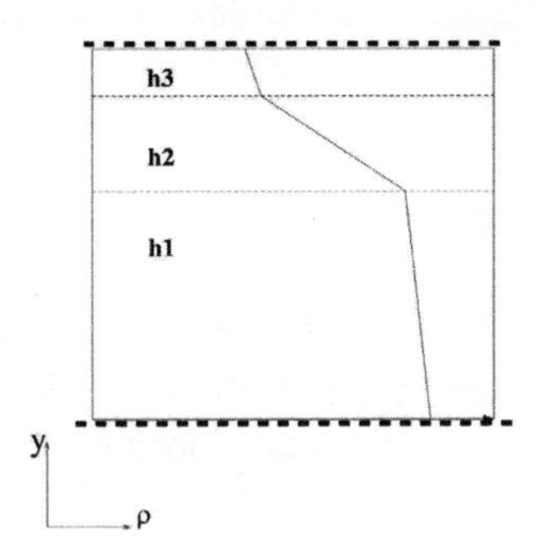

Figure 32. Density profile in three-layer model. Upper layer has thickness h_3, mid layer thickess h_2 and lower layer thickness h_1. Brunt-Väisälä frequency in the layers are N_3 (upper), N_2 (mid), N_1 (lower).

where $Y = K_2 h_2 = N_2 h_2 / c_0$. The longest wave mode is obtained for Y in the interval $(0, \pi)$.

Two-layer case. Very large internal solitary waves may propagate under conditions where the ocean has a top layer at rest of (almost) constant Brunt-Väisälä frequency and a lower layer without any significant stratification. Such a density profile is typical for observations made in e.g. the South China Sea, Sulu Sea and Norwegian Sea. This motion is modelled by assuming that N_∞ is constant in upper layer above lower layer of constant density, and is obtained by putting $h_3 = 0$ in (5.33). The speed becomes $c_0 = N_2 h_2 / (\pi/2)$ for $h_2/h_1 \to 0$ and $c_0 = N_2 h_2/\pi$ for $h_1/h_2 \to 0$.

5.6 Nonlinear three-layer wave motion. Solution by integral equations

In the case of nonlinear motion the field equation in each of the layers, assuming constant Brunt-Väisälä frequency in the layers at rest, reads

$$\nabla^2 \psi_j + \frac{N_j^2}{c^2} \psi_j = 0 \tag{5.34}$$

where c denotes the nonlinear wave speed.

An interface I localized at $y = \eta(x) - h_3$ separates the upper layer number three from the mid layer number two. Likewise, an interface $\hat{I}$ localized at $y = \hat{\eta}(x) - h_2 - h_3$ separates the mid layer number two from the lower layer number one. The values of η and $\hat{\eta}$ vanish for $x \to \pm\infty$. The wave motion is taking place between the rigid lids at the top and bottom boundaries of the fluid layer where the boundary conditions are $\psi_3 = 0$ at $y = 0$ and $\psi_1 = 0$ at $y = -H$.

The kinematic and dynamic boundary conditions at the separation line at $y = \eta(x) -$

h_3 gives that

$$\frac{\partial \psi_j}{\partial s} - c\frac{\partial \eta}{\partial s} = 0, \qquad j = 2, 3, \tag{5.35}$$

$$\frac{\partial \psi_2}{\partial n} = \frac{\partial \psi_3}{\partial n} \tag{5.36}$$

that both are satisfied at I, where s denotes the arclength along I and n the normal, pointing out of mid layer 2. Similar relations hold for $\psi_{1,2}$ at the lower boundary $\hat{I}$ with η replaced by $\hat{\eta}$.

The nonlinear wave problem is solved by means of integral equations. The relevant Green function satisfies the Helmholtz equation in each of the layers. For this purpose we introduce the function

$$Z_0(\alpha, \hat{x}) = Y_0(\hat{x}) + \alpha J_0(\hat{x}) \tag{5.37}$$

where J_0 and Y_0 denote Bessel functions of order zero, of first and second kind, respectively, and α a real constant to be choosen (Fructus and Grue, 2004). The importance of the nonsingular term $\alpha J_0(\hat{x})$ is indicated in eq. (5.46) below. $Z_0(\alpha, \hat{x})$ behaves like $\ln \hat{x}$ for $\hat{x} \to 0$.

In the upper layer the choice of Green function reads

$$G_3(x, y, x', y') = \frac{\pi}{2}\big[Z_0(\alpha, rN_3/c) - Z_0(\alpha, r_3 N_3/c)\big], \tag{5.38}$$

where $r = [(x - x')^2 + (y - y')]^{1/2}$, $r_3 = [(x - x')^2 + (y + y')^2]^{1/2}$. The function G_3 becomes zero at $y = 0$. In the mid layer the function

$$G_2(x, y, x', y') = \frac{\pi}{2} Z_0(\alpha, rN_2/c) \tag{5.39}$$

is used, and in the lower layer

$$G_1(x, y, x', y') = \frac{\pi}{2}\big[Z_0(\alpha, rN_1/c) - Z_0(\alpha, r_1 N_1/c)\big], \tag{5.40}$$

where $r_1 = [(x - x')^2 + (y + y' + 2H)^2]^{1/2}$. The function G_1 becomes zero at $y = -H = -(h_1 + h_2 + h_3)$.

The stream functions are determined by singularity distributions, i.e.

$$\psi_j = \int_I \sigma_j(s') G_j(x, y, x'(s'), y'(s'))\mathrm{d}s', \quad j = 1, 3 \tag{5.41}$$

$$\psi_2 = \int_I \sigma_2(s') G_2(x, y, x'(s'), y'(s'))\mathrm{d}s' + \int_{\hat{I}} \hat{\sigma}_2(s') G_2(x, y, x'(s'), y'(s'))\mathrm{d}s', \tag{5.42}$$

where σ_1, σ_2, $\hat{\sigma}_2$, σ_3 denote distributions to be determined. The kinematic boundary conditions (5.35) give, at I,

$$PV \int_I \sigma_3(s') \frac{\partial G_3}{\partial s}\mathrm{d}s' - c\frac{\partial \eta}{\partial s} = 0, \tag{5.43}$$

$$PV \int_I \sigma_2(s') \frac{\partial G_2}{\partial s}\mathrm{d}s' + \int_{\hat{I}} \hat{\sigma}_2(s') \frac{\partial G_2}{\partial s}\mathrm{d}s' - c\frac{\partial \eta}{\partial s} = 0, \tag{5.44}$$

where PV means principal value. The condition (5.36) gives

$$\pi[\sigma_2(s)+\sigma_3(s)]+\int_I\left(\sigma_2(s')\frac{\partial G_2}{\partial n}-\sigma_3(s')\frac{\partial G_3}{\partial n}\right)\mathrm{d}s'+\int_{\hat{I}}\hat{\sigma}_2(s')\frac{\partial G_2}{\partial n}\mathrm{d}s'=0\,. \tag{5.45}$$

The integral equations (5.43), (5.43), (5.45) are complemented by a set of similar equations at the lower boundary $\hat{I}$. The six equations determine the four unknown singularity distributions σ_1, σ_2, $\hat{\sigma}_2$, σ_3, and the profiles η and $\hat{\eta}$. The computations are initiated by weakly nonlinear KdV solution, and small increments in the wave speed c are specified. The linear part of the integral equation operator is inverted analytically by means of Fourier transform. This results in a set of equations where the transform of the derivative of the Green function appears in the following way (Fructus and Grue, 2004),

$$\mathcal{F}\left\{[Y_1(K|u|)+\alpha J_1(K|u|)]\frac{u}{|u|}\right\}=\begin{cases}\frac{-2\alpha ik}{K\sqrt{K^2-k^2}}, & |k|<K\\[2ex] \frac{2ik}{K\sqrt{k^2-K^2}}, & |k|>K,\end{cases} \tag{5.46}$$

where $\mathcal{F}$ denotes Fourier transform and J_1 and Y_1 Bessel functions of order one, of first and second kind, respectively. The inclusion of the nonsingular function, αJ_1 in the Green function means that the spectrum in Fourier space becomes complete.

5.7 Wave motion along a thick pycnocline

The mathematical method provides a useful basis for numerical implementation and computation of the waves. The computations are compared to experimental laboratory investigations of the large waves moving along the pycnocline. Experiments have the same set-up as sketched in figure 13, except that the pyncocline now has a substantial thickness. One purpose is to investigate the characteristics up to breaking that can be observed in experiments. While the theory provides highly accurate computations of the wave properties like speed, excursion of the individudal stream lines, wave induced velocity and vorticity, the latter are more cumbersome to extract with high precision in the experiments. The mathematical and numerical method enhances the calibration and outcome of the experiments.

Computations of the boundaries of the pycnocline are compared with laboratory observations in figure 33 which provides a snapshot of a solitary wave moving along the pycnocline. The wave induced velocity and stream lines within the pycnocline at the position of the wave crest are shown in figures 34 and 35b, respectively. The upper boundary of the pycnocline (the contour I) has a maximum excursion that slightly exceeds $3\frac{1}{2}$ times the upper layer thickness at rest. The large excursion is somewhat smaller than the nonbreaking waves observed in COPE, see the description and references in subsection 1.5. Numerical values of the Richardson number are evaluated by relation (5.25) and visualized in figure 35a. The minimal Ri evaluated from experiment and theory becomes 0.22 for this wave and is observed at maximum depression at vertical coordinate $y\simeq-1.5(h_2+h_3)$. The experiment exhibits a stable wave motion and indicates that shear instability may be excited for values of the Richardson number that are significantly lower

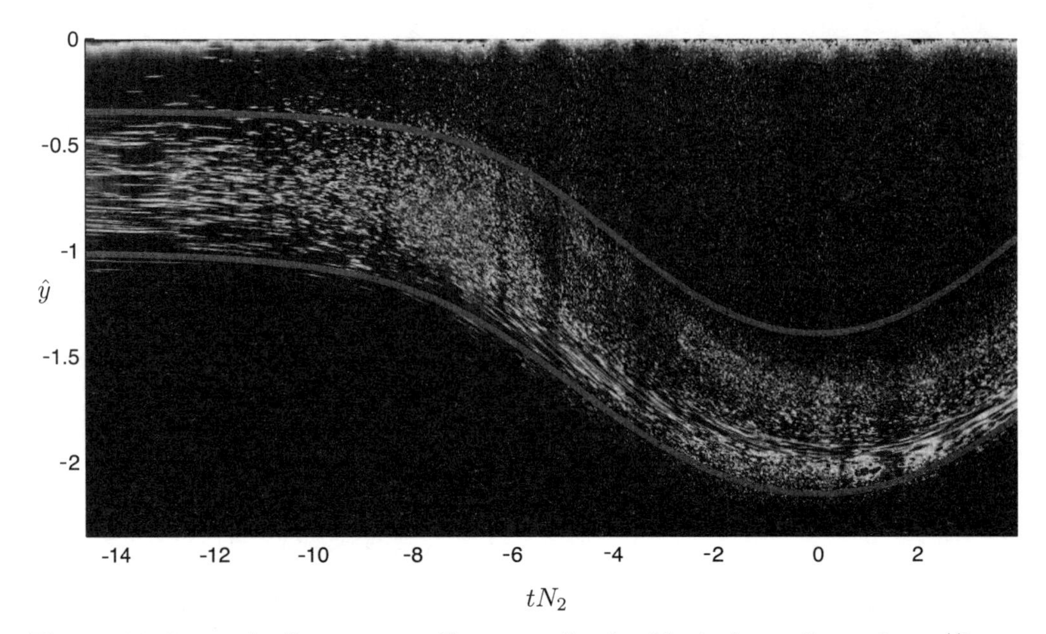

Figure 33. Internal solitary wave of large amplitude. Vertical coordinate $\hat{y} = y/(h_2 + h_3)$. Thick pycnocline. Experiment using Particle Image Velocimetry. Solid lines: fully nonlinear computations of the boundaries of pycnocline. The measured and computed wave speed 18 cm/s. Maximum excursion of $\hat{I}$ is $\hat{\eta} = 1.17(h_2 + h_3) = 3.51h_3$. $c/c_0 = 1.459$. Depth ratios $h_2/h_3 = 2$, $h_1/(h_1 + h_3) = 3.813$. Experimental $N_2 = 1.44\ \mathrm{s}^{-1}$.

than $\frac{1}{4}$. Computations reproduce experiment well. The relative difference between experiment and theory is measured by evaluating $\mathcal{E} = \sqrt{(u_{comp} - u_{exp})^2 + (v_{comp} - v_{exp})^2}/c$, which is typically 2–3% for the wave shown in figure 33 and similar waves.

Figure 36 shows two different theoretical velocity profiles. One is due to a wave moving along a relatively thick pycnocline with homogeneous layers above and below. The other profile is for a two-layer fluid with an upper layer that is linearly stratified and a lower layer that is homogeneous. In the first case the vertical variation of the horizontal velocity outside the pycnocline is very small. This implies that the horizontal fluid velocity is always less than the wave speed in cases with a homogeneous surface layer. This is easily seen from conservation of mass in the upper layer, see eq. (3.17), which gives that $\bar{u}_3/c = A_3/(h_3 + A_3) < 1$, where A_3 denotes the excursion of the upper boundary of the pycnocline and $\bar{u}_3$ the vertically averaged horizontal fluid velocity in the upper layer.

The situation is different when the upper layer is stratified all the way to the ocean surface. In that case the horizontal velocity has a strong vertical variation in the upper layer, where the computations of the horizontal fluid velocity may even exceed the wave speed in the upper part of the water column, indicating the formation of a trapped core in the wave. In the cases with moderate amplitude, the wave induced fluid velocity is always less than the wave speed. The computations are supported by experimental meas-

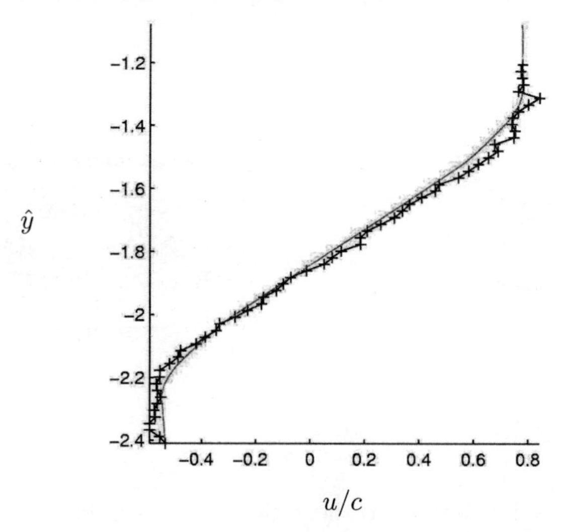

Figure 34. Measured (+ + +) and computed (——) velocity profile through the pycnocline at maximum depression. Same wave as in figure 33. $\hat{y} = y/(h_2 + h_3)$.

urements of the wave induced velocities, with good agreement for cases when $u/c < 1$. In cases of strong waves, where theory predicts fluid velocities that exceed the wave velocities at the boundaries ($u/c > 1$), the waves in the physical laboratory differ from the computations. The experimental waves are characterized by a core of breaking motion and velocity field $\mathbf{v} = -c\mathbf{i} + \mathbf{v}'$ where $\mathbf{v}'$, with $|\mathbf{v}'| << c$, represent small velocity fluctuations of the breaking core. Velocity profiles at maximum depression, with a breaking core, are exemplified in figure 37.

6 Concluding remarks

The main objective of this chapter has been to describe the fully nonlinear methods to compute the motion of very large oceanic internal waves. The waves are a spectacular phenomenon and are viewed from above through the wave induced signatures on the ocean surface. Their presence are well documented by observations during the past forty years. A typical feature is the very large vertical excursions of the wave motion.

We have here particularly discussed the mathematics and numerical implementation of a fully nonlinear and fully dispersive interface method. This has been found to be very useful in reproducing the nonlinear internal solitary waves of very large amplitude, including their generation and propagation along the pycnocline. We conclude from several numerical experiments, comparisons with observations in the field and measurements in the experimental laboratory that the interface method is highly useful in representing the internal wave motion in the ocean. This is true when the pycnocline is rather thin and is separating fluid layers that are homogeneous. Moreover, the results presented here show the relevance of the interface model also to situations where the pycnocline is not

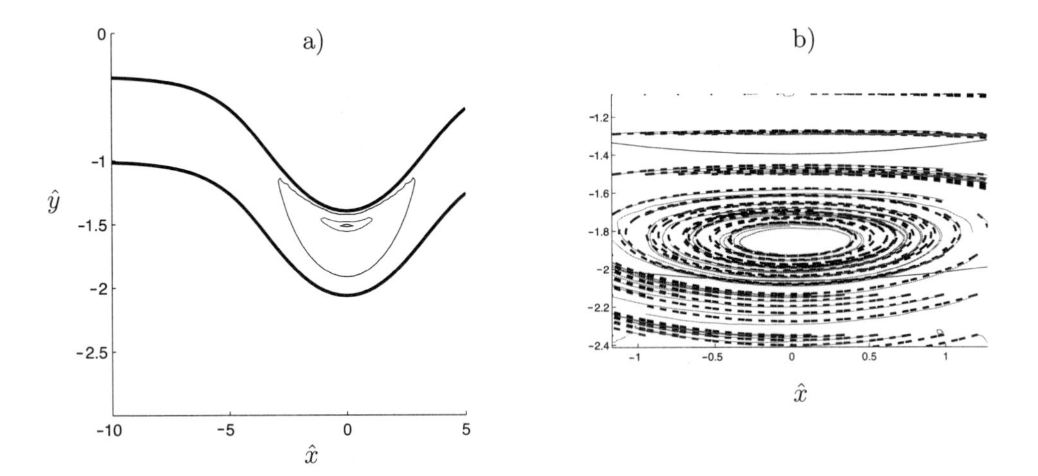

Figure 35. Same wave as in figure 33. a) theoretical wave with contours of constant values of $Ri = 0.22$ (inner), 0.25 (mid), 0.5 (outer). b) numerical (dashed) and experimental (solid) lines of constant ψ_2 within the wave core. Maximum value of stream function: $\psi_{2,max} = 1.71(h_1 + h_2)c_0$. $\hat{y} = y/(h_2 + h_3)$.

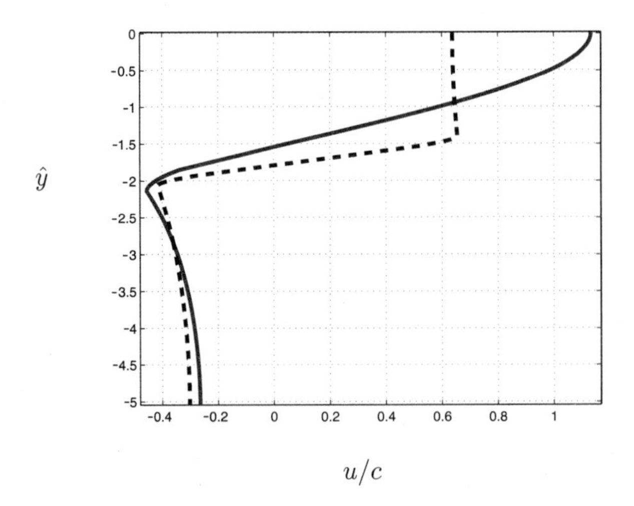

Figure 36. Theoretical velocity profiles at maximum depression. Three-layer model with $N_1 = N_3 = 0$, $N_2 \neq 0$, $h_2/h_3 = 0.87$, $h_1/(h_2 + h_3) = 4.13$, $a/(h_2 + h_3) = 1.22$ (- - -) $N_1 = 0$, $N_2 = N_3 \neq 0$, $h_1/(h_2+h_3) = 4.13$, $a/(h_2+h_3) = 1.22$ (——). $\hat{y} = y/(h_2+h_3)$.

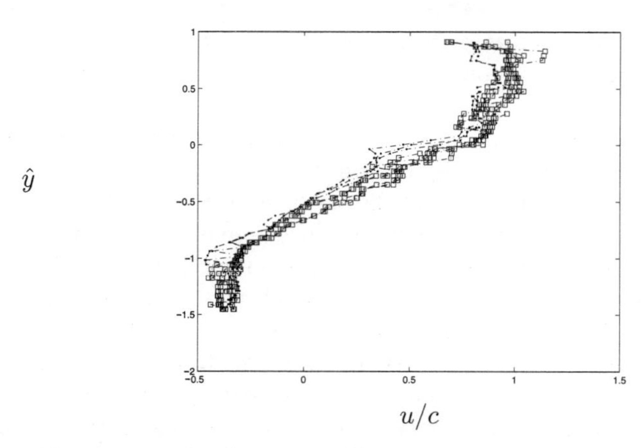

Figure 37. Experimental velocity profiles in strong waves. $N_1 = 0$, $N_2 = N_3 \neq 0$, $h_1/(h_2 + h_3) = 4.13$. $\hat{y} = y/(h_2 + h_3)$. From Grue et al. (2000). Reproduced by permission by J. Fluid Mech.

so thin, like in some of the laboratory experiments and e.g. the field measurements in COPE. The fully nonlinear model predictions tend to the asymptotes of the Korteweg-de Vries and Benjamin-Ono equations in the case when the wave amplitude is very small. The weakly nonlinear theories have been outlined here to provide references for the full model.

The usefulness of the interface method dramatically changes when it comes to modelling the motion within the pycnocline. The latter is needed when studying internal wave breaking caused by shear instability. This is also the case when the upper part of the ocean is stratified all the way to the sea surface. A model that can be justified assumes a constant Brunt-Väisälä frequency in the far field pycnocline or the upper layer of the ocean. Solution procedure involves the theory of integral equations. An advantage is that interpretation of the wave motion becomes relatively simple. Methods assuming continuous Brunt-Väisälä frequency in the vertical provide a much explored computational strategy of the waves, see e.g. Lamb (1994, 2004) and Cummins et al. (2003).

Further development of internal wave modelling should include new mathematical and numerical methods for the motion of internal waves interacting with the continental shelves and their slopes, and bottom topographies like submarine ridges. The motion of the pycnocline and internal waves at the shelf slope are excited by the wind and tide. These processes are characterized by the transfer of energy from motion at low frequencies, to high frequency motions on local scale. The motion is often strongly three-dimensional with essential variation along the vertical. The effect of bottom roughness is important. In other cases the motion is long crested, as in the transformation of long internal tides into trains of shorter solitary waves, a process that usually is only weakly three-dimensional. Improved understanding of the local breaking processes is fundamental. General background references include Legg and Adcroft (2003), Wunsch and Ferrari (2004), van Haren et al. (2004).

The modelling of strong bottom currents, motion of pycnoclines in the ocean, and internal waves interacting with the continental shelf should include be capable of resolving the motion on small scale. It is important that the modelling is highly nonlinear, highly dispersive as well as accounting for the effect of three-dimensionality. The earth's rotation and long and short scale features of the bottom topography should be accounted for. Ideally one should pursue two approaches of the mathematical and numerical description. This includes the approach where the wavenumber spectrum is fully resolved (section 4.5). Likewise, the essentially long wave methods used in ocean modelling should be improved with regards to representation of the dispersion. Work in this direction is in progress, see e.g. Alendal et al. (2005) and Berntsen and Furnes (2005). Modelling work should be guided by field measurements (see e.g. Moum et al., 2003; Hosegood and van Haren, 2004; Yttervik and Furnes, 2004) as well as precise laboratory experiments.

The international research community is seeking to understand how the internal wave induced breaking and mixing on local scale contribute to drive the slow ocean currents on global scale. These are complicated nonlinear processes taking place in the ocean. While the wind and internal tide are the driving mechansims of the energy cascade inducing the short-scale motions on local scale, the local motions provide, on the other hand, an important feedback mechanism on the motion on the global scale in the ocean. This in turn has implications on our understanding and improvement of the modelling of climate.

The author acknowledges with gratitude Prof. D. Howell Peregrine (University of Bristol), Dr. Henrik Kalisch (University of Bergen), Prof. Karsten Trulsen and Dr. Dorian Fructus for useful comments on the manuscript. Dr. Fructus is acknowledged for preparing figures 32-36, and Mr. Jon Reierstad for making the figure sketches. This work was funded by the Research Council of Norway through the Strategic University Program: Modelling of currents and waves for sea structures 2002-6 at the University of Oslo. The support is gratefully acknowledged.

A Inverse scattering theory. Lax pairs

A.1 Laboratory waves

The inverse scattering theory – valid for the integrable equations – is useful to estimate the number and size of the solitary waves that are generated from a weakly nonlinear initial state. Kao et al. (1985) tested the predictions by KdV theory in their experimental campaign. Their figure 8 shows that experiment and theory are relatively close in the case of an initial condition that is relatively long and with moderately small excursion from rest. By careful adjustment of the length and depth of the initial displacement, it is possible to generate a single solitary wave of large amplitude in the wave tank, however. In the latter case the motion is outside the range of validity of the weakly nonlinear model equations (Grue et al., 1999).

A.2 Brief history of solitons and inverse scattering theory

In 1834 the engineer John Scott Russell discovered the solitary wave of elevation in the Edinburgh channel. He provided a description of the wave speed and shape based on experiments. The Dutch hydrodynamicist, Korteweg published together with his student, de Vries a theoretical investigation in 1895 describing an equation for water wave motion where both nonlinear and dispersive effects were represented.

In 1965 – the same year as Perry and Schimke published their measurements of the large internal waves in the Andaman Sea – Zabusky (from Bell Telephone Lab. NJ) and Kruskal (from Princeton) published jointly a four pages long paper in Phys. Rev. Lett. describing an accurate procedure for the numerical integration of the KdV equation. More specifically they integrated $y_t + yy_x + \delta^2 y_{xxx} = 0$ on the interval $[0, 2\pi)$ with $\delta = 0.022$ and initial condition $y(x, t = 0) = \cos \pi x$. They found that the KdV equation had solitary pulse-like solutions with speed increasing with amplitude. The pulses conserved amplitude and shape during several nonlinear interactions and collisions. They noted that the solitary waves, similar to photons, shared the duality of wave and particle property, and introduced the term soliton for the waves.

Gardner, Green, Kruskal and Miura (research group at Princeton University) published in Nov. 1967 a 2 pages long paper in Phys. Rev. Lett. on the inverse scattering theory – referred to as a landmark paper. Working with the linear Schrödinger equation, they inserted $y = \lambda(t) + \psi_{xx}/\psi$ into the KdV equation $y_t - 6yy_x + y_{xxx} = 0$, finding that the eigenvalues $\lambda(t)$ were independent of time. They showed that the solution of the (nonlinear) KdV equation was obtained for all x, t by the initial state through the solution of a linear integral equation. The solitons were determined by a discrete set of eigenvalues. Peter Lax (at Courant Institute, New York University) submitted in Feb. 1968 a paper to Comm. Pure and Appl. Math. describing unitarian operators $U(t)$ that were such that $U(t)^{-1}L(t)U(t)$ became independent of time, where $L = \partial^2/\partial x^2 + \frac{1}{6}y$. He derived the equation $L_t = BL - LB = [B, L]$ where B is antisymmetric and $U_t = BU$. Lax derived sets of higher order KdV equations by using $y_t - [B_q, L] = 0,\ q = 1, 3, 5, ...$, where $B_1 = \partial/\partial x$, $B_3 = \partial^3/\partial x^3 + b_1 \partial/\partial x + \partial/\partial x\, b_1, \ldots$. Generalized Lax-pairs $[B, L]$ led to analytic integration of various nonlinear model equations, also termed integrable equations, and were investigated by other scientists, see e.g. the book on solitons by Drazin and Johnson (1989). Peter Lax became the Abel Laureate in 2005 for his achievements in mathematics. The Abel prize is issued by the Norwegian Academy of Science and Letters in Oslo, Norway

We note that the solitary waves obtained from solving the full Euler equations, discussed in sections 3-5 of the chapter, will attain the properties of the solitons due to the integrable equations only asymptotically as the wave amplitude tends to zero (see also figure 15). The fully nonlinear solitary waves only approximately conserve amplitude and shape during interactions and collisions. The waves are in American-British tradition not termed solitons, while in the Russian scientific community the notion (internal) soliton is often used in a generalized sense.

Bibliography

G. Alendal, J. Berntsen, E. Engum, G. K. Furnes, G. Kleiven and L. I. Eide (2005). Influence from 'Ocean Weather' on near seabed currents and events at Ormen Lange. Marine and Petroleum Geology Vol. 22, 21–31.

C. J. Amick and R. E. L. Turner (1986). A global theory of internal solitary waves in two-fluid systems. Trans. Am. Math. Soc., Vol. 298, 431-84.

J. R. Apel (1978). Observations of Internal-Wave Surface Signatures in ASTP Photographs. pp. 505-509, in: Apollo-Soyuz Test Project. Summary Science Report. Vol. II, NASA SP, Washington, Scientific and Technical Inf.

J. R. Apel, H. M. Byrne, J. R. Proni and R. L. Charnell (1975). Observations of oceanic internal and surface waves from the Earth Resources Technology Satellite. J. Geophys. Res. Vol. 80 (6), 865-81.

J. R. Apel, J. R. Holbrook, A. K. Liu and J. Tsai (1985). The Sulu Sea internal soliton experiment. J. Phys. Oceanogr. Vol. 15, 1625-51.

L. Armi and D. M. Farmer (1988). Flow of Mediterranean water through the Strait of Gibraltar. D. M. Farmer and L. Armi (1988). Flow of Atlantic water through the Strait of Gibraltar. Prog. in Oceanogr. Vol. 21, 1-105.

T. B. Benjamin (1967). Internal waves of permanent form in fluids of great depth. J. Fluid Mech., Vol. 29, 559-92.

D. J. Benney (1966). Long nonlinear waves in fluid flows. J. Math. Phys. Vol. 45, 52-63.

J. Berntsen and G. K. Furnes (2005). Internal pressure errors in sigma-coordinate ocean models – sensivity of the growth of the flow to the time stepping method and possible non-hydrostatic effets. Cont. Shelf Res. Vol. 25, 829-48.

V. I. Byshev, Yu. A. Ivanov and Ye. G. Morozov (1971). Study of temperature fluctuations in the frequency range of internal gravity waves. Izv. Atmos. Ocean. Phys. Vol. 7 (1), 25-30.

M. Carr and P. A. Davies (2006). The motion of an internal solitary wave of depression over a fixed bottom boundary in a shallow, two-layer fluid. Phys. Fluids Vol. 18, 016601-1–10.

W. Choi and R. Camassa (1999). Fully nonlinear internal waves in a two-fluid system. J. Fluid Mech. Vol. 396, 1-36.

P. Cummins, S. Vagle, L. Armi and D. M. Farmer (2003). Proc. R. Soc. Lond. A. Vol. 459, 1467-87.

A. Defant (1961). Physical Oceanography. Vol. II. Pergamon. 598 pp.

J. O. Dickey, P. L. Bender, J. E. Faller, X. X. Newhall, R. L. Ricklefs, J. G. Ries, P. J. Shelus, C. Veillet, A. L. Whipple, J. R. Wiant, J. G. Williams and C. F. Yoder (1994). Lunar Laser Ranging: A Continuing Legacy of the Apollo Program. Science, Vol. 265, 482-90.

J. W. Dold and D. H. Peregrine (1985), in: Numerical Methods for Fluid Dynamics, 2, Ed. K. W. Morton and M. J. Baines, Clarenden Press, Oxford, pp. 671-9.

P. G. Drazin and R. S. Johnson (1989). Solitons: an introduction. Camb. Univ. Press. 226 pp.

M. L. Dubreil-Jacotin (1932). Sur les ondes type permanent dans les liquides heterogenes. Atti della Reale Academia Nationale dei Lincei Vol. 15 (6), 814-9.

T. F. Duda and D. M. Farmer (1999), The 1998 WHOI/IOS/ONR Internal Solitary Wave Workshop: Contributed Papers, Woods Hole Oceanographic Inst. 247 pp.

V. W. Ekman (1904). On dead water: the Norwegian north polar expedition 1893-1896. Scientific Results. Edited by F. Nansen. Ch. XV, A. W. Brøgger. Christiania. 152 pp. and appendix with 17 pl.

A. Eliassen (1982). Vilhelm Bjerknes and his students. Ann. Rev. Fluid Mech. Vol. 14, 1-11.

W. A. Evans and M. J. Ford (1996). An integral equation approach to internal (2-layer) solitary waves. Phys. Fluids Vol. 8, 2032-47.

G. Ewing (1950). Slicks, surface films and internal waves. J. Mar. Res. Vol. 9 (3), 161-87.

D. M. Farmer and L. Armi (1999a). In: The 1998 WHOI/IOS/ONR Internal Solitary Wave Workshop: Contributed Papers, Woods Hole Oceanographic Inst. Ed. by T. F. Duda and D. M. Farmer, pp. 188-92.

D. M. Farmer and L. Armi (1999b). Stratified flow over topography: the role of small-scale entrainment and mixing in flow establishment. Proc. R. Soc. Lond. A. Vol. 455, 3221-58.

B. Franklin (1762). Behaviour of oil on water. Letter to John Pringle. Philadelphia. Dec. 1 1762. In: Experiments and observations on electricity, London 1769, 142-4.

D. Fructus and J. Grue (2004). Fully nonlinear solitary waves in a layered stratified fluid. J. Fluid Mech. Vol. 505, 323-47.

D. Fructus, D. Clamond, J. Grue and Ø. Kristiansen (2005). An efficient model for three-dimensional surface wave simulations. Part I Free space problems. J. Comp. Phys. Vol. 205, 665-85.

C. S. Gardner, J. M. Green, M. D. Kruskal and R. M. Miura (1967). Method for solving the Korteweg-de Vries equation and generalizations. Phys. Rev. Lett. 19, 1095-7.

A. E. Gargett and B. A. Hughes (1972). On the interaction of surface and internal waves. J. Fluid Mech. Vol. 52, 179-91.

R. D. Gaul (1961). Observations of internal waves near Hudson Canyon. J. Geophys. Res. Vol. 66 (11), 3821-30.

J. Grue (2002). On four highly nonlinear phenomena in wave theory and marine hydrodynamics. Appl. Ocean Res. Vol. 24, 261-74.

J. Grue (2004). Internal wave fields analyzed by imaging velocimetry. Chap. 6 in PIV and Water Waves., World Scientific, J. Grue and P. L.-F. Liu and G. K. Pedersen (Eds.), 239-78.

J. Grue (2005). Generation, propagation and breaking of internal solitary waves. Chaos Vol. 15, 037110-1–14.

J. Grue, H. A. Friis, E. Palm and P.-O. Rusås (1997). A method for computing unsteady fully nonlinear interfacial waves. J. Fluid Mech. Vol. 351, 223-52.

J. Grue, A. Jensen, P.-O. Rusås and J. K. Sveen (1999). Properties of large-amplitude internal waves. J. Fluid Mech. Vol. 380, 257-78.

J. Grue, A. Jensen, P.-O. Rusås and J. K. Sveen (2000). Breaking and broadening of internal solitary waves. J. Fluid Mech. Vol. 413, 181-217.

D. Halpern (1971). Observations on short-period internal waves in Massachusetts Bay. J. Mar. Res. Vol. 29, 116-33.

L. R. Haury, M. G. Briscoe and M. H. Orr (1979). Tidally generated internal wave packets in Massachusetts Bay. Nature Vol. 278, 312-7.

K. R. Helfrich (1992). Internal solitary wave breaking and run-up on a uniform slope. J. Fluid Mech. Vol. 243, 133-54.

K. R. Helfrich and W. K. Melville (2006). Long nonlinear internal waves. Annu. Rev. Fluid Mech. Vol. 38, 395-425.

P. E. Holloway, E. Pelinovsky, T. Talipova and B. Barnes (1997). A nonlinear model of internal tide transformation on the Australian north west shelf. J. Phys. Oceanogr. Vol. 27, 871-96.

P. Hosegood and H. van Haren (2004). Near-bed solibores over the continental slope in the Faeroe-Shetland Channel. Deep-Sea Res. II Vol. 51, 2943-71.

B. A. Hughes (1978). The effect of internal waves on surface wind waves. 2. Theoretical analysis. J. Geophys. Res. Vol. 83 (C1), 455-65.

B. A. Hughes and J. F. R. Gower (1983). SAR imagery and surface truth comparisons of internal waves in Georgia Strait, British Columbia, Canada J. Geophys. Res. Vol. 88 (C3), 1809-24.

B. A. Hughes and H. L. Grant (1978). The effect of internal waves on surface wind waves. 1. Experimental measurements, J. Geophys. Res. Vol. 83 (C1), 443-54.

The Internal Wave Atlas, "http://atlas.cms.udel.edu/search.html" (1998).

R. I. Joseph (1977). Solitary waves in a finite depth fluid. J. Phys. A: Math. Gen. Vol. 10, L225-7.

T. W. Kao, F.-S. Pan and D. Renouard (1985). Internal solitons on the pycnocline: generation, propagation, and shoaling and breaking over a slope. J. Fluid Mech. Vol. 159, 19-53.

G. H. Keulegan (1953). Characteristics of internal solitary waves. J. Res. Natl. Bureau of Standards Vol. 51, 133-40.

C. G. Koop and G. Butler (1981). An investigation of internal solitary waves in a two-fluid system. J. Fluid Mech. Vol. 112, 225-51.

D. J. Korteweg and G. de Vries (1895). On the change of form of long waves and advancing waves. Philos. Mag. Vol. 39 (5), 422-43.

R. A. Kropfli, L. A. Ostrovsky, T. P. Stanton, E. A. Skirta, A. N. Keane and V. Irosov (1999). Relationships between strong internal waves in the coastal zone and their radar and radiometric signatures. J. Geophys. Res. [Oceans] Vol. 104, 3133-48.

T. Kubota, D. R. S. Ko and L. D. Dobbs (1978). Weakly-nonlinear long internal waves in a stratified fluid of finite depth. J. Hydronautics Vol. 12, 157-65.

E. C. Lafond (1966). Internal waves. In: Encyclopedia of oceanography, New York: Reinold Publ., pp. 402-8.

K. G. Lamb (1994). Numerical experiments of internal wave generation by strong tidal flow across a finite amplitude bank edge. J. Geophys. Res. Vol. 99 (C1), 843-64.

K. G. Lamb (2004). On boundary-layer separation and internal wave generation at the Knight Inlet sill. Proc. R. Soc. Lond. A. Vol. 460, 2305-37.

K. G. Lamb and L. Yan (1996). The evolution of internal wave undular bores: comparison of a fully-nonlinear numerical model with weakly nonlinear theories. J. Phys. Oceanogr. Vol. 26, (2), 2712-34.

P. D. Lax (1968). Integrals of nonlinear equations of evolution and solitary waves. Comm. Pure Appl. Math. Vol. 21, 467-90.

C.-Y. Lee and R. C. Beardsley (1974). The generation of long internal waves in a weakly stratified shear flow. J. Geophys. Res. Vol. 79, (3), 453-62.

O. S. Lee (1961). Observations of internal waves in shallow water. Limnol. Oceanogr. Vol. 6 (3), 312-21.

S. Legg and A. Adcroft (2003). Internal wave breaking at concave and convex continental slopes. J. Phys. Oceanogr. Vol. 33, 2224-46.

C. E. Lennert-Cody and P. J. S. Franks (1999). In: The 1998 WHOI/IOS/ONR Internal Solitary Wave Workshop: Contributed Papers, Woods Hole Oceanographic Inst. Ed. by T. F. Duda and D. M. Farmer, pp. 69-72.

R. R. Long (1956). Solitary waves in one- and two-fluid systems. Tellus Vol. 8, 460-71.

M. S. Longuet-Higgins and R. W. Stewart (1961). The changes in amplitude of short gravity waves on steady non-uniform currents. J. Fluid Mech. Vol. 10, 529-49.

T. Maxworthy (1979). A note on the internal solitary waves produced by tidal flow over a three-dimensional ridge. J. Geophys. Res. Vol. 84 (C1), 338-46.

W. K. Melville and K. R. Helfrich (1987). Transcritical two-layer flow over topography. J. Fluid Mech. Vol. 178, 31-52.

J. W. Miles (1979). On internal solitary waves. Tellus Vol. 31, 456-62.

J. W. Miles (1981). On internal solitary waves II. Tellus Vol. 33, 397-401.

E. G. Morozov, K. Trulsen, M. G. Velarde and V. I. Vlasenko (2002). Internal tides in the Strait of Gibraltar. J. Phys. Oceanogr. Vol. 32, 3193-206.

J. N. Moum, D. M. Farmer, W. D. Smyth, L. Armi and S. Vagle (2003). Structure and generation of turbulence at interfaces strained by internal solitary waves propagating shoreward over the continental shelf. J. Phys. Oceangr. Vol. 33, 2093-112.

W. Munk and C. Wunsch (1998). Abyssal Recipes II: energetics of tidal and wind mixing. Deep Sea Res. Vol. 45, 1976-2009.

A. New, K. R. Dyer and R. E. Lewis (1987). Internal waves and intense mixing periods in a partially stratified estuary. Estuarine, Coastal and Shelf Science, Vol. 24, 15-33.

F. Nansen (1902). The oceanography of the North Polar Basin. Norwegian North Polar Expedition 1893-1896, Scientific Results, Vol. 3, no. 9, 427 pp. Christiania. Longmans, Green & Co. London.

H. Ono (1975). Algebraic solitary waves on stratified fluids. J. Phys. Soc. Japan Vol. 39, 1082-91.

A. R. Osborne and T. L. Burch (1980). Internal solitons in the Andaman Sea. Science Vol. 208 (4443), 451-60.

A. R. Osborne, T. L. Burch and R. I. Scarlet (1978). The influence of internal waves on deep-water drilling. J. Pet. Tech. Vol. 30, 1497-504.

L. A. Ostrovsky and J. Grue (2003). Evolution equations for strongly nonlinear internal waves. Phys. Fluids Vol. 15 (10), 2934-48.

L. A. Ostrovsky and Y. Stepanyants (2005). Internal solitons in laboratory experiments: Comparison with theoretical models. Chaos Vol. 15, 037111-1–28.

R. B. Perry and G. R. Schimke (1965). Large-amplitude internal waves observed off the Northwest Coast of Sumatra. J. Geophys. Res. Vol. 70 (10), 2319-24.

P. R. Pinet (1992). Oceanography. West Publ. Comp. 576 pp.

R. D. Pingree and G. T. Mardell (1985). Solitary internal waves in the Celtic Sea. Prog. Oceanogr. Vol. 14, 431-41.

D. I. Pullin and R. H. J. Grimshaw (1988). Finite amplitude solitary waves on the interface between two fluids. Phys. Fluids Vol. 31, 3350-9.

J. Roberts (1975). Internal gravity waves in the ocean. Marine science. Vol. 2. Marcel Dekker Inc. New York. 274 pp.

P.-O. Rusås and J. Grue (2002). Solitary waves and conjugate flows in a three-layer fluid. Eur. J. Mech.B/Fluids Vol. 21, 185-206.

J. S. Russell (1844). Report on waves. British Association Report.

H. Segur and J. L. Hammack (1982). Soliton models of long internal waves. J. Fluid Mech. Vol. 118, 285-304.

J. A. Shand (1953). Internal waves in the Georgia Strait. Eos. Trans. AGU, Vol. 34, 949-56.

T. P. Stanton and L. A. Ostrovsky (1998). Observations of highly nonlinear internal solitons over the continental shelf. Geophys. Res. Lett. Vol. 25, 2695-8.

M. Stastna and K. G. Lamb (2002). Vortex shedding and sediment resuspension associated with the interaction of an internal solitary wave and the bottom boundary layer. Geophys. Res. Lett. Vol. 29, 1512.

A. Svansson (2006). Otto Pettersson, oceanografen, kemisten, uppfinnaren. Tre Böcker Förlag AB, Göteborg. 375 pp.

M. P. Tulin, Y. Yao and P. Wang (2000). The generation and propagation of ship internal waves in a generally stratified ocean at high densimetric Froude numbers, including nonlinear effects. J. Ship Res. Vol. 44 (3), 197-227.

K.-K. Tung, T. F. Chan and T. Kubota (1982). Large amplitude internal waves of permanent form. Stud. Appl. Math. Vol. 66, 1-44.

B. Turkington, A. Eydeland and S. Wang (1991). A computational method for solitary waves in a continuously stratified fluid. Stud. Appl. Math. Vol. 85, 93-127.

R. E. L. Turner and J.-M. Vanden-Broeck (1988). Broadening of interfacial solitary waves. Phys. Fluids Vol. 31, 2486-90.

P. J. H. Unna (1942). Waves and tidal streams. Nature, Lond. Vol. 149, 219-20.

H. van Haren, L. St. Louis and D. Marshall (2004). Small and mesoscale processes and their impact on the large scale: an introduction. Deep-Sea Res. II Vol. 51, 2883-7.

C. Wunsch and R. Ferrari (2004). Vertical mixing, energy, and the general circulation of the oceans. Annu. Rev. Fluid Mech. Vol. 36, 281-314.

C.-S. Yih (1960). Exact solutions for steady two-dimensional flow of a stratified fluid. J. Fluid Mech. Vol. 9. 161-74.

R. Yttervik and G. K. Furnes (2004). Current measurements on the continental slope west of Norway in an area with a pronounced two-layer density profile. Deep Sea Res. I Vol. 52, 161-78.

N. J. Zabusky and M. D. Kruskal (1965). Interactions of 'solitons' in a collisionless plasma and the recurrence of initial states. Phys. Rev. Lett. 15, 240-3.

Internal Tides. Global Field of Internal Tides and Mixing Caused by Internal Tides

Eugene Morozov

Department of Internal Waves, Shirshov Institute of Oceanology
Russian Academy of Sciences
Moscow, Russia

Abstract Different approaches to the study of internal waves in the ocean are analyzed. Generation of internal tides over submarine ridges is considered on the basis of numerical models and measurements in the ocean. Energy fluxes from submarine ridges exceed many times the fluxes from continental slopes because the dominating part of the tidal flow is directed parallel to the coastline. Submarine ridges if normal to the tidal flow form an obstacle that can cause generation of large internal waves. Internal tides are extreme when the depth of the ridge crest is comparatively small with respect to the surrounding depths. Energy fluxes from most submarine ridges were estimated. They account for approximately one fourth of the total energy loss from the barotropic tides. Model estimates were compared with the measurements on moorings at 30 study regions in the oceans. Combined calculations and measurements result in a map of global distribution of internal tide amplitudes. The study is extended to the Arctic region. Extreme internal tides were recorded near the Mascarene Ridge in the Indian Ocean, Mid-Atlantic Ridge in the South Atlantic, Great Meteor bank and the Strait of Gibraltar.

A high correlation has been found between bottom topography and vertical wavenumber spectra of vertical displacements calculated from potential temperature profiles measured by CTD instruments. Increased spectral densities of vertical wavenumber spectra are related to the presence of fine structure. The latter is caused by vertical motions in the ocean, which lead to mixing. Hence, vertical wavenumber spectra are an integral characteristic of many processes, which induce mixing: breaking of internal waves, intrusions, upwelling, frontal dynamics, etc. Spectral densities (*dropped spectra*) near submarine ridges are several times greater than in the regions far from abrupt topography where they are close to the background spectra described by the Garrett-Munk model. We found regions of enhanced mixing near the bottom in the deep Equatorial and Vema channels. This method also indicated strong mixing of Mediterranean and Atlantic waters west of the Strait of Gibraltar, mixing of North Atlantic Deep Water with Antarctic Intermediate and Antarctic Bottom waters, and mixing at the front of the North Atlantic Current.

1 Global Field of Internal Tides

Internal tides with a semidiurnal frequency are recorded almost everywhere in the ocean. Their amplitudes are generally much greater than the background level characterized by the Garrett-Munk (1972, 1975) models. Intense field measurements of internal waves, theoretical study, and numerical modeling brought us to the level of such understanding of the process that it became possible to study the mechanisms governing semidiurnal internal wave energy distribution in the ocean. The generation of internal tides occurs mainly due to the interaction of barotropic tides with bottom topography. Tidal currents flowing over the features of bottom topography obtain a vertical component. This leads to periodical vertical displacement of the isopycnals and generation of tidal internal waves.

Various methods have been used for the calculation of the generation of internal tides over bottom topography. The analysis was made for such features as continental slopes and uneven topography in the abyssal depths. These models involved either continuous stratification or a few layers. Both the ray theory approach and normal modes were used in the models. The present work is based on the model developed by P. Baines (1982) and on the measurements from moored current and temperature meters carried out from the ships of the Shirshov Institute of Oceanology of the Russian Academy of Sciences. The discussion in this sub-chapter follows the publication by Morozov (1995).

1.1 The model

We applied the Baines' (1982) model to calculate the energy of the internal waves generated over submarine ridges. The approach made by P. Baines is as follows. The linear equations governing the barotropic tide are subtracted from the equations of internal tide. The remaining equations for internal motion can be rewritten so that the fluctuations are driven by a body force **F**:

$$\frac{\partial \mathbf{u}_i}{\partial t} + \mathbf{f} \times \mathbf{u}_i + \frac{1}{\overline{\rho}_0}\nabla p_i + \frac{\overline{\rho} g \mathbf{k}}{\overline{\rho}_0} = \mathbf{F}, \tag{1.1}$$

$$\rho_{B_t} = -w_B \frac{d\rho_0}{dz}, \tag{1.2}$$

$$\mathbf{F} = -\frac{g\rho_B}{\overline{\rho}_0}\mathbf{k} = -\frac{QN^2\, zh'(x)}{\omega h^2}\mathbf{k}\sin\omega t, \tag{1.3}$$

where g is the acceleration due to gravity, p, ρ are the perturbations due to the wave motion, $p = p_B + p_i$, $\rho = \rho_B + \overline{\rho}$, $\overline{\rho}_0$ is the mean density of the ocean, ρ_B, ρ_i are density perturbations caused by the barotropic motion and internal motion, respectively; $\rho_0 = \rho_0(z)$ is the vertical distribution of density in the state of equilibrium, u is the fluid velocity, f is the Coriolis parameter, Q is the mass or more exactly volume flux, N is the Brunt-Väisala frequency, $z = -h(x)$ is the equation for the rigid ocean bottom, ω is tidal frequency, and t is time. Further, symbols in bold mean vectors and **k** the unit vector along the vertical.

The force is non-zero only in the regions of varying depth due to the presence of the first derivative of depth in the equation for the force (1.3). The equations are solved by introducing a stream function Ψ for the internal wave field:

$$u_i = -\Psi_z, \quad w_i = -\Psi_x. \tag{1.4}$$

For the solution with tidal frequency

$$\psi = \Psi(x, z)e^{-i\omega t} \tag{1.5}$$

the final equation is

$$(N^2 - \omega^2)\Psi_{xx} - (\omega^2 - f^2)\Psi_{zz} = N^2 Qz(1/h)_{xx}. \tag{1.6}$$

N is the stratification given by

$$\begin{aligned} N^2 &= 0, \qquad -d < z < 0, \\ N^2 &= N_0^2(z) + \frac{g\Delta\rho_0}{\rho_0}\delta(z+d), \qquad -h < z < -d \end{aligned} \tag{1.7}$$

where δ is the delta Dirac function, $\Delta\rho_0$ is the density difference across the seasonal thermocline $z = -d$; N_0 is constant (the value of Brunt-Väisala frequency below the upper homogeneous layer).

The Ψ function can be written as the sum $\Psi = \Psi_A + \Psi_B$, where Ψ_A is defined to have a two-layer interfacial mode and Ψ_B denotes the remainder of the internal tidal motion over the entire depth.

The boundary conditions for both functions are zero at the surface $z = 0$ and at the sea bottom $z = -h$.

Solutions are found for the two layer interfacial mode and for continuous deep stratification. The cases of flat and steep topography are studied separately. The waves induced on the continental slope propagate into the open ocean. The energy flux per meter of the shelf or ridge length is given by:

$$E_f = \int pu\,\mathrm{d}x \qquad \text{or} \qquad E_f = C_g E, \tag{1.8}$$

where E_f is the vertically integrated energy density and C_g is the group velocity. Baines (1982) calculated the energy fluxes for internal tidal motion. For steep slopes (where the inclination of the slope is close to the inclination of the curve given by the characteristic equation) the energy flux is expressed by:

$$E_f = K\rho_0 Q^2 N_0 (1 - f^2/\omega^2)^{1/2}. \tag{1.9}$$

The coefficient K is complex depending upon the stratification, the depth of the seasonal thermocline, geometry of the slope, and the depth of the shelf-break or ridge crest. The energy flux also depends on the mass flux. The shallower the shelf, the deeper the surrounding waters, the greater the energy flux and amplitude of the internal tide. The most significant contribution to the energy flux is made by the mass or volume flux, Q. For shelf areas, Baines (1982) used the coastal data of tidal amplitudes and estimated the mass flux as $Q = \omega a_0 l$, where a_0 is the amplitude of the surface elevation and l is the shelf width.

The estimates of the mass flux are needed for the calculation of energy fluxes of the internal tides generated in the open ocean over submarine ridges. It is possible to use

Table 1. Mass fluxes of tidal currents and energy fluxes of semidiurnal internal waves over several submarine ridges per one linear meter of ridge length

Ridge	Mass flux, (m^2/s)	Energy flux, (J m/s)
North Atlantic	137	6853
Mendocino	115	4741
Macquarie	126	2988
Emperor	134	5343
Mascarene	174	16507

the satellite measurements of surface elevation in the open ocean. However, we made different estimates. The mass flux was estimated as $Q = \omega HS$, where H is ocean depth, S is the horizontal displacement of a water particle by the tide during half of the tidal period. $S = 2u_0 \sin \theta \ / \ \omega$, where θ is the angle between the direction of the tide and the ridge. In order to estimate the amplitudes of tidal velocities we used the results of tide numerical modeling (Accad and Pekeris, 1978; Gordeyev et al., 1975) and also the measurements from moored current meters where detailed vertical structure of the currents was available.

The velocity time series were band filtered so that only the tidal motions remained. The time series were then vertically averaged to obtain the velocities of the barotropic tide and to eliminate the tidal internal waves. In those areas, where we had current measurements, the results of numerical modeling and averaging of current measurements did not differ greatly. The estimates of the velocities were taken in the range 1-2 cm/s depending on the site of measurements.

The bottom characteristics of the ridges were taken from navigation charts and ETOPO bottom topography data set in the Internet. The stratification parameters were taken from the World Ocean Atlas (1975, 1977) and WOCE data set. Table 1 gives the mass fluxes and energy fluxes for five ridges, near which we had mooring arrays: North Atlantic (30°N); Mendocino (west of California); Macquarie (south of New Zealand); Emperor seamounts (northwest Pacific); Mascaren Ridge (western Indian Ocean).

The amplitudes of internal tides can be estimated from the known energy flux. Using equation $E_f = C_g E$ we can calculate the energy density. Here, C_g is the group velocity. Actually, the group velocity is the partial derivative of the frequency to the wavenumber. It can be estimated from the following equation, which is a simplification of Eq. (2) from Garrett and Munk (1979).

$$C_g = \frac{kN}{\omega k_z^2}. \tag{1.10}$$

The horizontal wavenumber k was calculated by integrating the internal wave equation using the measured Brunt-Väisala frequency distribution with depth:

$$\frac{d^2 w}{dz^2} + \frac{N^2(z) - \omega^2}{\omega^2 - f^2} w k^2 = 0. \tag{1.11}$$

Here $N^2(z)$ is the squared Brunt-Väisala frequency, ω is the semidiurnal tidal frequency, and f is the Coriolis parameter. The vertical velocity (the boundary conditions) at the surface and bottom is zero.

The vertical wavenumber k_z was calculated from the ray trajectories

$$\frac{dx}{dz} = \frac{k_z}{k} = \sqrt{\frac{N^2(z) - \omega^2}{\omega^2 - f^2}}. \tag{1.12}$$

The energy density is given by: $E = 0.5\rho_0 V^2$, the vertical velocities far from the surface and bottom are $w = v/\cos\phi$, where $\cos\phi = \omega/N$. If we make a vertical integration of energy, then:

$$E = \tfrac{1}{2}\rho v^2 H \tag{1.13}$$

where v is vertically integrated velocity, and $w = v/\cos\phi$ the vertically-averaged vertical velocity. Thus, the amplitudes of vertical displacements are $\zeta = 2w\omega$.

The total energy fluxes for the oceans are as follows: 4×10^{11} W for the Atlantic; 4×10^{11} W for the Pacific; 3×10^{11} W for the Indian Ocean. The total amount for the World Ocean (11×10^{11} W) is about one-fourth of the barotropic tide dissipation 4.3×10^{12} W (Cartwright, 1977). The recent estimate of tidal dissipation from astronomy is equal to $(2.50 \pm 0.05) \times 10^{12}$ W (TW) from M2 only. Previous estimates allocated this dissipation in the turbulent bottom boundary layers of marginal seas. Recent TOPEX/POSEIDON altimetric estimates combined with dynamical models suggest that 0.6-0.9 TW may occur due to abyssal mixing (Munk and Wunsch, 1998). Baines (1982) estimated that less than 1% of tidal energy is transferred into the energy of tidal internal waves over continental slopes. Bell (1975) estimated that about 10% of the tidal energy is spent on internal tidal generation over uneven bottom topography in deep basins. All values are uncertain.

The energy fluxes of internal waves generated over submarine ridges in general exceed several times those internal waves generated over continental slopes. The currents of the barotropic tidal wave in most parts of the ocean are parallel to the shores, and only a small part of the mass flux crosses the slope line. Many submarine ridges are perpendicular to the tidal currents and form an obstacle that provides an intense internal wave generation. Nevertheless, there is evidence of large internal waves amplitudes generated over continental slopes. Pingree and New (1991) recorded high amplitudes in the Bay of Biscay, Boyce (1962) measured large amplitudes in the Strait of Gibraltar. These intense tidal internal waves are generated, when a great part of the tidal flow crosses the continental slope. Baines (1982) predicted high amplitudes in the Bay of Biscay.

1.2 Measurements

Let us assume that the submarine ridges are the areas of intense internal wave generation and let us analyze the properties of tidal internal waves in relation to the position of the experimental areas to the nearest submarine ridge following Morozov (1995). The data analyzed here are mostly temperature measurements from moored buoys. The data of approximately 1000 moorings in 30 various regions of the ocean were processed. Amplitudes, directions, and wavelengths of semidiurnal internal tidal waves were calculated.

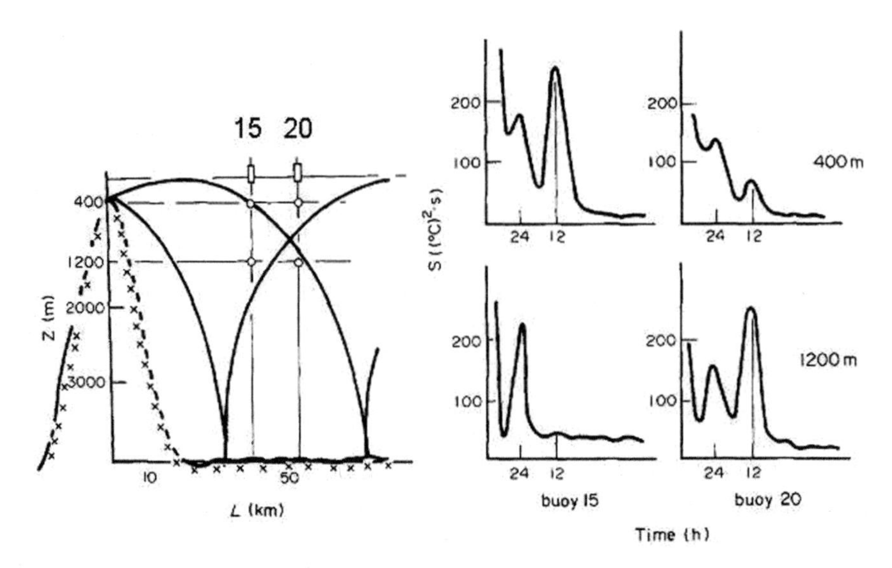

Figure 1. Scheme of ray trajectories for semidiurnal oscillations with positions of moorings 15 and 20 near Henderson seamount in the Eastern Pacific and the temperature spectra measured at levels of 400 and 1200 m on buoys 15 and 20. Printed with permission from Elsevier.

The amplitudes were calculated by dividing the semidiurnal temperature amplitudes by the mean vertical temperature gradient. The direction and wavelength were estimated by calculating spatiotemporal spectra for semidiurnal frequency (Barber, 1963) in the regions where we had clusters of moorings with temperature meters at one level on several moored buoys. Let us briefly describe the measurements carried out in different regions.

Henderson seamount in the Eastern Pacific (25°N, 119°W). This is a solitary seamount in a deep basin with its summit located approximately 400 m below the surface. Ray propagation of internal tidal fluctuations has been observed here. The trajectories of the rays can be calculated using equation (1.12). The instruments located on the ray trajectories recorded intense temperature fluctuations, which resulted in the calculation of sharp spectral peaks. The trajectory with the most intense fluctuations starts from the very top of the seamount (Baines, 1973). Instruments not located on the trajectories did not record any tidal peaks. The spectra for buoy 15 (figure 1), level 400, and buoy 20, level 1200 m located on the ray path have intense peaks, whereas the 12-hour peaks for buoy 15, level 1200 m and buoy 20, level 400 m are weak. The slant coherence between the time series characterized by intense semidiurnal peaks is 0.74. The coherences between other pairs of time series are low.

Mascarene Ridge in the western Indian Ocean. Very intense tidal internal waves were measured in this region. A chart of the region with the location of moorings is shown

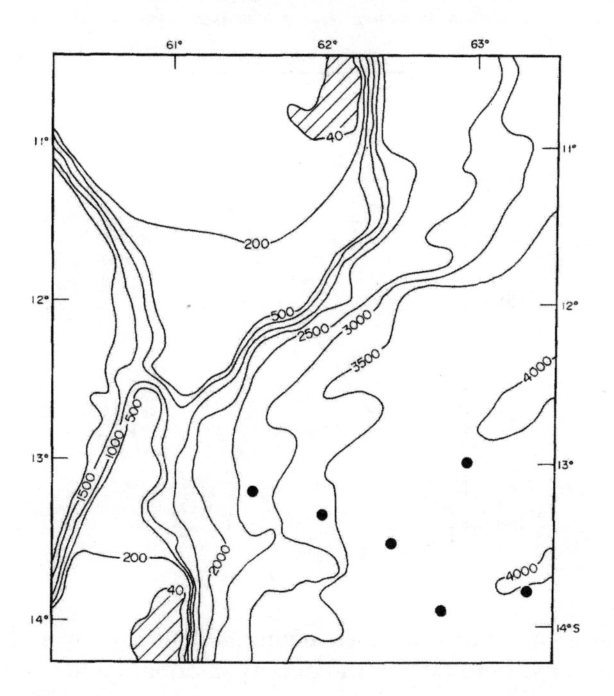

Figure 2. Chart of moorings (dots) at the Mascarene Ridge. Printed with permission from Elsevier.

in figure 2. The largest vertical displacements associated with internal tides were equal to 150-160 m. The measured amplitudes at several levels together with the eigen function for the first mode show that the first mode approximation for the internal tide motion is quite satisfactory (figure 3). Calculations give 90-95% as the estimate of the contribution of the first mode to the total amount of the wave energy. The amplitudes are affected by the spring neap cycle and vary from 60 to 150 m (double amplitudes). Horizontal coherences are high and exceed the 95% confidence level (figure 4a). The spring neap 14-day variability occurs on the eastern buoy with a phase delay of one day later than on the western buoy. This gives an estimate of 2.3 m/s for the group velocity of the waves. The amplitudes of the internal tide are very large because the currents of the barotropic tide are normal to the ridge. The amplitudes of the currents associated with the barotropic tide are also large, and the geometry of the slope provides intense internal wave generation. The shallow banks (40 m), the shallow channel between the banks (mean depth is 400 m), and deep waters around the ridge (4000 m) form a geometry appropriate for the generation of intense internal waves. These factors combined provide exclusive conditions for the generation of internal tides of high amplitude.

Region, 600 km south of the Mendocino Ridge, 700 km west of San Francisco. The detected waves propagate south from the ridge, and there are no waves recorded that propagate from the continental shelf. The amplitudes are approximately equal to 20 m.

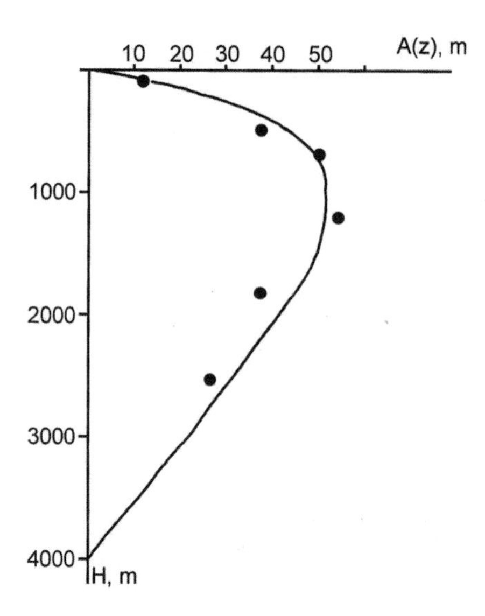

Figure 3. First mode eigen function for semidiurnal internal waves. The Brunt-Väisala frequency was calculated from the vertical density distribution in the Mascarene region. The solid circles show the amplitudes of waves measured by moored temperature meters. Printed with permission from Elsevier.

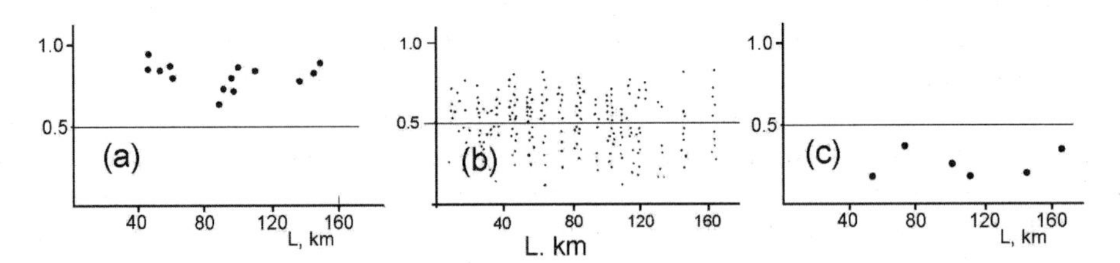

Figure 4. Horizontal coherence as function of the horizontal separation between moorings: (a) measurements near the Mascarene Ridge (200 m); (b) measurements in the central Atlantic, Polygon-70 experiment (200 m). (c) measurements near Kerguelen Island (300 m). Printed with permission from Elsevier.

These measurements became a turning point in our understanding of the tidal internal wave generation. The waves were expected to propagate from the continental shelf, but actually the strongest internal tide was generated over the Mendocino Ridge.

East of Macquarie Island and south of New Zealand. This is an example of recording of two wave systems of internal tide. One system of waves was recorded propagating from the Macquarie Ridge with amplitudes of 15-20 m, the other propagated from the South Pacific Ridge with smaller amplitudes.

The North Atlantic (29°N), east of the Mid-Atlantic Ridge (MAR). This is another example of recording two wave systems. The two wave systems were recorded propagating from two ridges of the MAR system with amplitudes of 30-40 m.

Four sites near the equator of the Indian Ocean: 85°E, 75°E, 65°E, and 55°E. Large amplitudes of internal tides were recorded near the Mascarene (55°E) and Maldivian (75°E) ridges with amplitudes reaching 40 m. The amplitudes in the study areas near the Arabian-Indian Ridge (65°E) and the Eastern Indian Ridge (90°E) are smaller and do not exceed 25 m. The crests of the two latter ridges are much deeper.

The South Atlantic (21°S) near Brazil, Trindadi and Martin Vaz Islands. Vertical displacements of 50 m seaward of the banks were recorded in this region. It is noteworthy that the barotropic tide in the South Atlantic is very strong. Thus, an obstacle to the currents of the barotropic tide (ridges and islands) cause intense generation of the internal tide.

Kusu-Palau Ridge south of Japan (26°N). This is an example of strong internal tide generation caused by strong barotropic tide and steep submarine ridge. Vertical displacements caused by the internal tide were equal to 60 m.

South of Iceland (54°N, 27°W). This is another example of strong generation of internal tide over steep slopes of the ridge. Vertical displacements were equal to 60 m. The buoy was deployed at a distance of 300 km from the Reykjanes Ridge.

Region east of the Great Meteor banks in the North Atlantic (31°N, 26°W). Steep slopes of the Great Meteor banks with flat summits at a depth of 300 m together with strong barotropic tide cause intense generation of internal tide. The maximal vertical displacements were equal to 100 m. The wave of the first mode propagated east from the banks, where the mooring array was deployed. The wavelength was estimated as 160 km. The cause of intense internal tide generation was the same as near the Mascarene Ridge. The moorings in the last three sites were set after the work with the model was completed and high amplitudes of the internal tide in these regions were predicted by the calculations.

Northwestern Pacific region. This is a region where many moorings were deployed, which made it possible to estimate the direction of the internal tide propagation with high accuracy. Five moorings were deployed in 1981, and 188 buoys in 1987 (Megapolygon). The distance from the Emperor Ridge was 500 km. The waves propagating from the Emperor seamounts were recorded in both experiments. The amplitudes were approximately equal to 20 m.

Atlantic Polygon-70 with 17 buoys deployed in 1970 and Mesopolygon with 70 buoys deployed in 1985 almost in the same region 16-20°N, 33-37°W. The experiments were carried out in the Cabo Verde Basin far from the Mid-Atlantic Ridge and far from the continental slopes. These regions represent the examples of bottom internal tide generation. The waves were generated over an uneven bottom: many hills 500-1000 m high above the ocean floor serve as the spread source area. The measurements were made just over the generation area. This explains why several modes of the internal tide were recorded. Usually high modes of the internal tide decay rapidly. The amplitudes were moderate (20 m). Due to the presence of the multi-modal structure the coherences are random (figure 4b).

Madagascar Basin. The experiment site was located far from the ridges. Extraordinary variability was recorded at six moorings 10 km apart from one another. Intense peaks were recorded on different levels of each mooring with almost complete absence of peaks at the same level on other moorings. A ray trajectory of 12-hour oscillation can be traced from a seamount 1000 m above the bottom explaining the semidiurnal oscillations measured by several instruments located on this trajectory, while the other meters set out of the trajectory measured only the background spectra (figure 5).

Sargasso Sea, POLYMODE, Array-1, and Array-2. The measurements were carried out on many moorings during a long time period. Measurements were carried out in the deep basin far from slopes and ridges. Very weak and unstable waves were recorded. The amplitudes of the internal tide were close to the background level.

Crozet Basin north of Kerguelen Island. A linear mooring array was deployed not perpendicular to the ridge as usual but almost parallel to the ridge. Very weak 12-hour oscillations were recorded. Most of the power spectra had no peaks. The coherences were insignificant (figure 4c). Barotropic tidal currents are parallel to the "flat" ridge and thus, cannot cause significant internal wave generation.

The power spectra of the vertical displacements caused by the internal tide in different regions are plotted together in one graph (figure 6). One can see significant difference between the intense and weak internal tides. The coherences also differ greatly (figure 4).

1.3 Discussion about the Global Field of Internal Tides

The observations in many study regions of the ocean indicate that internal waves with a period of 12.4 h caused by the M2 component of the barotropic wave are the most

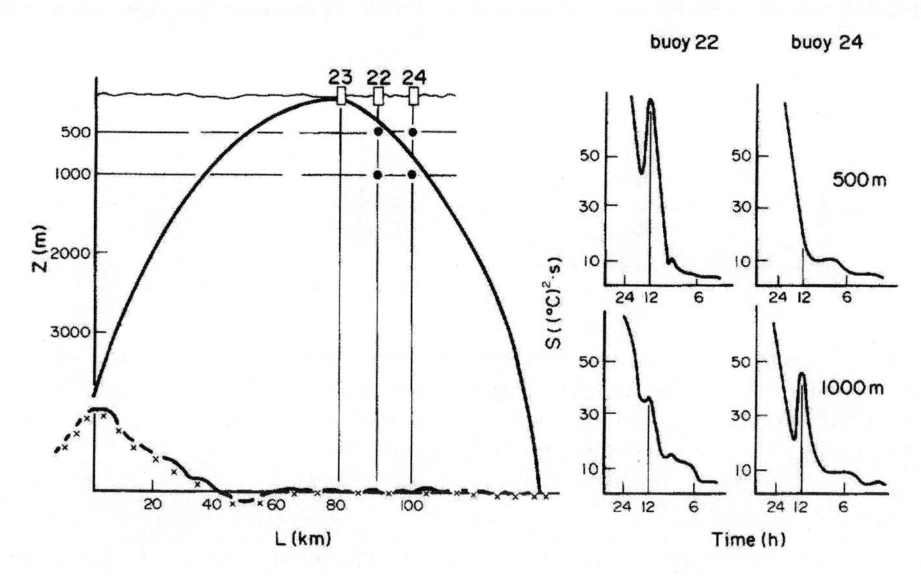

Figure 5. Scheme of ray trajectories in the Madagascar region on moorings 23, 22, and 24. Temperature spectra were measured at levels 500 and 1000 m. Printed with permission from Elsevier.

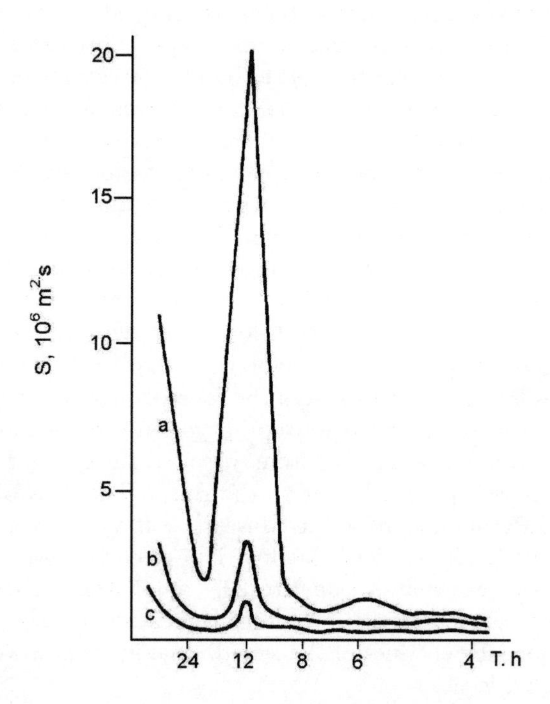

Figure 6. Spectra of vertical displacements in Mascarene (a), Polygon-70 (b), and Kerguelen (c) regions. Printed with permission from Elsevier.

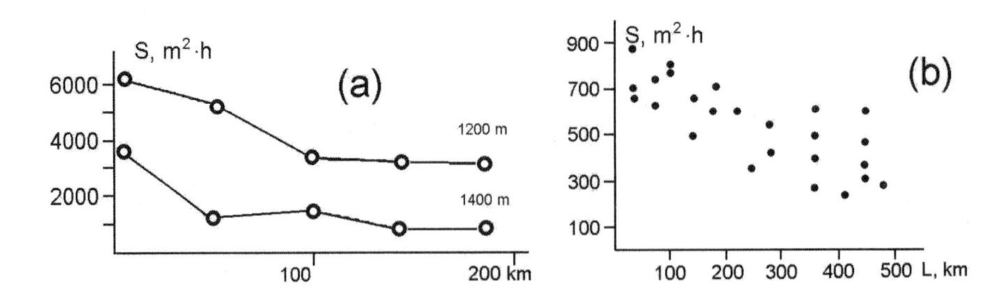

Figure 7. Horizontal decay of semidiurnal internal oscillations. Estimates of vertical displacement spectra (m2h) are plotted as function of distance from the ridge: Mascarene region (1000 and 1200 m) (a) and northwestern Pacific (1200 m), as distance from the Emperor Ridge (b). Printed with permission from Elsevier.

energetic ones. Let us return back to the model. The energy fluxes were estimated for most of submarine ridges in the ocean. The total amount of energy was estimated as one-fourth of the barotropic tide loss. Internal wave amplitudes were calculated on the basis of energy fluxes. The greatest amplitudes were found close to the ridges where the surface elevations in the open ocean due to the barotropic tide are large. The main regions of the generation of internal tides are submarine ridges, continental slopes, and isolated seamounts. In the case when the currents of the barotropic tide are normal to a submarine ridge, their efficiency to generate the internal tide is maximal. The geometrical form of the bottom topography is the other factor facilitating intense generation of internal tides. If the crest of the ridge is shallow and the surrounding waters are deep the generation of internal tides increases. The maximal generation of internal tides takes place when the slopes of the ridge are close to the inclination of the characteristic of the internal wave determined from equation (1.12). Near the ridges, the first mode is formed. In the course of the wave propagation and due to additional wave generation over uneven topography, the oscillations caused by internal wave transform from deterministic to random.

One of the main objectives of this sub-chapter was to map the internal tide amplitudes in the ocean. Besides the energy fluxes from the source areas, an estimate of the internal wave energy decay in the course of the wave propagation is needed for this purpose. The estimate of the wave decay was evaluated from those measurements where the long arrays of moorings were deployed perpendicular to the ridge. The data of the measurements in the northwestern Pacific and near the Mascarene Ridge were processed (figure 7). The estimate assumed in the model is 30-40% loss of energy over the distance of one wavelength, which is approximately equal to 20% amplitude decrease. Of course, this estimate is very rough. Near the ridge, the decay is greater. In deep basins far away from the areas of generation, the energies of the semidiurnal internal waves tend to approach the natural background level.

Thus, it became possible to make a map of the internal tide amplitudes in the ocean (figure 8). The map combines calculations and measurements. The greatest amplitudes were recorded near the ridges, where the natural conditions favor the intense internal

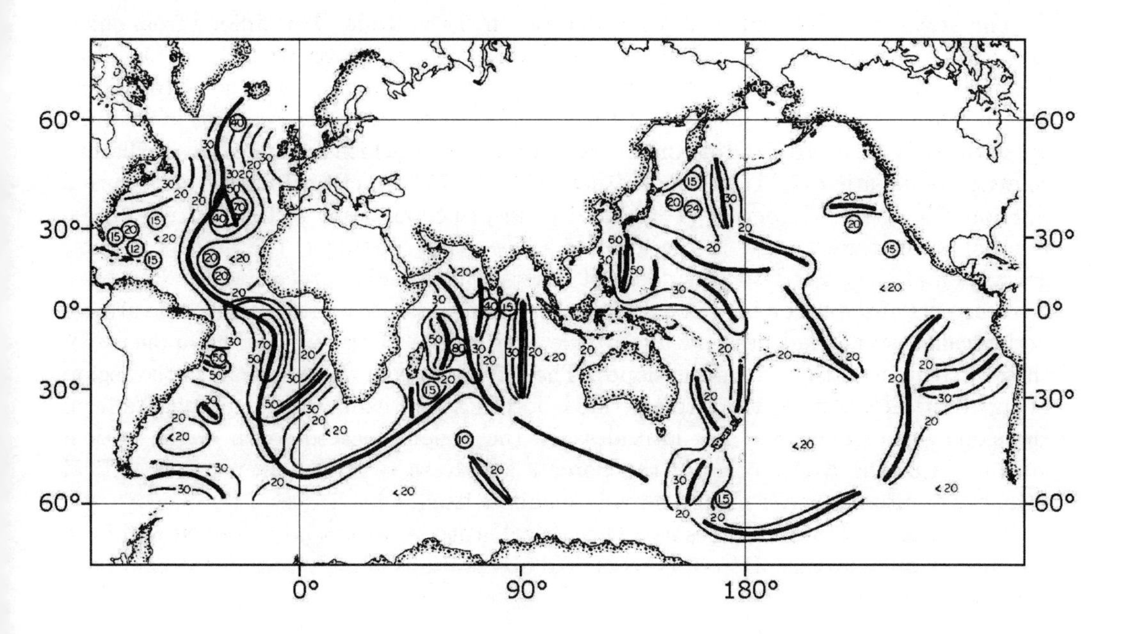

Figure 8. Map of semidiurnal internal waves double amplitudes (meters). Wide lines show the ridges, thin solid lines are the calculated contour lines of double amplitudes of semidiurnal internal waves. The numerals in the circles not associated with the contour lines are the measured values on moored buoys. Printed with permission from Elsevier.

tide generation. Such ridges include the Mascarene and Kusu Palau ridges, the Great Meteor banks, and obviously the Mid-Atlantic Ridge in the South Atlantic and Walvis Ridge.

Of course the map is very rough as it presents only the general outline of the amplitude distribution, which should be understood as "large", "moderate", and "small". The mapped amplitudes must be regarded as "vertically averaged" and also "temporally averaged" with a half month (spring-neap) time scale.

2 Internal Tide at High Latitudes

In the previous section we demonstrated that internal waves of the tidal period transform the velocity field of the barotropic tide so that the relatively low velocities of the barotropic tide form a field with greater velocities in certain regions of the sea associated with abrupt topography. The objective of this sub-chapter is to extend the previous results to polar latitudes and determine the spatial variability of the internal tide at high latitude sea on the basis of the generation of the internal tide resulting from the interaction of the barotropic tide with bottom topography. The discussion in this sub-chapter follows the publications in (Morozov and Pisarev, 2002; Morozov, Neiman et al., 2003)

The properties of the internal waves depend on the latitude. This follows from equation (1.11) through the dependence of the Coriolis parameter on latitude. This effect becomes stronger at high latitudes. With the approach to the polar latitudes, the difference between the tidal frequency and the Coriolis parameter decreases, and at latitude approximately equal to 75° this difference becomes zero, which excludes the oscillatory solutions of equation (1.11). Thus, latitudes close to 75° are critical for the semidiurnal internal tide of the M2 period ($T = 12.4$ h). In principle, north of this latitude, these internal waves cannot exist in the form of a free wave. However, the process of their generation at the polar latitudes does not differ from a similar process at low latitudes. The currents of the barotropic tide flow over submarine slopes and the isopycnal surfaces periodically ascend and descend due to the appearance of the vertical components of these currents near the submarine slopes. The difference is in the fact that, in the region of supercritical latitudes, the internal tide is forced and it cannot freely propagate from the slope as in the case of low latitudes. In the regions located south of the critical latitude, near the Arctic coast of the Barents and Kara seas, the effect of the critical behavior of internal waves should already manifest itself.

The behavior of internal tides near the critical latitude has not been studied well from the theoretical approach. Descriptions of field observations of high latitude internal tides are also not numerous.

2.1 Numerical model

We used a two-dimensional model of internal wave generation and propagation. The coordinate axes in the model were directed normal to the coastline and downward. Internal waves are considered flat and propagating normal to the coast. The assumption that the wave processes are two-dimensional almost does not change the characteristics of the internal wave field in the shelf and continental slope regions since the variability of the internal waves in the direction normal to the coast is usually much stronger than their variability along the coastline. The application of a three-dimensional approach would require powerful computers and developing a three-dimensional model.

The study is carried out using the nonlinear model developed by Vlasenko (1991). The latest development of this model has been already used to calculate internal wave dynamics in the ocean (Morozov et al., 2002; Vlasenko and Hutter, 2002). We consider a two-dimensional (x, z) flow in a continuously stratified rotating ocean of variable depth. Internal waves are described by the following system of equations:

$$\begin{aligned} &\Omega_t + J(\Omega, \Psi) - fV_z = \frac{g\rho_x}{\rho_0} + K(x)\Omega_{xx} + K\Omega_{zz} + (K\Psi_{zz})_z, \\ &V_t + J(V, \Psi) + f\Psi_z = K(x)V_{xx} + (KV_z)_z, \\ &\rho_t + J(\rho, \Psi) + \frac{\rho_0 N^2(z)}{g}\Psi_x = R(x)\rho_{xx} + (R\rho_z)_z + (R\rho_{0_z})_z, \end{aligned} \tag{2.1}$$

where Ψ is the stream function, $\Psi_z = U$, $\Psi_x = -W$, $\Omega = \Psi_{xx} + \Psi_{zz}$ is the vorticity, (U, V, W) is the velocity vector, N is the Brunt-Väisala frequency, ρ is the density disturbance due to the wave motion, ρ_0 is the mean density, f is the Coriolis parameter,

$K, K(x), R$, and $R(x)$ are the vertical and horizontal coefficients of turbulent viscosity and mass diffusivity, respectively, J is the Jacobian and g is the acceleration due to gravity.

The model is two dimensional. However, we introduce the equation for the V-component of velocity normal to the x, z plane to account for the effects of rotation. Here, the V-component varies only in x, z plane and does not change along the transversal y-axis. The boundary conditions at the surface, $z = 0$ are:

$$\rho_z = 0, \ \ \Omega = 0, \ \ \Psi = 0. \tag{2.2}$$

We assume zero vertical motion and no tangential stresses at the surface.

Zero vorticity means the absence of tangential stresses at the free surface. Zero stream function determines zero vertical motion at $z = 0$. Zero density variations by the vertical at the sea surface follow from the assumption that heat and salt transport across the surface are zero. This assumption presupposes that precipitation, evaporation, and ice formation are negligible for internal wave processes.

At the bottom, we assume zero heat and mass transport:

$$\text{at} \ \ z = -H(x), \ \ \partial\rho/\partial n = 0, \ \ \Psi = \Psi_0 \sin \omega t, \tag{2.3}$$

where ω is the tidal frequency, n denotes the unit normal vector to the bottom and Ψ_0 is the amplitude of the mass transport in a barotropic tidal current. The boundary condition for vorticity at the bottom is calculated using equation $\Omega = \Delta\Psi$ with the value of the Ψ field obtained at the previous time step.

The wave perturbations of vorticity, stream function, and density are assumed zero at the lateral boundaries located far from the bottom irregularities at the submarine ridge. We stop the calculations when the wave perturbations reach the lateral boundaries. Stratification was specified in each of the 20 layers. The bottom topography was specified on the basis of navigation charts.

The calculation starts from the state of rest:

$$\text{at} \ \ t = 0: \ \ \Omega = 0, \ \ \rho = 0, \ \ \Psi = 0 \tag{2.4}$$

We specify the density field, which is not disturbed by internal waves with the corresponding distribution of the Brunt-Väisala frequency. A long tidal barotropic wave propagates from the open part of the sea to the continental slope. The tidal currents related to this wave obtain a vertical component when they approach the continental slope, which is an obstacle in the path of their propagation. The vertical components, which change periodically, cause vertical oscillations of isopycnal surfaces, and the internal waves of the tidal period are generated due to this mechanism. Thus, the problem of calculating the parameters of the baroclinic wave perturbations is solved on the basis of specified parameters of the incident wave, stratification, and bottom topography.

2.2 Numerical experiments to study internal tides

The numerical model allows us to analyze two-dimensional distributions of internal tide properties. In order to evaluate the three-dimensional distribution of amplitudes we used the approach based on the strip-theory. This theory is used in aerodynamics to calculate the airflow around an airplane wing. This approach allows one to determine the three-dimensional field pattern on the basis of calculations using a two-dimensional model if the variations of the properties along the transversal coordinate are significantly smaller than along the longitudinal coordinate. In our case, the longitudinal coordinate is directed normal to the slope. We divide the sea into strips (20-50 km wide) with approximately the same properties across the strips so that the variations parallel to the coastline in bottom topography, stratification, and barotropic tide are not strong. Thus, the strips for the calculations are located normal to the isobaths. Since the main variability associated with internal waves is found normal to the slopes, the entire basin of the sea was divided into 110 strips. The calculations of the internal tide generation and propagation were made in each of the strips. Within each strip we specified the model topography of the bottom, characteristic stratification, and currents of the barotropic tide.

Water transport by the currents of the barotropic tide in each zone was specified by recalculating the currents of the barotropic tide into the stream function. (Model simulations of the barotropic tide in the Barents Sea has been presented by Gjevik et al., 1994.) The currents of the barotropic tide were calculated from the TOPEX/POSEIDON satellite data from the NASA database at Oregon State University (USA). We used the TPXO.5.1 global inverse tidal model (http://www.oce.orst.edu/po/research/tide/) with satellite data assimilation. The model was developed on the basis of the OTIS software (Oregon State University Tidal Inversion Software Model) (Egbert and Erofeeva, 2002). The ellipses of the tidal currents at characteristic points of the basin are shown in figure 9. The sea bottom was specified on the basis the ETOPO database. The typical parameters of the stratification were specified on the basis of the Russian-US Oceanographic Atlas of the Polar Ocean. We used the climatic data averaged over approximately 50 years of observations.

The barotropic tidal currents flowing across this slope obtain a vertical component. The periodical vertical component causes displacements of the isopycnal surfaces, thus generating internal waves. Since the eigen function for the vertical oscillations induced by internal waves is zero at the surface and at the bottom and its maximum is located approximately in the middle of the water column, the maximum vertical velocities should be found precisely in the middle of the column. The eigen function for the horizontal velocities has maxima at the bottom and at the surface; therefore, internal waves should cause relatively high near-bottom and surface currents in certain regions of the shelf and slope.

Strong stratification causes large amplitudes of the internal waves and associated currents and increases the length of the internal wave. The calculations carried out with the numerical model allow us to estimate the amplitude of the internal waves and internal wave energy fluxes in the immediate vicinity of the generation region. Tidal currents flowing over the rough features of steep bottom slopes attain a vertical component, which leads to the vertical displacements of density contour lines. As a result, the internal wave is generated with the frequency of the main perturbation, that is, the semidiurnal tide frequency.

For the model calculations, we used a work field 80 km long with a horizontal step

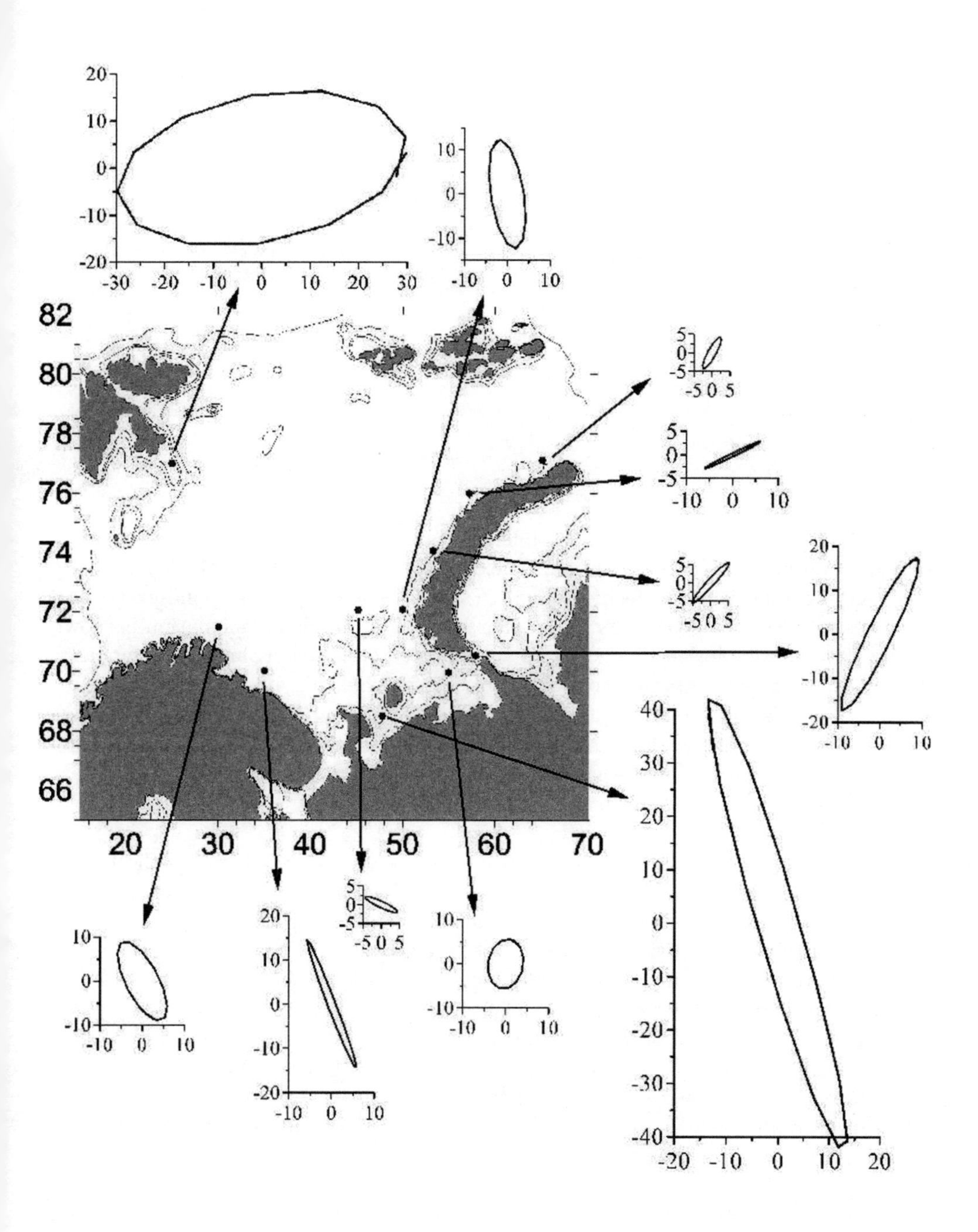

Figure 9. Ellipses of the barotropic tide currents at certain typical points of the Barents Sea obtained from the TOPEX/POSEIDON satellite data. Isobaths of 25, 50, 100, and 500 m are shown with thin lines. Velocity scale of the ellipses are cm/s. Reproduced with permission by Nauka.

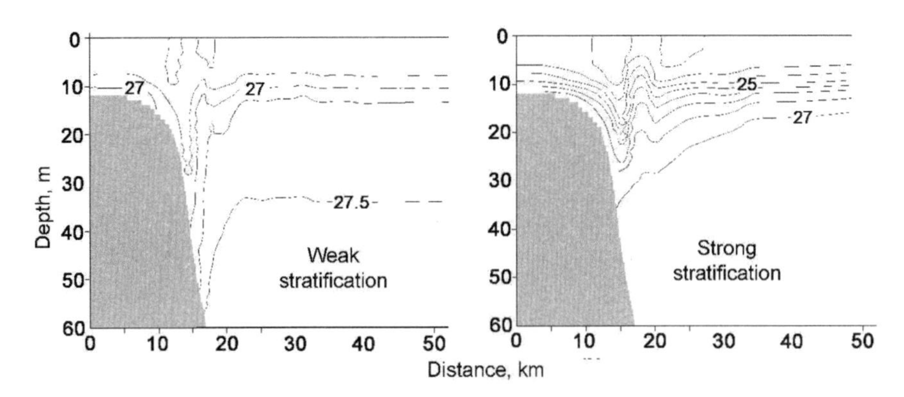

Figure 10. Field of the isopycnal surfaces for strong and weak stratification. Reproduced with permission by Nauka.

of 100 m and 20 vertical levels. The time step was equal to 2 s. These parameters satisfy the Courant-Friedrichs condition. The coefficients of the horizontal eddy viscosity and density diffusivity were specified equal to 1.5 m^2/s and the corresponding vertical coefficients were equal to 0.0001 m^2/s. A characteristic example of calculation is shown in figure 10. The calculations in each strip were plotted on the chart and then smoothed to obtain a pattern with gradual spatial changes in internal tide amplitudes.

The decay of internal tide amplitudes north of the critical latitude is stronger than in the southern part of the Barents Sea. In the region of critical latitudes, the internal tide decays at a distance smaller than one wavelength. It is noteworthy that at moderate latitudes the internal tide can propagate over distances exceeding 10 wavelengths (~1500 km).

The frequency range for the existence of internal waves extends between the Brunt-Väisala frequency and the local inertial frequency. The geographical location of the Arctic shelf at latitudes greater than 70°N leads to the following effect. The latitude of the Arctic shelf is close to critical with respect to the frequency of the M2 wave; therefore, the wave cannot exist here in non-degenerate form. However, it should be generated over the continental slope, and in a certain vicinity of the slope it would exist as a forced wave. The wave decays when it propagates either toward the coast or toward the ocean. The decay of the wave in the shelf zone would generate packets of nonlinear short-periodic internal waves, which should produce intensive bottom currents.

As a result of dividing the sea into many calculation areas and using the data for the vertical stratification and barotropic currents in each of the areas we obtained a pattern of the distribution of internal tide amplitudes over the basin of the sea during the summer period. The coefficients of turbulent viscosity and diffusion used in the model calculation were chosen so that the resulting amplitudes would be very close to the measurements of internal tide at a few moorings in the open sea and in the Kara Strait between the Barents and Kara seas. The obtained chart is shown in figure 11. Different gradation of gray color in the sea basin corresponds to the amplitude of vertical displacement in

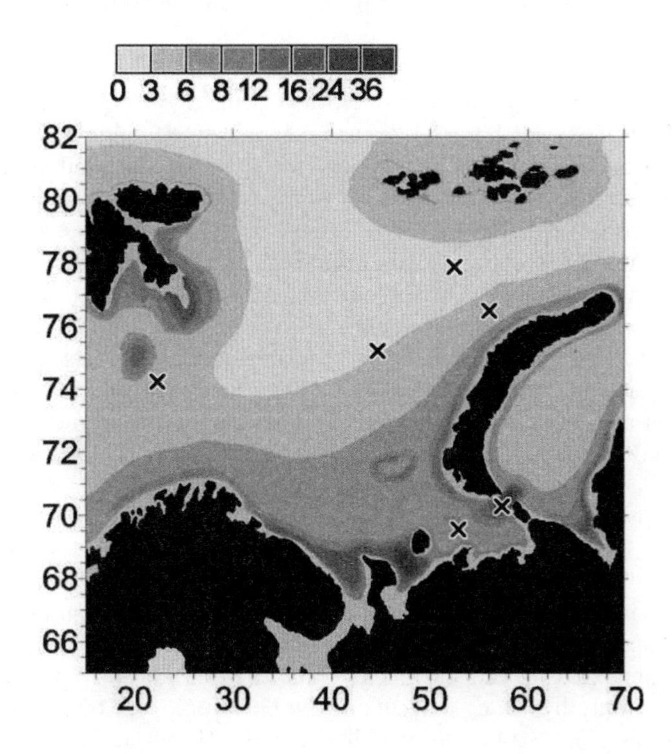

Figure 11. Chart of internal tide amplitudes in the Barents Sea for the summer season (in meters). Gradations of amplitudes (0-36 m) are shown in the upper part of the figure. Crosses indicate locations of moorings. Reproduced with permission by Nauka.

meters.

The amplitudes obtained from model calculations were compared with measurements on moorings. Six sites of mooring data were available. Their locations are shown in figure 11 with crosses. The difference between measured and calculated amplitudes were within 15%, which can be assumed as the accuracy of the chart.

One can distinguish a few regions with large amplitudes of the internal tides. Extreme amplitudes are observed in the Kara Strait between the Barents and Kara seas. The depth of the sea here is about 120-150 m, while the amplitudes can reach 40-50 m, and the corresponding wave heights (double amplitudes) are almost equal to the sea depth. Thus, internal tides occupy almost the total depth from surface to bottom. This effect is observed in some other straits, for example, in the Strait of Gibraltar.

The amplitudes of internal tides (about 10 m) along the northern slope of the Kola Peninsula (69°N, 39°E) are related to the steep slopes along the coast and strong barotropic tide. Large amplitudes of internal tides between the Kola and Kanin peninsulas (69°N, 42°E) and around the Kanin Peninsula are caused by significant currents of the barotropic tide despite the fact that the slope of the bottom is flatter than in the northern part of the Kola Peninsula.

Large amplitudes of the internal tide found in the region of the Geese Bank (Gusinaya Banka) (71°N, 46°E) and western coast of Novaya Zemlya (74°N, 53°E) are determined by the joint effect of strong barotropic currents and steep submarine slopes. Large amplitudes in the northern part of the sea over moderately steep submarine slopes are caused by the physical mechanism of internal tide generation at critical and supercritical latitudes where a very steep slope is not necessary for internal tide generation.

Interesting results were obtained comparing the chart of internal tide amplitudes and the chart of water turbidity in the surface layer based on the satellite SeaWiFS (Sea-viewing Wide Field-of-view Sensor) ocean color scanner data. The chart of water turbidity (figure 12) was obtained using a special algorithm for determining the concentration of suspension matter in the Barents Sea waters on the basis of the SeaWiFS data (Burenkov et al., 2001). A comparison of these charts demonstrates similar features in the distribution of internal tide amplitudes and suspension matter concentration in shallow waters. The correlation between the internal tide amplitudes with water turbidity at shallow depths is explained by the fact that strong bottom currents induced by the internal waves facilitate stirring up bottom sediments and their mixing in the entire water column (see also Gjevik et al., 1994). The general features, which are well seen on the chart, are increased turbidity in the region of the Kanin Peninsula, where the currents of the barotropic tide and the amplitudes of the internal tide are large. In the shallow water regions close to the strait connecting the Barents and White seas increased turbidity is caused only by strong currents of the barotropic tide over shallow depths of the strait. Near the Kara Strait the turbidity is high. However, the strait was covered with ice and we could not get satellite data on turbidity. High turbidity is also observed along the northwestern coast of Novaya Zemlya where we estimate large amplitudes of the internal tide. The observed correlation between the charts is well justified in the shallow regions. In the center of the sea, both the suspended matter concentration and amplitudes of internal tides are small. Here the concentration of suspension matter can be determined by other factors.

Finally we can make the following conclusions related to internal tide in the Arctic region. In the region of the Arctic coast, the barotropic tide generates an internal tide, whose influence most strongly manifests itself near different features of uneven bottom topography. The internal tide, even in the regions located south of the critical latitude, practically does not develop into a free propagating internal wave. The fluctuations of the isopycnal surfaces with significant amplitudes are observed only near bottom slopes. It is likely that here internal waves lose their energy, which is transferred to the energy of short-period waves and into the potential energy of the mixed water column. This should be confirmed in the distribution of the hydrological properties. More mixed layers should be located near submarine slopes.

The calculations made with the numerical model indicate that the bottom slopes characteristic of the region under study can lead to the generation of a forced internal tide. Since the stratification and bottom slopes are irregular, the internal wave field should also have a patchy structure. The maximum amplitudes would be observed at the places where the inclination of the characteristic lines for the internal tides would

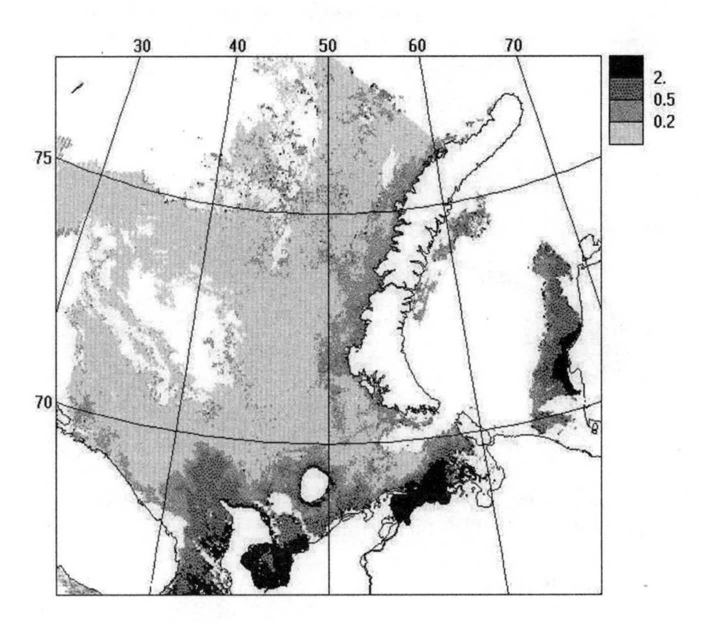

Figure 12. Distribution of the suspended matter in the Barents Sea based on the satellite SeaWiFS color scanner data in July-August 1998. Gradations of the suspended matter concentration (2, 1, and 0 mg/l) are shown in the right-hand part of the figure. White color denotes missing data. Reproduced with permission by Nauka.

randomly coincide with the bottom slope. In order to form large amplitudes of the internal tide, the bottom slopes should not necessarily be as steep as they should be in the regions of low and mid-latitudes. Due to the fact that the inclination of the curves of the characteristic equation for the internal waves in the Arctic latitudes is significantly smaller, large amplitudes of the internal tide can be generated even at small bottom inclinations. The proximity of the critical latitude for the internal tide, which is determined not only by the latitude of the place (approximately 75°N) but also by the vorticity of the currents, will play a significant role here.

3 Internal tides in the Kara Strait

The Kara Strait (figure 13) connects the Barents and Kara seas in the Arctic Ocean. The Barents Sea is a relatively warm sea because the warm water from the Atlantic Ocean is transported into the sea, and the southern part of the sea usually does not freeze in winter. On the contrary, the Kara Sea is a very cold sea. The ice cover in the southern part of the sea disappears only for a short period of time in summer. A surface flow of warm water exists from the Barents Sea into the Kara Sea. The discussion in this sub-chapter follows the publication in (Morozov, Parrilla, et al., 2003).

Internal waves were poorly studied in this region and we do not possess any reliable

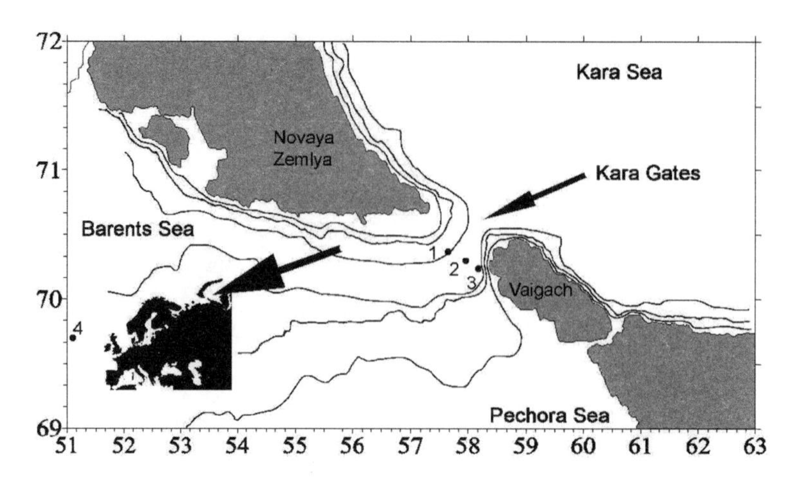

Figure 13. Position of three moorings in the Kara Strait (Kara Gates). The fourth mooring is in the Barents Sea. The 25 m, 50 m, and 100 m depth isobaths are shown. Printed with permission from Elsevier.

historical oceanographic field measurements of short periodic internal waves here. According to the theoretical considerations, the generation of internal tides occurs on the steep slopes of the sill, which crosses the Kara Strait, as well as at the shelf edges (Hibia, 1990). In September-October, 1997, four moorings were set for a few days in the Kara Strait. The positions of moorings are shown in figure 13. Time sampling was equal to 15 min.

Mean currents of the flow directed to the Kara Sea with a velocity of about 6 cm/s were recorded in the surface layer on the mooring closest to the Novaya Zemlya Islands. The tidal component exceeded the mean currents; therefore reversing flow was periodically observed. In the middle of the strait, the mooring recorded a stable water transport from the Barents Sea with a mean velocity of 18 cm/s at a depth of 90 m.

Temperature time series measured on moorings show that temperature fluctuations in the strait are very strong. A temperature time series measured at 115 m on mooring 3 is shown in figure 14. An elliptic band filter (Parks and Burrus, 1987) was used to separate the component of the temperature fluctuations with the M2 period (12.42 h). In order to isolate the fluctuations with M2 frequency we provided a frequency band between $1/12.37\ h^{-1}$ and $1/12.47\ h^{-1}$ tuned to a frequency of $1/12.42\ h^{-1}$. The data at the beginning and end of the time series are not cut off, which is extremely important for such short time series. The temperature time series after band filtering is shown in the upper part of figure 14. Temperature fluctuations with the M2 period reach 0.25°C.

The data at depths of 65, 115, and 215 m we used to plot the displacements of isotherms. The interpolation of the locations of the isotherms was carried out onto a grid with a 5 meter depth interval and 15 minute sampling interval of measurements. The displacement of the 0.4°C isotherm during 64 hours from October 4 to October 7, 1997 is shown in figure 15. The displacement of the minus 0.4°C isotherm ranges

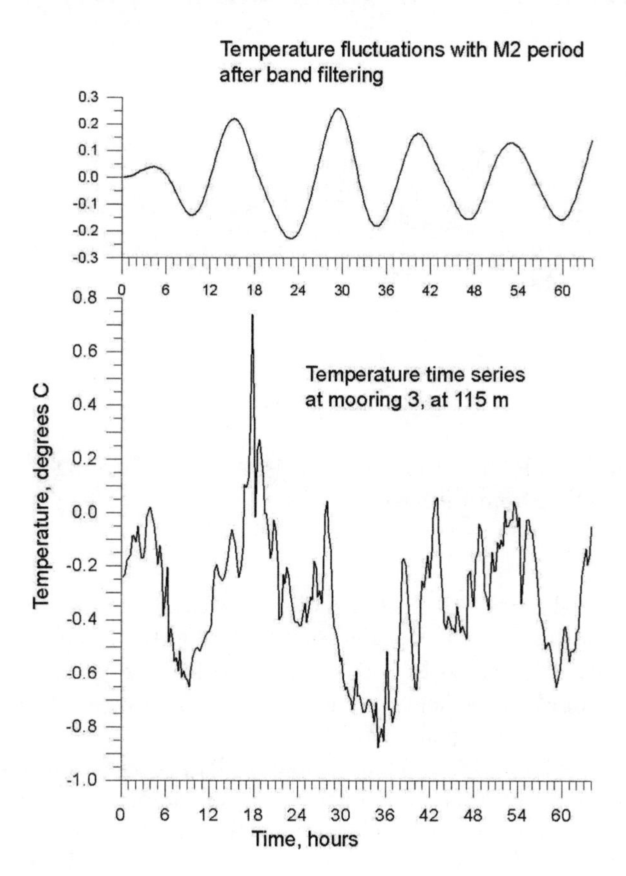

Figure 14. Temperature time series at 115 m on mooring 3 in the Kara Strait (lower panel). Band filtered time series to separate the fluctuations with M2 period are shown in the upper panel. Printed with permission from Elsevier.

approximately from 110 to 170 m. This is a very significant span of oscillations with respect to the maximum sea bottom depth (230 m) in the narrow channel. The average surrounding depths in the region are equal to 120-160 m. The most intense vertical motion occurs in the lower part of the water column. We interpret this as a forced motion due to the influence of the barotropic tide currents (depth averaged currents) flowing over steep bottom topography.

In order to determine the velocity of the barotropic tide currents we used a band pass filter to separate the component with a 12.4-hour period. The ellipse of the M2 tidal currents obtained at mooring 3 is shown in figure 16 with a thick line. The amplitude of the barotropic tide current according to these calculations was equal to 19 cm/s. This is in good agreement with the calculation of tides, which were carried out by S. Erofeeva in the Oregon State University (USA) using the TOPEX/POSEIDON data (personal communication). The maximum velocity of the barotropic tide on October 1,

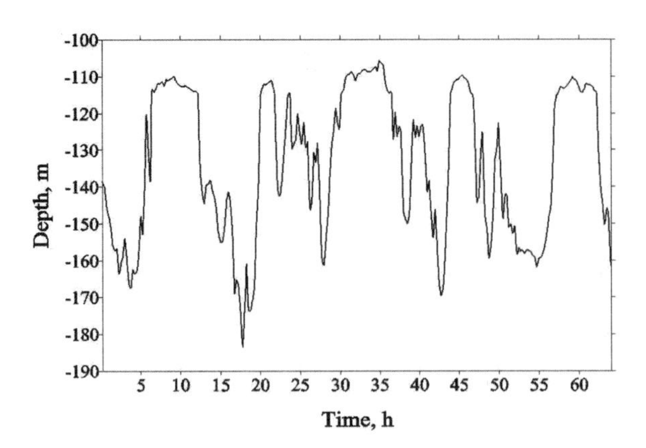

Figure 15. Vertical displacement of the 0.4°C isotherm at mooring 3 in the Kara Strait. Printed with permission from Elsevier.

1997 calculated from the satellite data of surface elevations was equal to 19.5 cm/s. The direction of the main tidal axis in both cases was normal to the sill.

Due to the floating ice there is a high risk of losing surface moorings in the Kara Strait so that they could not be kept for a longer time. Southwest of the strait a mooring was deployed in the ice-free region. This made possible to carry out the measurements during 18 days. A spectrum of temperature fluctuations measured at a depth of 40 m at this station is shown in figure 17. The fluctuations of temperature are characterized by one reliable peak at a frequency of the semidiurnal tide. The amplitude of the vertical displacements at this mooring was five times smaller than in the Kara Strait. Thus, the amplitude of the wave decreased at least five times while it propagated over a distance of 200 km. Since other sources of internal tide generation in the region could also contribute to the wave amplitude, the actual decrease of the amplitude could be even greater.

We estimated the wavelength of the internal tide from the dispersion relation. We assumed that the bottom is flat beyond the generation region and no forcing exists beyond the slopes of the sill. To a first approximation we can estimate the wavelength of internal tide by numerical integration of equation (1.11) for the vertical velocity (w) induced by internal waves. The vertical velocity at the surface and bottom is zero. This is an eigen value problem with depth-dependent Brunt-Väisala frequency. The summer stratification used in the calculation is very strong: the upper layer temperature reached 7°C while the temperature in the lower layers was below zero. We integrated this equation with a vertical step of 10 m for the typical stratification. The wavenumbers obtained in the solution correspond to the modes of the internal tides. Mode one has no zero crossings (only zero values at the bottom and surface), mode two has one zero crossing, etc. The graph of the eigen function is shown in figure 18.

We assumed that the mean depth in the region west of the Kara Strait was 160 m. The calculated wavelength of the semidiurnal internal tide (mode one) was 48 km. This result corresponds to the case of zero mean current. A vertically uniform current of

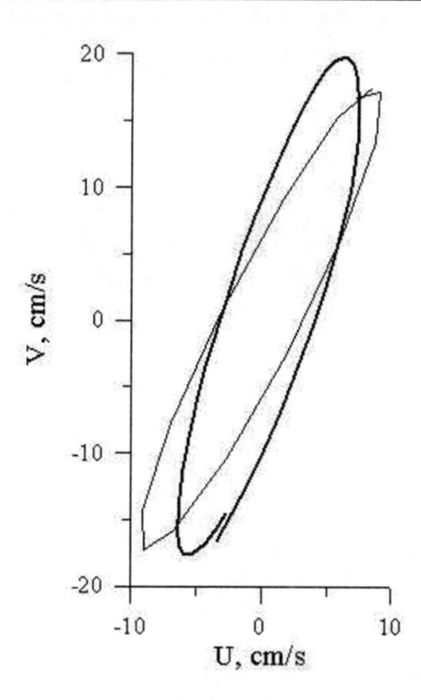

Figure 16. Ellipse of tidal currents at the M2 frequency on mooring 3 calculated from the measurements is outlined with a thick line. Ellipse of tidal currents calculated on the basis of the TOPEX/POSEIDON data is shown with a thin line. Printed with permission from Elsevier.

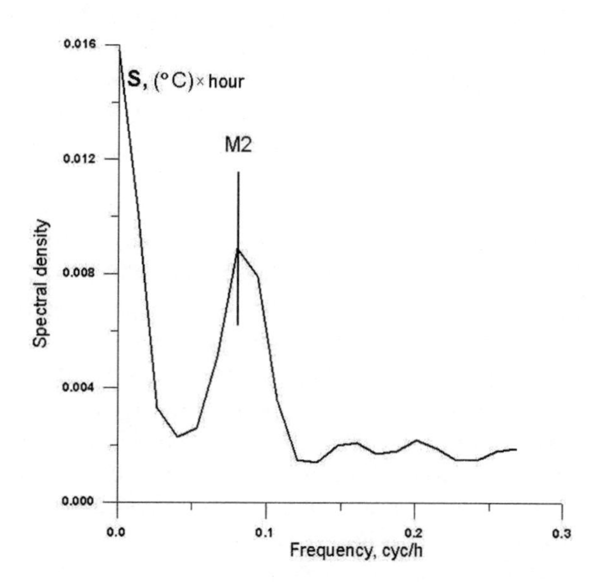

Figure 17. Spectrum of temperature fluctuations, measured at 40 m depth on mooring 4. The depth of the sea is 80 m. Printed with permission from Elsevier.

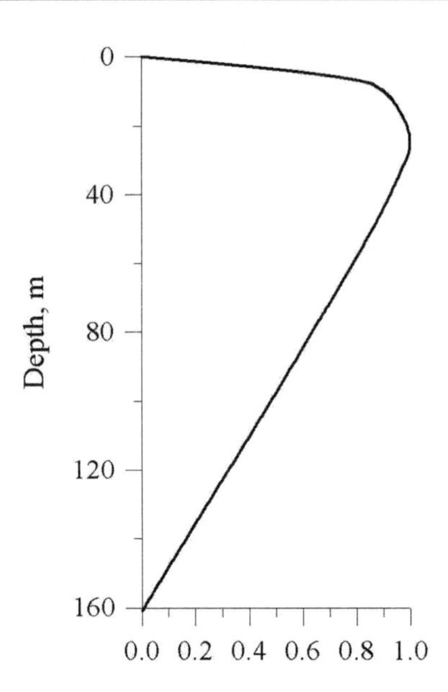

Figure 18. Vertical displacement induced by internal tide (first mode) normalized by the maximum value. Printed with permission from Elsevier.

19 cm/s directed opposite to the wave decreases the wavelength by 3 km and increases it approximately by the same amount when the wave propagates in the same direction as the flow. Surprisingly long wavelength for the Kara Strait region is explained by the close location of the region to the critical latitudes, where the M2 period becomes equal to the inertial period. Due to a small denominator the coefficient in the second term of equation (1.11) becomes large, which should be compensated by decreasing of wavenumber k. At latitudes close to 74.5°N, the inertial period becomes equal to the M2 period. In this case, equation (1.11) has a singularity and requires special methods for solution. The corresponding physical processes are analyzed in Nøst (1994), Gjevik et al. (1994), Furevik and Foldvik (1996) and Robertson (2001).

Numerical modeling using the model described above was performed to study the influence of the currents in the Kara Strait on internal tides.

Calculations were carried out using a domain 150 km long with a horizontal step of 50 m and 20 vertical levels. The time step was approximately equal to 2.2 seconds. We specified the coefficients of horizontal eddy viscosity and density diffusivity as 5 m^2/s over the ridge and 1.2 m^2/s beyond the ridge over the flat bottom in the model. The coefficients of vertical turbulent viscosity and density diffusion were set to 0.001 m^2/s over the ridge and zero beyond the ridge. A small horizontal step in these calculations allows us to increase the non-linearity, which suppresses the dispersion due to strong rotation at high latitudes.

We introduced in the model calculations the flow from the Barents Sea, which is directed to the northeast and occupies the entire depth. In the model (figure 19) the

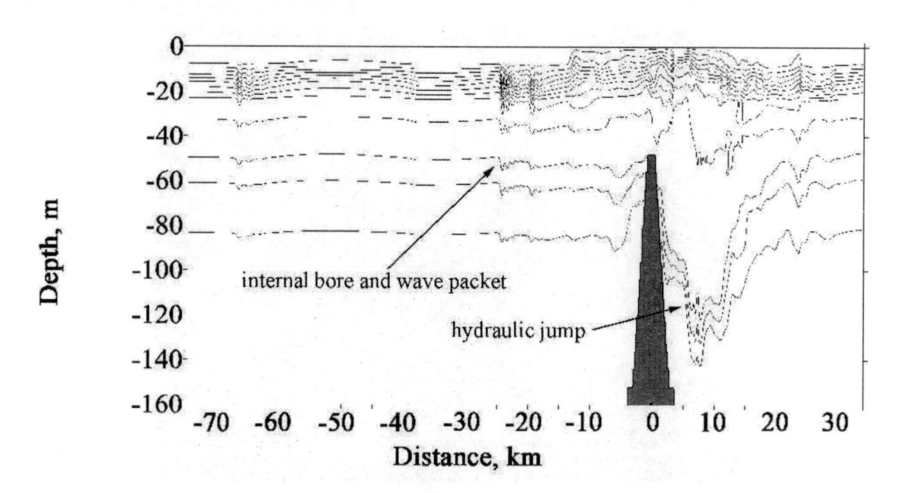

Figure 19. Numerical calculation of internal waves in the Kara Strait. The contour lines of density between 25.00 and 29.00 are shown with an interval of 0.4. Additional contour lines of 29.05 and 29.1 are shown in the deep layers. Gray colored pattern shows bottom topography. Printed with permission from Elsevier.

direction of the flow is from left to right. Its mean velocity was 9 cm/s. A periodical barotropic tidal flow was superimposed on this current.

Periodical changes in the tidal horizontal flow induce an internal wave propagating in both directions from the sill. The perturbations of the density field induced by the propagating internal tide after three tidal periods of calculation are shown in figure 19.

The fluctuations of the density field are not symmetrical relative to the sill that crosses the strait. The internal tide propagating in the southwestern direction opposite to the mean flow is stronger than internal tide propagating to the northeast. Due to the presence of the flow, the wavelength in the southwestern direction is approximately equal to 44 km, while in the northeastern direction it is equal to 51 km. The leading edge of the wave is flat and the trailing edge is steep. Owing to strong non-linear effects and the mean current opposite to the wave propagation internal bore is formed. The isopycnals sharply deepen to 20 m forming a bore. A packet of short-period waves follows the bore. These smaller scale waves and the bore induce vertical motions, which manifest themselves at the surface.

The model calculations yield that the wavelength of the wave propagating to the southwest is 44 km. This agrees with the calculations made on the basis of the dispersion relation. The introduction of the current from the Barents Sea intensifies the internal bore in the southwestern part of the strait. This is seen on a satellite image of the sea surface.

Satellite photos of internal waves in the Kara Strait show the surface manifestation of internal tides (figure 20). The surface manifestation of internal tide is observed in the southwestern part of the strait. The northeastward flow from the Barents Sea is most important here. The internal tide propagating to southwest is intensified by this flow.

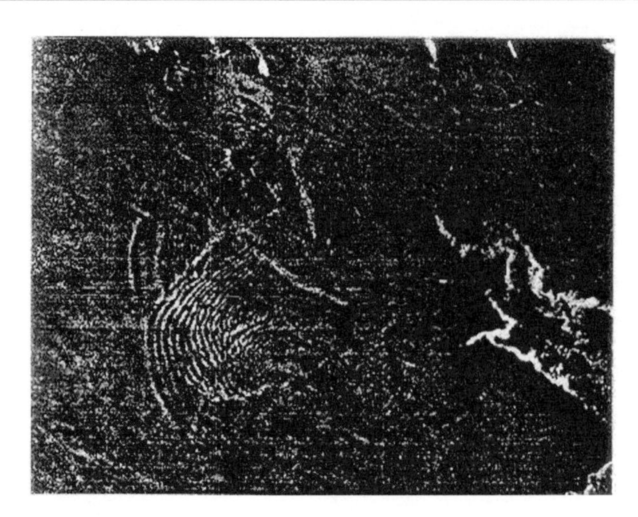

Figure 20. Satellite photos of internal waves in the Kara Strait (Almaz satellite, 1991). Printed with permission from Elsevier.

4 Internal Tides in the Strait of Gibraltar

The Strait of Gibraltar is the only connection between the Atlantic Ocean and Mediterranean Sea. The flow in the strait is characterized by a two-layer system of opposite currents resulting from the difference in sea level and water density (salinity and temperature) between the Atlantic Ocean and the Mediterranean Sea. Intense evaporation in the Mediterranean, which is estimated as 50 cm of water column per year, is compensated by a strong surface current of relatively fresher water from the Atlantic Ocean. A deep-water current of more saline Mediterranean water flows into the ocean owing to the difference in water density. A barotropic tidal wave with mean velocities in the range 70-80 cm/s propagates through the strait. The barotropic tide generates an internal tidal wave, when the tidal currents flow over uneven topography in the Strait, especially over the Camarinal Sill (Armi and Farmer, 1988) The discussion in this subchapter follows the publication (Morozov, Trulsen et al., 2002).

Our analysis (Morozov, Trulsen et al., 2002) shows that extreme tidal internal waves are observed only over the Camarinal Sill. A strong internal bore is observed in the eastern part of the strait caused by the interaction of the internal tide with the currents. The mixing caused by the decay of internal waves makes the interface between the inflowing and outflowing waters in the western part of the Strait thicker than in its eastern part.

Moorings deployed in the Strait of Gibraltar during the 1985-1986 Gibraltar experiment are shown in the scheme in figure 21. The data from buoy 2 were used to study the vertical structure of the internal tidal oscillations. The instruments measuring velocity and temperature were located at depths: 91, 112, 135, 182, 235, and 302 m. Due to the strong currents in the strait the instruments were subject to large vertical displacements exceeding 60 m. The instruments were supplied with pressure sensors. This made it possible to calculate the real location of the isotherms. The analysis shows that the

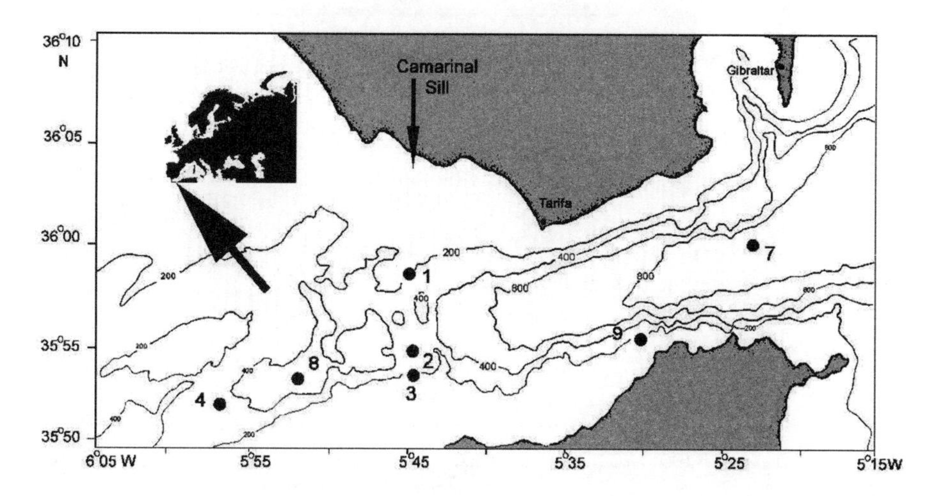

Figure 21. Locations of moorings during the Gibraltar experiment in 1985-1986. Reproduced with permission of American Meteorological Society (AMS).

displacement of the 13°C isotherm gives the best illustration pattern because this temperature value always remains within the depth interval of measurements. The vertical displacement of this isotherm ranges from 100 to 300 m. Superposition of diurnal and semidiurnal tidal components is clearly seen in figure 22.

We used an elliptic band filter (Parks and Burrus, 1987) to separate four tidal components, which form the internal oscillations over the sill shown by the displacement of the 13°C isotherm. The graphs of fluctuations with the frequencies of the main tidal components are shown in figure 23. The mean amplitudes of internal fluctuations are equal to 12.7, 68.8, 22.5, and 18.9 m for the components of internal tide S2 (12.0 h), M2 (12.42 h), K1 (23.93 h), and O1 (25.82 h), respectively. The internal tidal oscillations are observed over the sill exactly at the source of their generation; hence we conclude that these waves are forced internal oscillations, which transform into a free propagating wave, east and west of the sill. Internal tidal waves rapidly loose their energy and at a distance of approximately 50 kilometers from the sill their amplitude becomes three times smaller.

In order to estimate the wavelength of the semidiurnal internal tide we used the antenna method (Barber, 1963) developed for the arbitrary position of the sensors. Assuming that the main generation of tidal internal waves occurs at the Camarinal Sill, the moorings in the Strait were divided into two groups: buoys 1, 2, 3, 8, 4 for the internal tide propagating west of the sill and buoys 2, 3, 9, 7 for the easterly propagating internal tide.

The wavelength was calculated on the basis of 10 different combinations of the time series duration for the western group. The resulting wavelength ranges from 45 to 60 kilometers, and the direction of the wave varies from 210 to 280 degrees. The wavelength of the internal tide propagating to the east was estimated from 16 different combinations

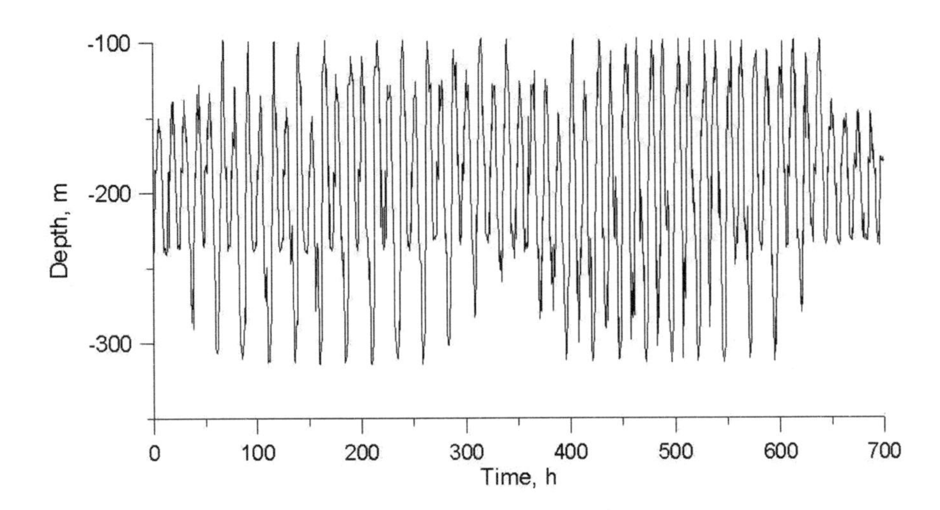

Figure 22. Depth variation of 13°C isotherm from July 2 to July 31, 1986. Reproduced with permission of AMS.

of moorings. The wavelength ranges from 90 to 140 km, while the direction of propagation varies between 90 and 120 degrees. The wavelength of the internal tide propagating in the eastern direction is greater and the difference is not surprising because the depths east of the Camarinal Sill are greater than in the western part of the Strait.

The wavelength of the internal tide estimated on the basis of the dispersion relation east of the sill was 94 km, and west of the sill the wavelength was equal to 59 km. We assumed the mean depth east of the sill equal to 800 m, and west of the sill equal to 500 m.

We applied a nonlinear numerical model to analyze the water motion in a continuously stratified rotating ocean of variable depth. The area of calculation was a domain 300 km long with a horizontal step of 200 m and 20 vertical levels. The time step was equal to 7 seconds. The main objective of the model investigation was to study the influence of the currents on internal tide propagation in the Strait of Gibraltar. The first approximation of the model calculation was zero mean current. We introduced only a periodical barotropic tidal flow by periodical increasing and decreasing the stream function at the surface.

The periodical changes in the horizontal flow with amplitude of 80 cm/s induce an internal wave propagating in both directions from the Camarinal Sill located in the middle of the computation area. The results of the calculation after four tidal periods presented as perturbations of the density field induced by propagating internal tide are shown in figure 24a. The fluctuations of the density field are not symmetrical with respect to the position of the Camarinal Sill because the bottom topography east and west of the sill is different. A calculation, when the bottom topography was modeled by a symmetric sill gave a symmetric pattern of propagating fluctuations in both directions from the sill.

An internal bore is formed on the trailing edge of the wave, which is steeper than the leading edge. A packet of shorter internal waves follows the bore. The bore and wave

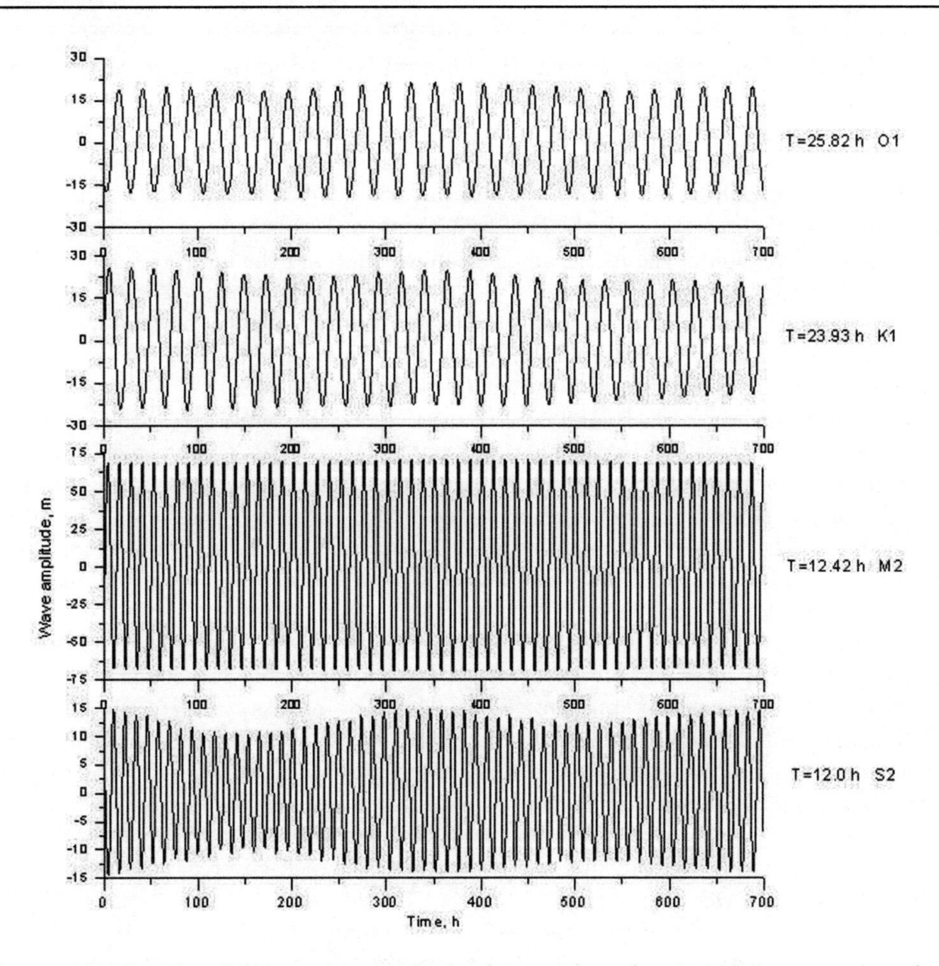

Figure 23. Graphs of fluctuations with the frequencies of main tidal components from July 2 to July 31, 1986, after band filtering. Reproduced with permission of AMS.

train of shorter internal waves is less manifested in the western part compared to the eastern part of the basin.

In the next step of the calculation we introduced a steady westward current in the entire water column of the strait. The velocity of the mean current was 30 cm/s. Similarly to the first version of calculations we superimposed a periodical barotropic tidal current on the mean current. The amplitude of the barotropic tidal velocity was also equal to 80 cm/s.

The westward flow changes the internal wave field (figure 24b). In the eastern part, where the internal tide propagates opposite to the current, we observe a well pronounced internal bore followed by a train of short period internal waves. The mean current opposite to the wave makes the wavelength shorter. The slopes of the internal tide are steeper, which leads to wave breaking and formation of a packet of shorter internal waves. For a uniform westerly current the bore is observed in the entire water column. An internal bore is also formed west of the sill, but only a weak packet of shorter waves appears. Internal tidal waves occupy the entire water column. So far as the outflowing current in the lower part of the water column is thicker than the surface inflowing current, and the maximum of the first mode of the internal tide is located in the outflowing current, the interaction of the wave with the outflowing current is more important.

In the third version of the model calculation we analyzed the internal tide developing in the strait with two opposite currents modeling the real situation. The eastward flow with mean velocities of 50 cm/s occupies the upper 250-300 m layer and the lower current with a vertically average velocity of 25 cm/s occupies the rest of the water column. These two opposite flows yield a strong shear at 200 m depth.

We superimpose the same tidal flow as in the two previous versions of the calculations. The wavelength of the easterly propagating wave is 74 km. The leading edge of the wave is flatter than the trailing edge, which is very steep. This leads to the formation of an internal bore. A train of shorter internal waves follows a sharp depression of the density contour lines. The westerly propagating wave is shorter. Its wavelength is 52 km, as the depth west of the Camarinal Sill is smaller. The structure of the wave is approximately the same except for the fact that the internal bore is lesser expressed in the western part than in the eastern part of the basin. The presence of the two opposite currents with a shear intensifies the internal bore in the upper layer at depths of 100-200 m, while in the deeper water it becomes less apparent (figure 24c).

Numerical calculations of isopycnal fluctuations in the main thermocline demonstrating the formation and propagation of internal bore are shown in figure 25. The direction of the barotropic tidal currents and time moments of low and high water are indicated on the right side of the wave patterns. We present seven sequential snapshots of the location of isopycnals with an interval of two hours, which cover a tidal period. The regime of internal waves propagating to the west (in the same direction as the outflowing current) and those propagating to the east differs significantly. Internal waves propagating eastward have greater amplitude. The waves propagate in the direction opposite to the currents as a periodical pulse of internal tidal perturbations, which rapidly increases its amplitude and then breaks into a packet of shorter internal waves.

Initially, the amplitude of the vertical displacements caused by internal waves exceeded 100 m. It decreases while the wave propagates away from the sill and small-scale internal waves developing at the steep slopes of isopycnals associated with undular internal bores are generated. These short internal waves with a wavelength of about 7-8 km develop on the trailing edge of the propagating tidal internal wave. They absorb energy from the internal tide and propagate in the same direction with approximately the same speed as the main wave carrier. A packet of short period internal waves starts to form east of the sill before the high water, when the easterly-directed barotropic current decreases. After the high water, the barotropic tidal current turns to the west and decreases the velocity of easterly propagating internal tide. This compresses the wave packet and intensifies it. Several hours later after low water, when the direction of the current coincides with the propagation of the wave packet, the latter becomes longer, the amplitudes of the waves in the packet decrease and its leading edge propagates faster. The speed of the leading edge changes from 0.8 to 1.6 m/s depending on the relative

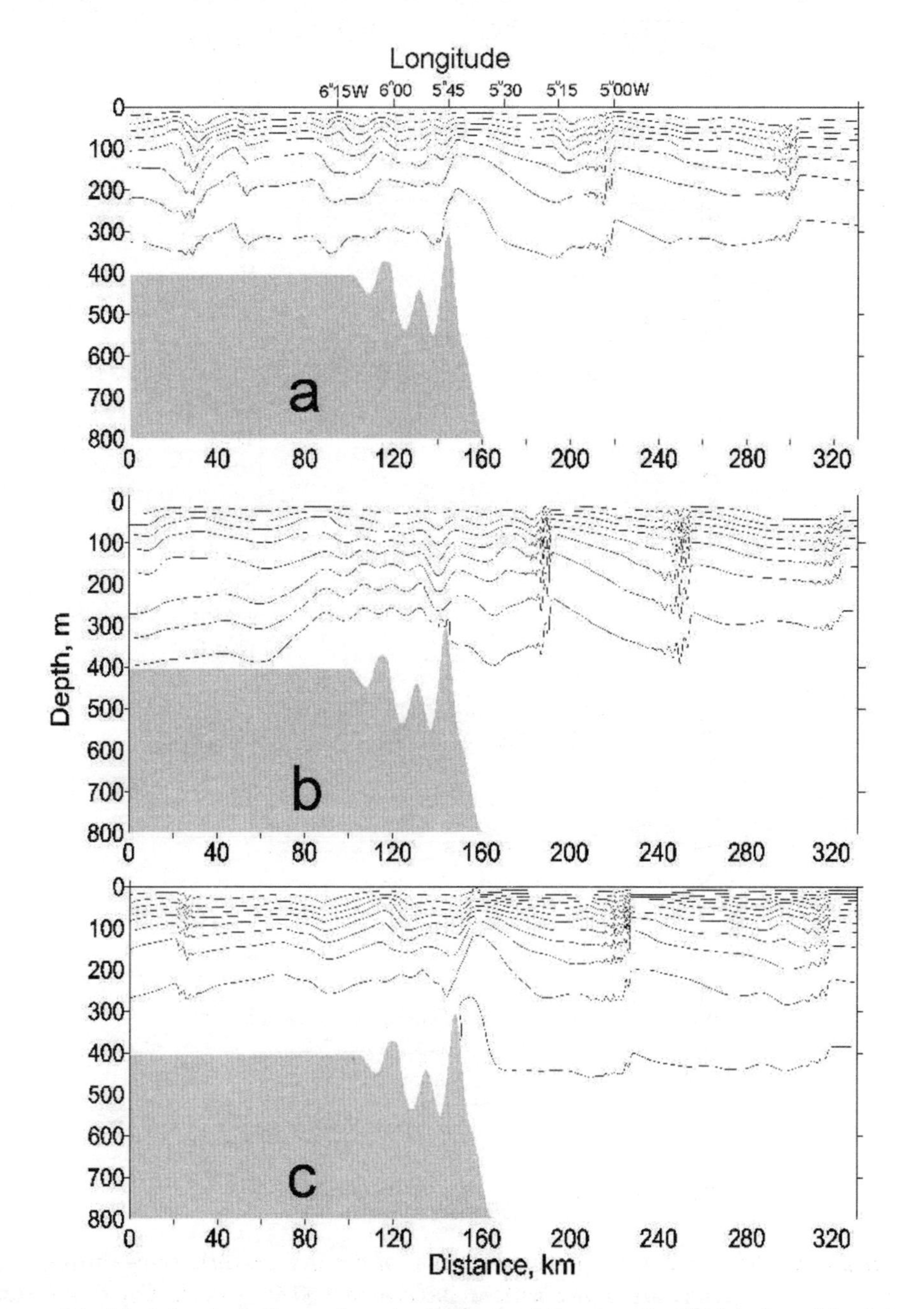

Figure 24. Model calculations of perturbations of the density field induced by internal tide. The contour lines are shown with an interval of 0.00025 g/cm3. The perturbations develop on the background of different currents: (a) Zero mean current (b) Westward barotropic current (c) Oppositely directed inflowing and outflowing currents. Reproduced with permission of AMS.

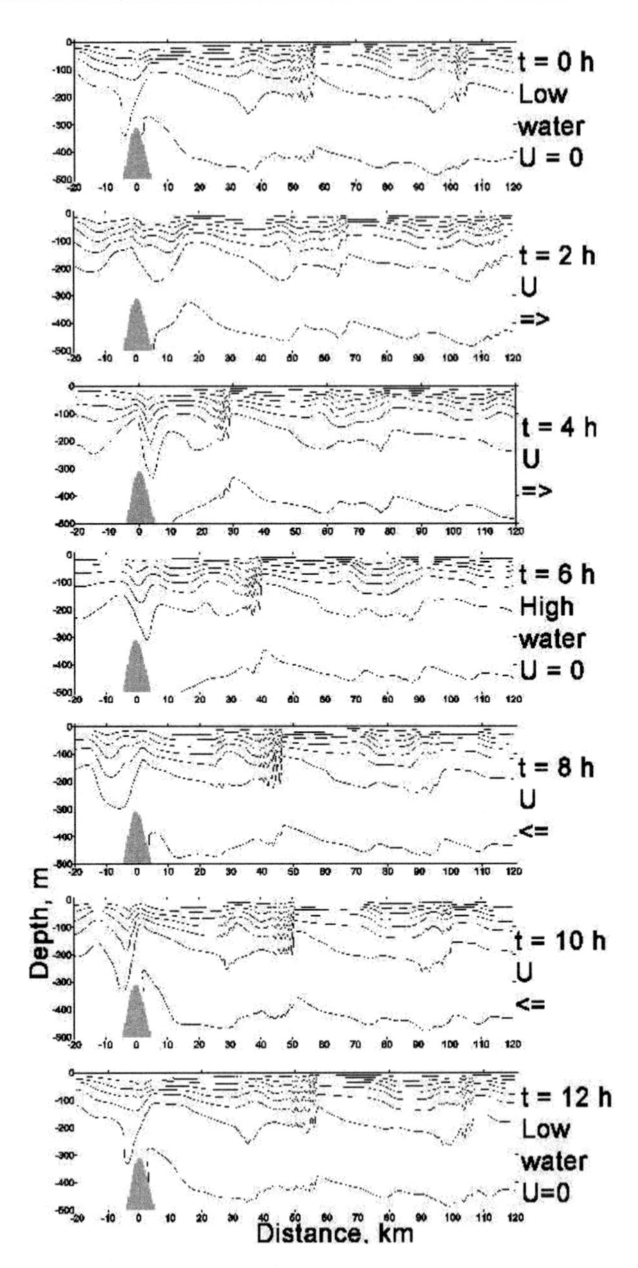

Figure 25. Model calculations of the evolution of density perturbations during a tidal period. The contour lines are shown with an interval of 0.00025 g/cm^3. The time intervals are given with respect to the tidal period ($T = 12.4$ h). Initial time ($t = 0$) corresponds to zero velocities of the barotropic tide along the strait. The time moments of low and high water and the direction of the barotropic tide are shown near the patterns.

velocity of the barotropic flow. One can observe that an undular bore also propagates to the west, but here it is weaker than on the eastern side.

5 Application of WOCE sections to a global view of mixing in the Atlantic Ocean

The cold water that sinks at polar latitudes is transported equatorwards by the deeper limb of the Meridional Overturning Circulation (MOC); its upper limb transports polewards the water heated in the tropical regions. These waters are mixed in the ocean thus maintaining the existing oceanic stratification. Munk (1997) and Munk and Wunsch (1998) proposed a mechanism to explain the role of the bottom topography in global ocean mixing. The interaction of topography with the barotropic tide generates the baroclinic tide. Thus, tidal energy spreading from the ocean topography is a key mechanism for ocean mixing (Munk and Wunsch, 1998).

Special experiments are needed to study the turbulent energy dissipation and to determine the values of diffusitives in the ocean. A special experiment was undertaken near Hawaii (Hawaii Ocean Mixing Experiment: HOME) to study the problem of topographic influence on mixing (Pinkel et al., 2000). We suggest a new method to find the regions of intense mixing in the ocean (the so-called *hot spots*) where the energy of the barotropic tide and other motions are transferred to small-scale processes and mixing. We use the ocean scale measurements as means to recognize the regions of possible enhanced mixing, which are usually associated with strong vertical motions.

New approaches to find the regions of enhanced mixing can be based on the data archived in the databases of the oceanographic projects. In this sub-section we use WOCE (World Ocean Circulation Experiment) CTD (Conductivity-Temperature-Depth) measurements to find the regions of strong mixing and estimate the intensity of mixing. For this purpose we have chosen the Atlantic Ocean where a significant amount of hydrographic measurements has been accumulated during the WOCE.

Strong mixing is unambiguously associated with vertical motions that are more intense near abrupt slopes where the barotropic tide obtains a vertical component. This process leads to vertical displacement of isopycnals thus inducing tidal internal waves. The main objective is to find a correlation between the bottom topography and the fluctuations of potential temperature on CTD profiles recalculated into vertical displacements. The intensity of the fluctuations is expressed numerically using the vertical wavenumber spectra of the temperature and vertical displacements compared to the level of the background GM spectra. From the WOCE hydrographic sections data in the Atlantic we identify the regions where mixing is strong, as well as regions of weaker mixing. We also point out other regions of enhanced mixing, which are not associated with submarine ridges.

5.1 Dropped spectra of CTD profiles

Garrett and Munk (1972, 1975) contrived a universal spectrum for internal waves in the entire ocean. The dimensionless energy spectrum of internal waves, which they composed, depends on the frequency (ω) and wavenumber (k_x, k_y, k_z). If we calculate the spectrum of the fluctuations from vertical profiles of temperature and salinity, these

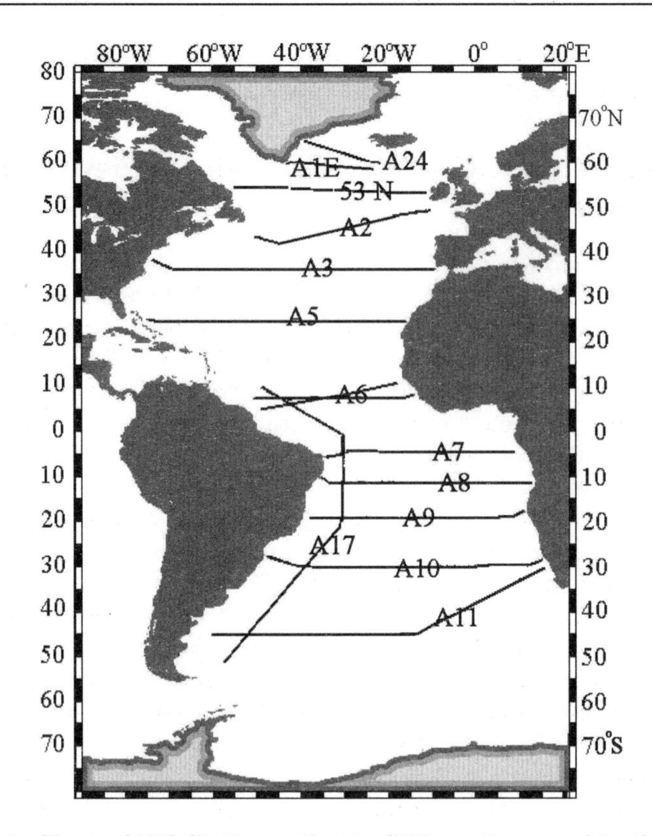

Figure 26. Chart of WOCE transatlantic CTD sections used in the analysis.

spectra are termed *dropped spectra* since they are derived from the data of instruments lowered from a ship. The physical meaning of the differences between *dropped* spectra calculated from different measurements is based on the different level of fine-structure inhomogeneities on the vertical profile. The homogeneous layers and steps with sharp vertical gradients on the vertical profiles of temperature and salinity may appear due to many processes: breaking of internal waves, diffusion, convection, salt fingers, winter cooling, etc. All of them are associated with vertical motion and mixing in the ocean.

5.2 Analysis of data

To analyze the variability of "dropped" spectra we processed all latitudinal transatlantic WOCE sections (figure 26). Usually, the stations on the sections have an average separation of approximately 30 nautical miles (1 nautical mile is 1852 m). The vertical spacing of the measured data is 2 dbar, which is approximately equal to 2 m.

The spectra were calculated from the CTD cast data with a 2-dbar interval as series for spectral analysis. Preliminary high filtering of the series was applied to filter out fluctuations with vertical wavenumbers smaller than 0.01 cyc/m. In order to compare the results with the GM spectrum we calculated the spectra of the vertical displacements that were found as deviations of temperature from the filtered profile normalized by the corresponding vertical temperature gradients. The vertical temperature gradients in the layers were calculated as temperature difference between the upper and lower levels divided by the layer depth. The variability of spectral densities was evaluated by averaging spectral densities in the wavenumber range from 0.028 to 0.055 cyc/m (the corresponding vertical scale is 36 - 18 m). We have chosen this vertical scale of fluctuations for the analysis because the maximum of this interval should be at least ten times smaller than the thickness of the layer after we filtered out long fluctuations (36 dbar is 16 times smaller than 600 dbar) and its minimum should be several times greater than the vertical interval between data samples (18 dbar is 9 times greater than 2 dbar). The average value of the GM dropped spectra was also calculated for the same band of wavenumbers. The exact values of the limiting frequencies for the range are not very important because there are no dominating peaks on the spectra. The ratio of spectral density of the measured displacements to the GM dropped spectra (E_{displ}/E_{GM}) was analyzed over the length of the WOCE sections.

5.3 Topographic influence on vertical wavenumber spectra

Let us show that the vertical temperature profiles are affected by their position relative to topographic features. This influence can be found from the statistical analysis of the vertical profiles of temperature and salinity and their vertical gradients. If we calculate the spectral densities of the temperature or of its vertical gradient fluctuations, the level of the spectral densities would be higher for the profiles taken in the regions with strong vertical motions.

Usually, the spectral densities calculated from profiles taken near submarine ridges are greater than those calculated from profiles far away from the ridges. To illustrate the differences we show two vertical temperature profiles taken on section A10 at 30°S. Vertical temperature profiles at station 64 (longitude 7°20'W, left curve) and at station 69 (longitude 3°12'W, close to the Walvis Ridge, right curve) are shown in figure 27a. The profile at station 64 is smoother than the one at station 69, which is clearly seen on the graphs. A high level of inhomogeneities on the profile at station 69 is caused by a large number of steps and layers with high vertical gradients of temperature that are to a great extent induced by internal wave strain, breaking internal waves, and other processes, which can induce mixing over the slopes of the Walvis Ridge.

The spectra of vertical displacements shown in figure 27b are appropriate to compare with the GM model. The upper curve shows the spectrum at station 69 located at 3°12'W near the Walvis Ridge. The lower curve is the average spectrum at stations 59-64, located between 11°01'W and 7°20'W at such a distance from the ridge where the spectrum may be considered as the background. The straight line corresponds to the Garrett-Munk model. This figure clearly illustrates that the level of spectral densities at station 69 near the ridge exceeds significantly the background level. This is caused by the inhomogeneities on the vertical profile due to the presence of fine structure and intrusions characteristic of the regions with intense vertical motion.

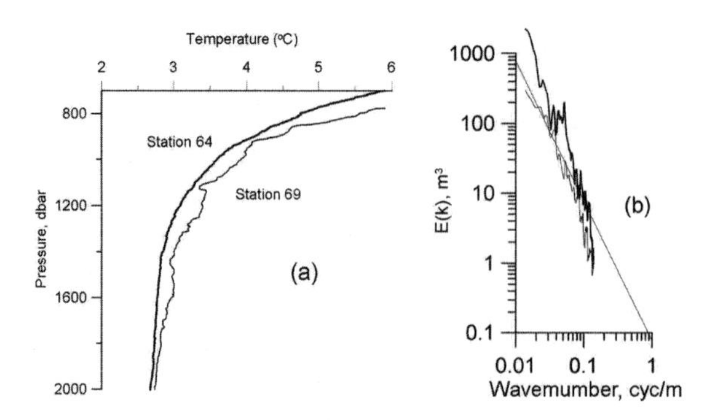

Figure 27. (a) Profiles of temperature on section A10 at station 69 (30°00S, 3°12W, right thin curve) near the Walvis Ridge and at station 64 (30°00S, 7°20W, left thick curve) at a distance of 200 km from the ridge. (b) Spectra of vertical displacements compared with the GM model. Thick curve: station 69 (section A10) near the Walvis Ridge at 3.2°W. Thin curve: averaged spectrum at stations 59, 60, 61, 62, 63, 64 (section A10) between 11.1°W and 7.33°W west of the Walvis Ridge. Straight line shows the GM spectrum.

5.4 Topographic influence of submarine ridges in the water column 600 dbar above the bottom

Since we are interested in the influence of bottom topography on mixing we shall analyze the spectral densities in the layer 600 dbar over the bottom. The spatial distribution of the ratio of spectral densities of vertical displacements to the value of GM dropped spectra in the wavenumber range from 0.028 to 0.055 cyc/m along section A8 is shown in figure 28. High spectral densities of vertical displacements are found over the slopes of the Mid-Atlantic Ridge. The ratio of vertical displacement spectra to the GM spectra in the deep basins east and west of the Mid-Atlantic Ridge are below unity, whereas over the ridge their values reach 2.5. The correlation between bottom topography and spectral densities in the 600 dbar bottom layer is found over the A9 section (figure 29, middle panel) and almost at each transatlantic section.

We find similar increase of spectral densities in the Northern Hemisphere at section A2, which was occupied three times in 1994, 1997, and 1999 (figure 30 a-c). The patterns differ from one year to another but the general feature of greater spectral densities over the slopes of the Mid-Atlantic Ridge remains.

In 1993 and 2000, two A6 sections were carried out north of the equator. They also indicate that high spectral densities of temperature and vertical displacements are found near the sloping bottom. The spatial variability of the spectral density of temperature fluctuations and spectral density of vertical displacements demonstrate approximately the same features. The increase in the spectral densities near 37-42°W in the region of the Equatorial Channel (figure 31) is not associated with any ridge. However, it is very unusual and important. We shall discuss this effect below.

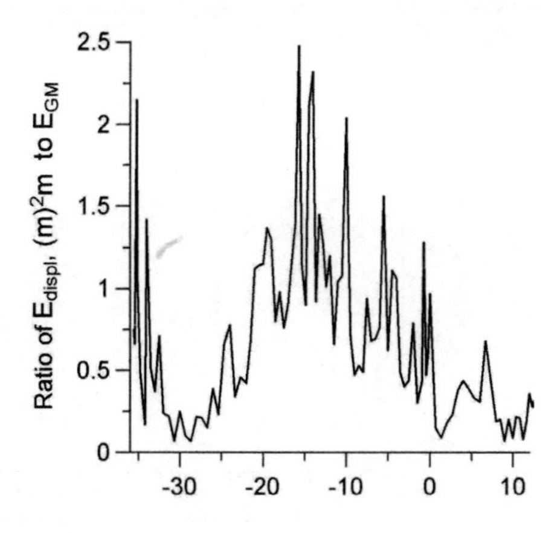

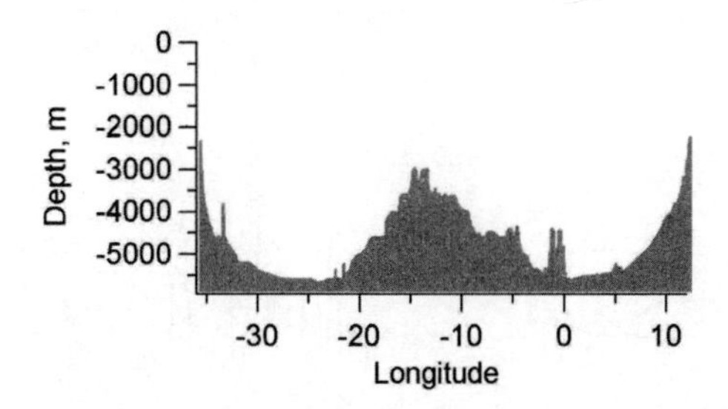

Figure 28. Section A8 approximately along 11°S. Longitudinal variability of the ratio of the spectra of vertical displacements calculated from temperature fluctuations to the GM dropped spectra in the wavenumber range from 0.028 to 0.055 cyc/m (upper panel) and bottom topography (lower panel); (600 dbar bottom layer). Reproduced with permission by Nauka.

5.5 Topographic influence of submarine ridges in the water column between 2000 and 3000 dbar

In the previous section we found an increase in spectral densities of temperature and vertical displacements in the near bottom layer near submarine ridges, which is caused by mixing processes induced by strong vertical motions. At the same time, when we demonstrate this increase over the slopes, we have to keep in mind that the depths of the layers over the slopes and over deep basins are different. While analyzing a 600 dbar layer over the slopes of a ridge we should remember that its actual depth is somewhat between 2000 and 3000 dbar. If we analyze a layer of the same thickness in a deep basin it can be located within an interval from 4000 to 5000 dbar, where all motions are less energetic. In order to make a comparison between the layers located at the same depths over different bathymetry we made calculations in the layers located at the same depth along the section within the interval from 2000 to 3200 dbar.

In figure 29 (upper panel), we show the spectral densities within the layer between

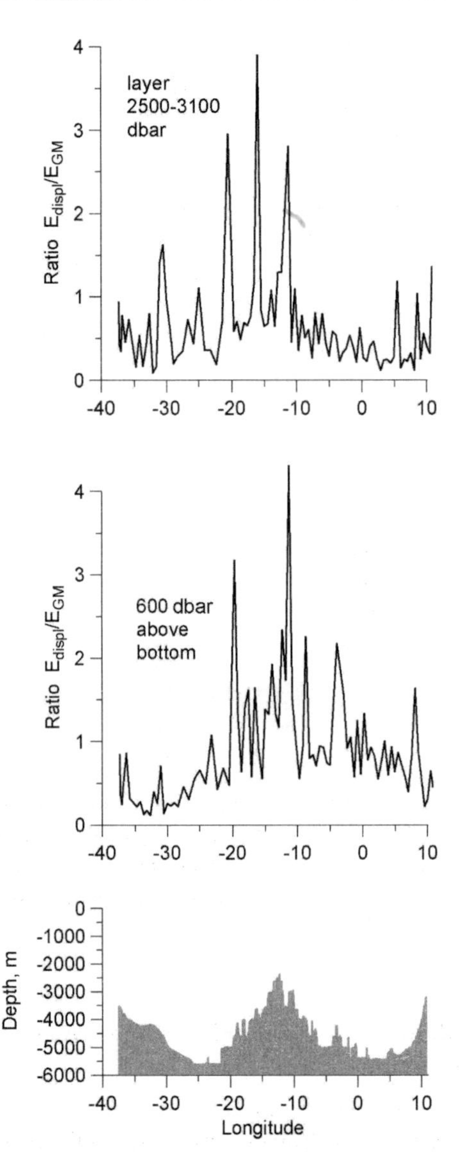

Figure 29. Section A9 approximately along 19°S. Same as in figure 28 but for the 2500 - 3100 dbar layer (upper panel) and 600 dbar layer over the bottom (middle panel). Reproduced with permission by Nauka.

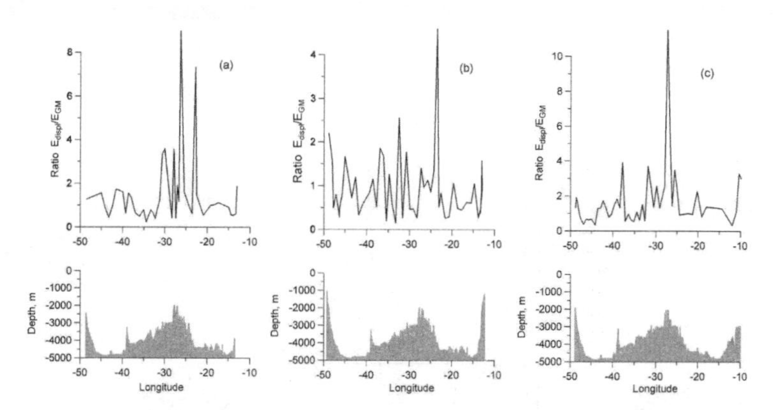

Figure 30. Section A2 approximately along 48°N. Same as in figure 28 but for the measurements made in different years: 1994 (a), 1997 (b), and 1999 (c).

2500 and 3100 dbar on section A9 in the South Atlantic. This graph is very similar to figure 29 (middle panel), which shows the variation of E_{displ}/E_{GM} on the A9 section in the 600-dbar bottom layer. The correlation with bottom topography is usually greater if we analyze a depth interval located at greater depths.

The spatial variability of the ratio of spectral density of vertical displacements to the GM spectrum on section A6 (2000) is shown in figure 31. The spectra were calculated from the vertical profile from 2000 dbar to 2600 dbar (figure 31, upper panel). Calculations of wavenumber spectra on this section (A6 2000) in the depth range between 2000 and 2600 dbar demonstrate increased spectral densities over the Mid-Atlantic Ridge similar to the 600-dbar bottom layer (figure 31, middle panel), but do not reveal any significant increase in spectral densities at 37°-42°W longitudes, which was found in the calculations of spectra from the data in the 600 dbar level over the bottom (figure 31, middle panel). We shall discuss this effect in the next section.

5.6 Spreading of Antarctic Bottom Water in the Vema and Equatorial channels

We can use the method of calculating one-dimensional spatial spectra from CTD profiles to localize enhanced mixing not only in the regions of submarine ridges but also in the abyssal depths. The measurements made at the A6 2000 section revealed an increase in the spectral densities near 37-42°W in the region of the narrow deep Equatorial Channel, where Antarctic Bottom Water flows from the Brazil Basin (figure 31, middle panel). It is noteworthy that no significant increase in spectral densities is found in the layer between 2000 and 2600 dbar (figure 31, upper panel) above the channel. A chart of the region with the sections analyzed here is shown in figure 32.

Antarctic Bottom Water (AABW) is formed in the Weddell Sea and propagates to the north in the Argentine Basin. It penetrates to the Brazil Basin through the narrow Vema Channel (31°15S, 39°20W) and partly through the Hunter Channel and spreads in the deeper part of the Brazil Basin. In the northern part of the basin, AABW enters the

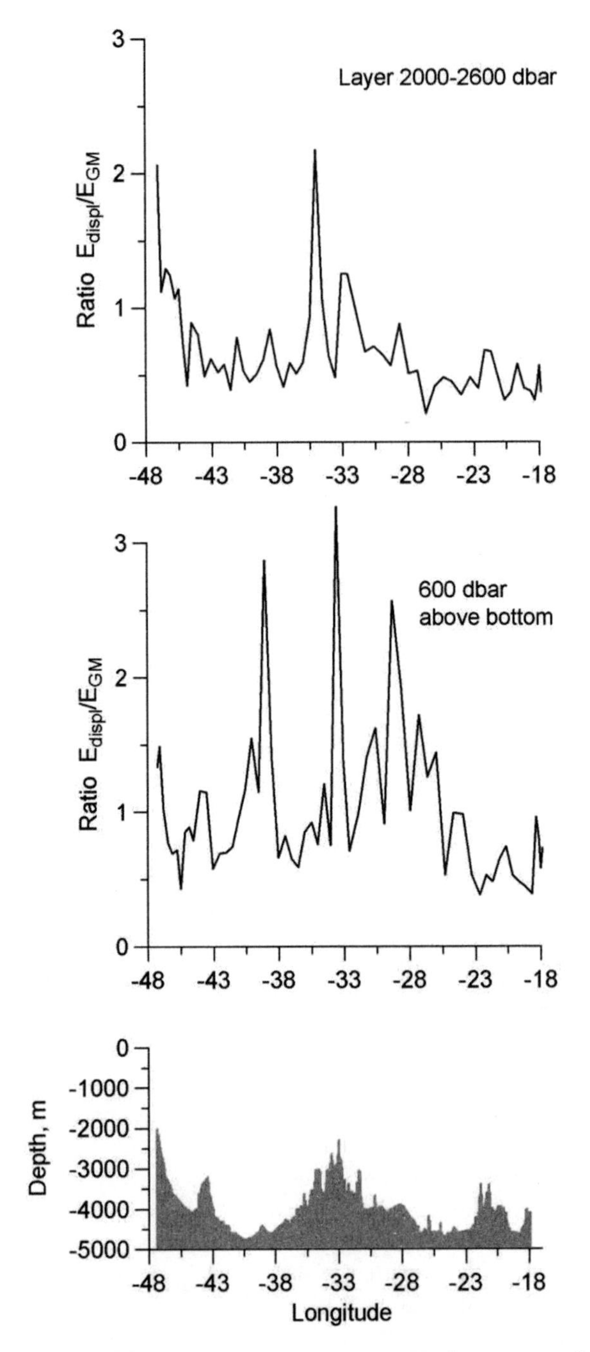

Figure 31. Section A6 (2000) approximately along 8°N. Same as in figure 29 but for the 2000 - 2600 dbar layer (upper panel) and 600 dbar layer over the bottom (middle panel).

Equatorial Channel (2°S, 34°W) propagating to the western North Atlantic basin. Here the flow accelerates while the bottom and lateral friction near the slopes of the channel induce strong mixing with the North Atlantic Deep Water (NADW) located above AABW (Sandoval and Weatherly, 2001). The NADW and AABW flow in opposite directions, and the shear results in additional mixing. The mixing process increases the temperature of AABW (figure 33, lower panel). The potential temperature of AABW slowly increases while it spreads in the Brazil Basin, and as soon as the water enters the Equatorial Channel its temperature increases very rapidly. In this region we find an increase in the spectral densities of vertical displacements (figure 33, upper panel). A sharp increase in the values of E_{displ}/E_{GM} ratio is observed north of 2°S, just when AABW enters the Equatorial Channel. The near bottom vertical profiles of temperature confirm the strong mixing in the benthic layers of the Equatorial Channel. A homogeneous benthic layer of about 100 meter thick is observed at station 175 in the Equatorial Channel (latitude 0°15N) (figure 34a). At station 160 located at 4°30S, the temperature decreases gradually to the bottom because the bottom mixing is weaker here due to the slower velocity of the flow.

It is important that the E_{displ}/E_{GM} ratio also increases in the interval between 28-32°S. The Vema Channel is located here, through which Antarctic Bottom Water spreads from the Argentine Basin to the Brazil Basin. The channel is only 20-50 km wide and its bottom is located at 4700-4900 m below the plateau at 4200 m. Increased velocities and bottom and lateral friction in the Vema Channel cause strong mixing.

5.7 Frontal zone of the North Atlantic Current

The western end of section A2 crossed the North Atlantic Current. It is clear that strong mixing occurs near the frontal zone of this current. Our measurements show increased values of the E_{displ}/E_{GM} ratio. In figure 35 (upper panel) we show the spectral densities of vertical displacements in the layer between 2300 and 2800 dbar. The spectral densities at stations located near the front (44°-45°W) of the current are very high. In this layer, the effect of the Mid-Atlantic Ridge, which we discussed above, is negligible compared with the fluctuations on profiles in the zone of the front of the North Atlantic Current. Vertical profiles at these stations contain intrusions with anomalies reaching 1°C with vertical scales of 20-30 m. These anomalies dominate over all other temperature fluctuations on the entire section. The spectrum calculated in the deeper interval between 2600 and 3200 dbar (figure 35, middle panel) demonstrates that anomalies below the front of the North Atlantic Current became less manifested in deeper layers, and the fluctuations associated with fine structure near the Mid-Atlantic became greater.

5.8 Influence of the Mediterranean outflow in the Atlantic Ocean

In figure 36 we show the ratio of spectral densities of vertical displacements to GM spectrum within three layers on section A3. The eastern end of this section is located near the Strait of Gibraltar. The outflowing current from the Mediterranean Sea transports warm and saline water. This water spreads in a thick layer located at a depth interval between 800-1600 dbar and mixes with the Atlantic waters. This effect is clearly seen in figure 36 (upper panel) by increased spectral densities in the 1000-1600 dbar in the

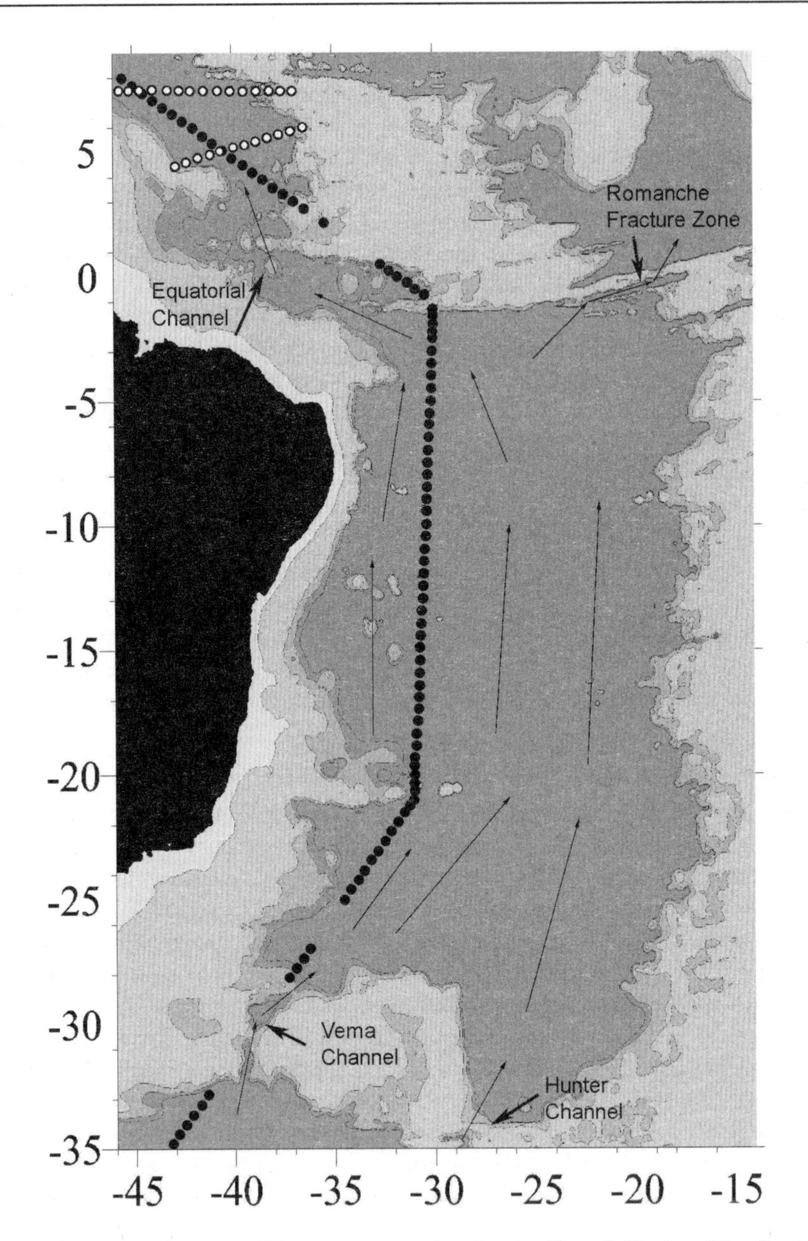

Figure 32. Antarctic Bottom Water propagation in the Brazil Basin. The South American continent is shown in black color. The depths deeper than 4300 m are shown in dark gray color. The locations of stations on the A17 section taken for the analysis are shown by black dots. Open circles show the position of stations on the A6 (1994) and A6 (2000) sections taken for the analysis. We analyzed only the deep stations, which reached Antarctic Bottom Water. Thin arrows show the propagation of Antarctic Bottom Water. Thick arrows show the locations of the Vema Channel, Hunter Channel, Equatorial Channel, and Romanche Fracture Zone.

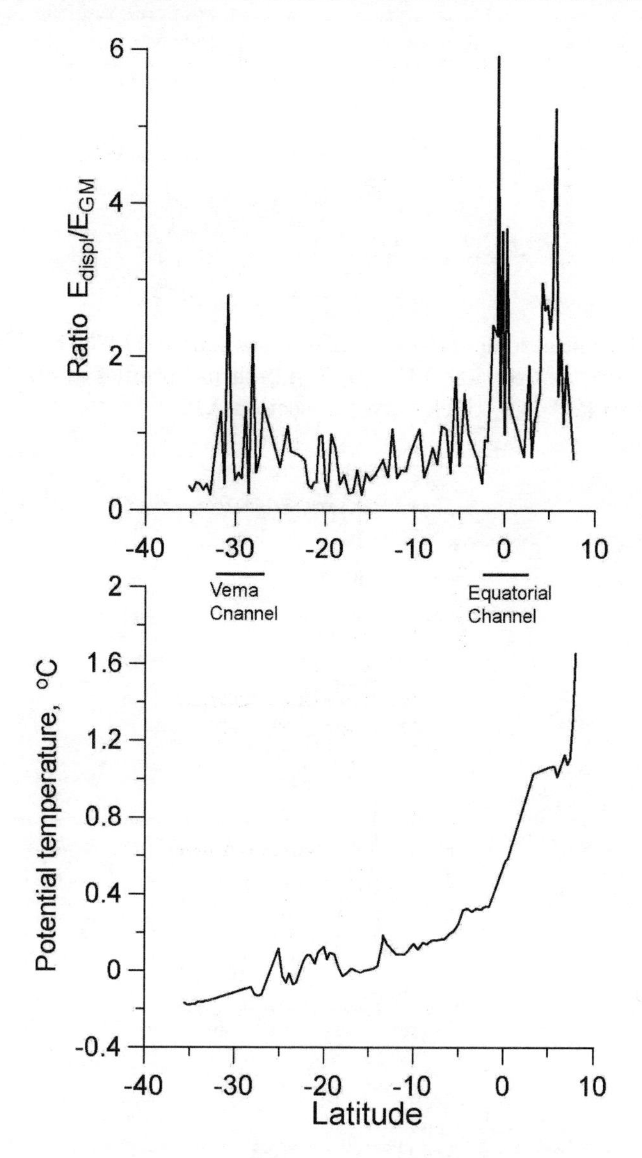

Figure 33. Quasi-meridional section A17. Latitudinal variability of the ratio of the spectra of vertical displacements to the GM dropped spectra in the wavenumber range from 0.028 to 0.055 cyc/m; (600 db bottom layer) (upper panel). A sharp increase in the values of E_{displ}/E_{GM} ratio is observed north of 2°S, just when AABW enters the Equatorial Channel. This leads to an increase in the potential temperature, which is shown in the lower panel: a graph of near bottom potential temperatures from section A17 that approximately follows the flow of AABW. The locations of A17 stations are shown in figure 32. Locations of the Vema and Equatorial channels are shown with horizontal lines below the latitude axis on the upper panel.

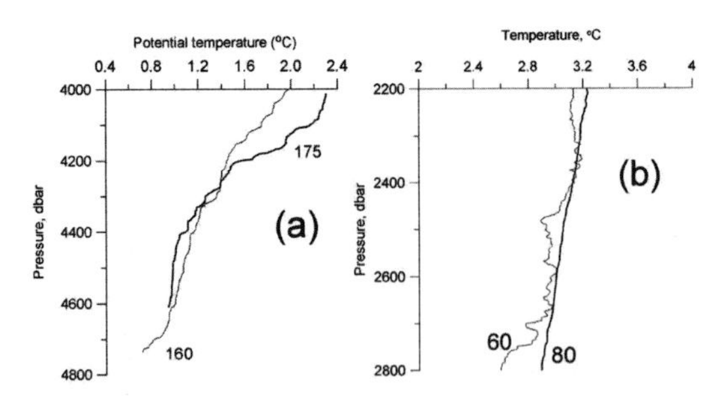

Figure 34. (a) Potential temperature profiles at stations 160 (4°30'S, thin curve) and 175 (0°15'N), thick curve) of section A17. (b) Temperature profiles at stations 60 (35°36'S, thin curve) and 80 (27°44'S, thick curve) of section A17.

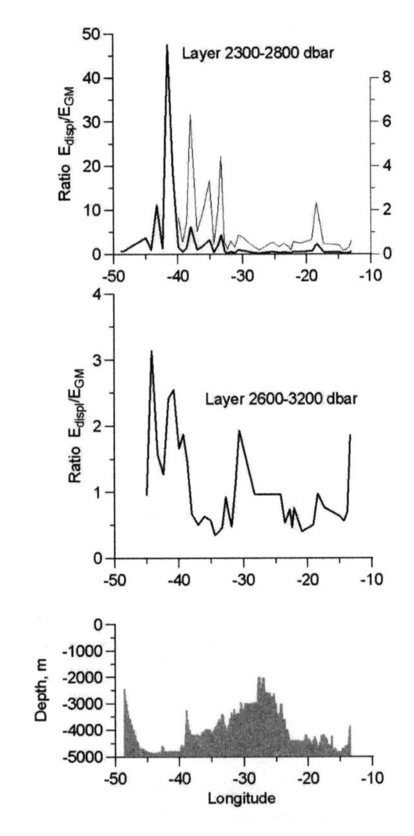

Figure 35. Section A2 (1994) approximately along 48°N. Same as in figure 29 but for the 2300 - 2800 dbar layer (upper panel) and 2600 - 3200 dbar layer (middle panel). The thin line on the graph in the upper panel is the same as the thick line but on the other vertical scale (right scale) and limited in the interval east of 40°W. The western end of the section crosses the North Atlantic Current.

eastern part of the section. An increase in spectral densities over the Mid-Atlantic ridge in this layer is smaller. It increases in the lower layers, and becomes strong in the 2000-2600 dbar layer, while the effect of the Mediterranean water inflow disappears below 1600 dbar figure 36 (lower panels).

5.9 Spreading of North Atlantic Deep Water

North Atlantic Deep Water (NADW) formed at high latitudes of the Atlantic propagates to the south as a thick layer of high salinity between 1500 and 3500 dbar. Antarctic Intermediate Water (AAIW) propagates to the north above NADW in a layer between 500 and 1500 dbar. Antarctic Bottom Water (AABW) flows to the north between NADW and the ocean bottom. The A17 salinity section (1994) demonstrating the location of these layers is shown in figure 37, lower panel. Antarctic waters are characterized by increased concentration of silicates, whereas the concentration of silicates in NADW is low. The silicates concentration in the core of NADW (level of salinity maximum) as function of latitude is shown in figure 37, upper panel. North of 35°S, the concentration of silicates is below 30 μg-atom/kg, and only slightly changes as NADW spreads to the south without mixing with Antarctic waters. Approximately south of 35°S the layer of NADW becomes thinner due to mixing with overlying AAIW and underlying AABW and the concentration of silicates supplied from Antarctic waters increases. In figure 37 (middle panel) we show the ratio of spectral densities of vertical displacements to GM spectrum within a layer between 2200 and 2800 dbar on section A17. Increased spectral densities of vertical displacements within this layer are found south of 35°S (at 45°W), exactly in the region where the NADW layer becomes thinner and concentration of silicates transported by Antarctic waters increases. The increase in silicates indicates the region of mixing between North Atlantic and Antarctic waters because there is no other source for silicates except for their supply with the Antarctic waters. Temperature profiles at two stations 60 (35°36'S) and 80 (27°44'S) are shown in figure 34. It is our opinion that uneven profile at station 60 is caused by horizontal intrusions of waters of different origin rather than by internal waves. This region of mixing is a transoceanic front in the deep ocean layers. Similar features were found on the A16 section, which follows the 25°W meridian.

Finally we can make the following conclusions related to the mixing in the ocean. The analysis of data from CTD-profiling at different WOCE sections in the Atlantic shows that the changes in the vertical structure of the temperature and salinity fields are well pronounced near topographic features. These changes can be caused by various processes, which induce vertical motions (internal tides, breaking internal waves, upwelling, intrusions near fronts, etc.) generated near the submarine ridges or by bottom and lateral friction of near bottom flows. The changes due to internal tides are caused by internal tide breaking, energy transfer to the waves of higher frequency, and formation of the fine vertical structure.

Changes in the vertical structure are reflected in the one-dimensional spatial spectra of temperature, and salinity, which are calculated from the vertical profiles. The spectra were recalculated to the spectra of vertical displacements to compare them with the GM model. In the regions far from the wave source over the ridge, the spectral densities

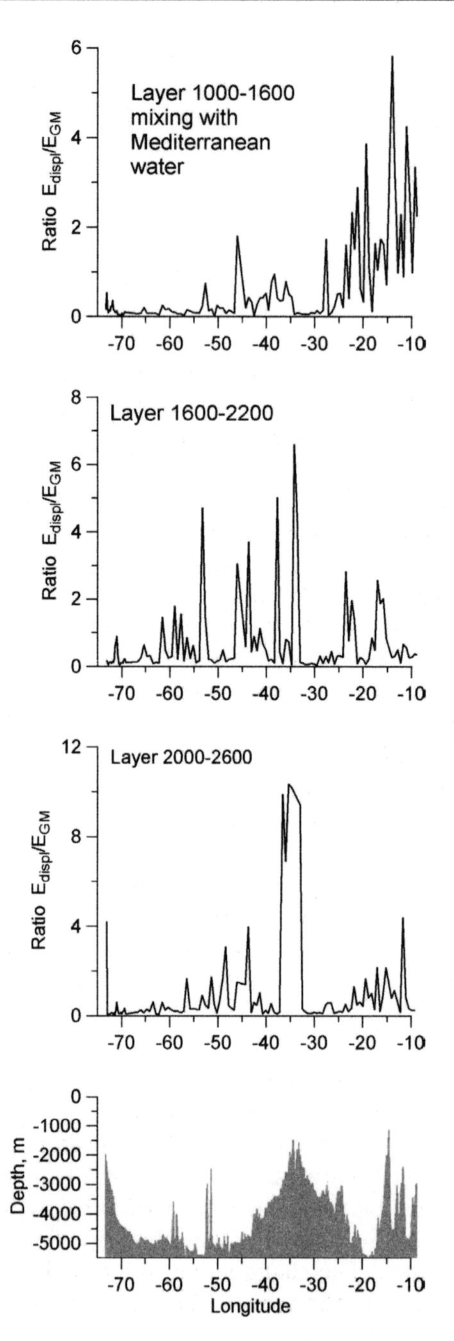

Figure 36. Section A3 approximately along 35°N. Same as in figure 29 but for the 1000 - 1600 dbar layer (upper panel); 1600 - 2200 dbar layer (first middle panel); 2000 - 26000 dbar layer (second middle panel). The eastern end of the section is located west of the Strait of Gibraltar.

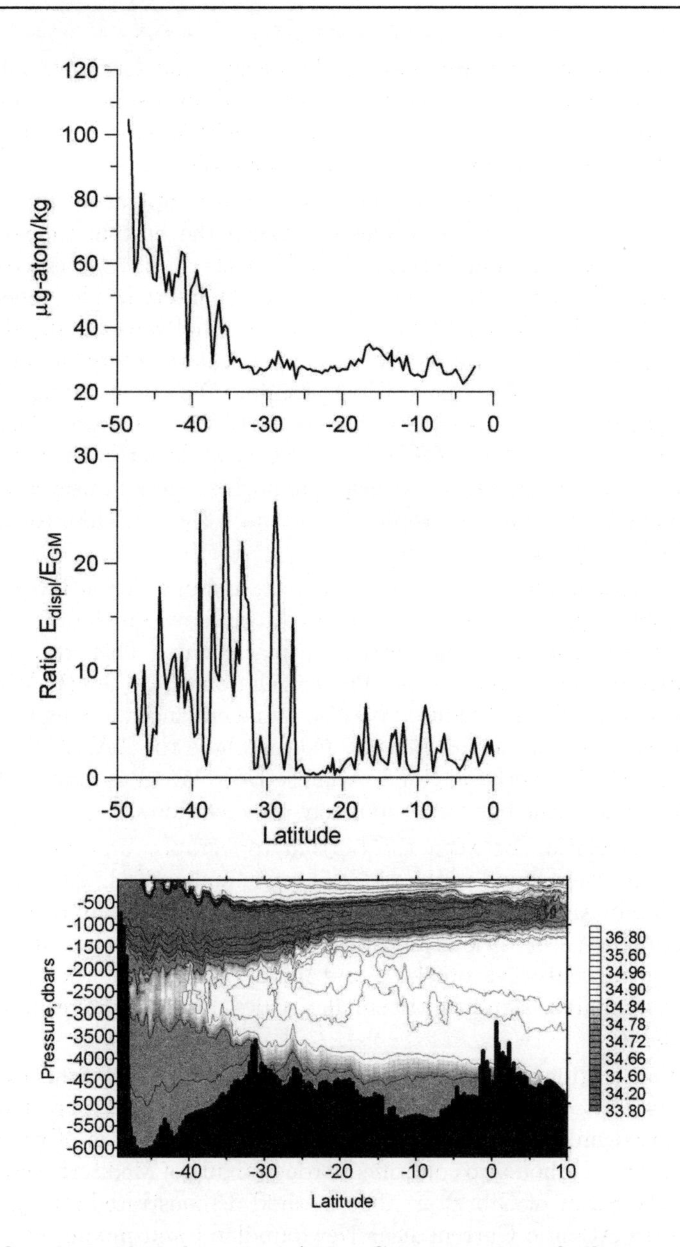

Figure 37. Quasi-meridional section A17. Concentration of silicates in the core of NADW (maximum salinity) versus latitude (upper panel). Latitudinal variability of the ratio of the spectra of vertical displacements to the GM dropped spectra in the wavenumber range from 0.028 to 0.055 cyc/m; (layer between 2200 and 2800 dbar) (middle panel). Salinity section A17 (lower panel). Antarctic waters of low salinity are shown in gray color, whereas NADW and upper layer of high salinity are in white color. A sharp increase in the values of E_{displ}/E_{GM} ratio is observed south of 35°S where enhanced mixing between Antarctic and North Atlantic waters occurs.

depending on the wavenumber are close to the background Garret-Munk model spectra, while those near the ridges exceed the background level sometimes by one order of magnitude. Strong mixing in the ocean is associated with intense vertical motion, which most likely develops near the slopes of bottom topography.

The analysis of the vertical wavenumber spectra of temperature profiles from CTD data indicates that a high correlation exists between the bottom topography and the fluctuations of temperature and vertical displacements in the water column at fixed wavenumber range. These fluctuations can be caused by internal tides generated over the slopes of submarine ridges and by high frequency internal waves induced when intense internal tide breaks. Internal waves strain the vertical structure of the hydrographic fields, which can be recorded by usual CTD-profiling. The fluctuations of temperature and salinity and the fine structure observed in the vertical profiles are reflected in the one-dimensional spatial spectra of the CTD series. Spectral densities calculated from these profiles (the so-called dropped spectra) near submarine ridges are several times greater than in the regions far from abrupt topography, where they are close to the background spectra described by the Garrett-Munk model.

The analysis made in the 600-dbar bottom layer, and in different layers between 1000 and 3200 dbar, indicates that there is high correlation between the bottom topography and fluctuations of temperature and vertical displacements. This correlation manifests itself most clearly in the bottom layers. The correlation in the South Atlantic is better pronounced than in the North Atlantic. We also found enhanced mixing in the Equatorial Channel in the western part of the Atlantic Ocean where the AABW flow narrows and a strong mixing with the overlying North Atlantic Deep Water is induced by lateral and boundary friction and shear between oppositely directed flows.

The highest correlation between the bottom topography and spectral densities of wavenumber spectra were found in the range of wavenumbers from 0.028 to 0.055 cyc/m. There are no dominating peaks on the spectra at these wavenumbers. We conclude that fine structure steps of various thicknesses within this length scale can be formed in the process of internal wave breaking and mixing. The data with 2 dbar sampling do not allow us to analyze significantly smaller scales, because the results would not be statistically reliable.

The Mid-Atlantic Ridge especially in the South Atlantic is distinguished as the region of strong mixing. The Equatorial Channel and Vema Channel were found to be the regions of enhanced mixing due to the bottom and lateral friction of the bottom water flows. The suggested method also confirmed strong mixing of Mediterranean and Atlantic waters west of the Strait of Gibraltar. The method demonstrated strong mixing at the front of the North Atlantic Current near Newfoundland and mixing of North Atlantic Deep Water with Antarctic Intermediate and Antarctic Bottom waters south of 35°S.

6 Several Approaches to the Investigation of Tidal Internal Waves in the Northern Part of the Pacific Ocean

According to the present-day concept about the balance of tidal energy in the ocean the dissipation of tidal internal waves is one of the main sources for mixing in the ocean

(Munk and Wunsch 1998), which maintains the balance between the inflow of cold waters from the polar regions into the deep layers of the ocean and warming of water in the upper layers at low latitudes. Traditionally, the investigation of tidal internal waves in the ocean is carried out by analyzing the data of measurements at moored buoy stations. Setting of moorings in the ocean is a very expensive experiment, and despite the permanent growth of the number of moorings, the amount of available data is not very large. Many regions in the ocean remain *blank spots* in the sense of investigating the oceanic processes with direct instrumental measurements of currents and temperature on moored buoys. At present, numerous CTD-profiling measurements are carried out in the ocean, XBT-temperature sections are taken from the ships of opportunity, neutral buoyancy floats and surface drifters are launched, which give enormous information about the variability of temperature and salinity fields in the ocean. Carrying out these measurements requires significantly smaller material outlay. A vast amount of these measurements has been accumulated already. So far as the propagation of tidal internal waves with amplitudes of tens of meters, and sometimes reaching 100 m distorts the structure of all fields in the ocean we suggest a study of those manifestations of tidal internal waves, which can be recorded by means of these mass measurements in the ocean. We have chosen the northern part of the Pacific Ocean as the region for the experiments. This basin is of significant interest for the countries in the region. Intensive oceanographic measurements have been carried out here and a large amount of various data has been accumulated. The measurements, which we have carried out during the MEGAPOLYGON and other experiments show that the tidal internal wave field in this region is determined by their generation near the Emperor Ridge due to the interaction of the barotropic tide currents with uneven topography on the slopes of the ridge. Wave amplitudes in the region of generation reach almost one hundred meters. During their propagation from the submarine ridge the waves loose their energy and decay. We investigated the waves in this region using several approaches. The combination of traditional and unusual approaches in the region, where many different measurements have been made, allowed us to obtain comparative characteristics of the results obtained.

6.1 Moored data analysis

This is a traditional method of investigating internal waves. The classification of Garrett-Munk (1975) refers the spectral characteristics of internal waves, obtained from these measurements, to the concept of *moored spectra*. The major amount of information in this region has been obtained from the MEGAPOLYGON experiment data, carried out in 1987. A schematic of moorings location is shown in figure 38. More than 170 moorings have been set in a square of 460 km x 520 km. Temperature meters at 1200 m depth were set on moorings in the northwestern (14 buoys) and southeastern (15 buoys) parts of the study area.

The calculation of spatiotemporal spectral estimates has been carried out according to the method suggested by Barber (1963).

The location of moorings at the study area was carried out at an almost regular grid, thus both groups of moorings with temperature meters (northwestern and southeastern) unambiguously resolve the waves in the range from 77 to 400 km, which is appropriate

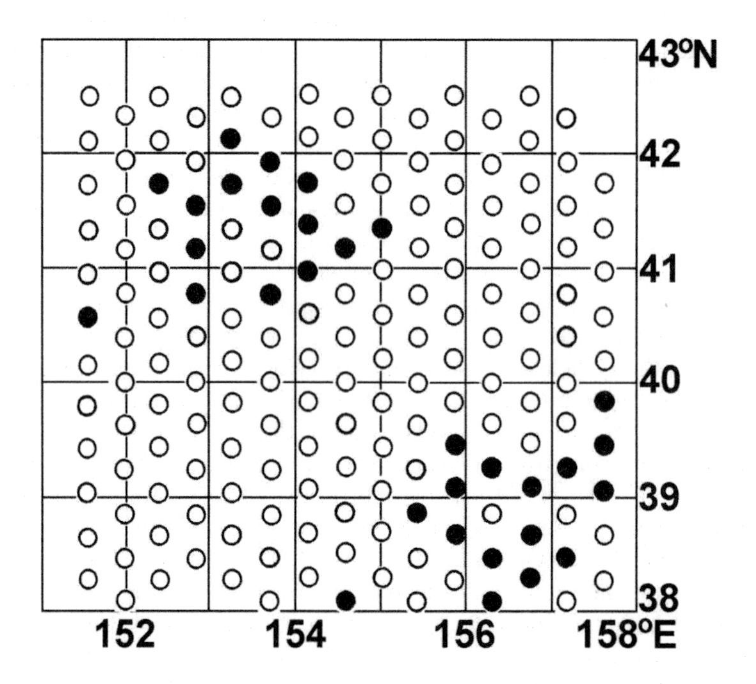

Figure 38. Schematic of location of moorings at the MEGAPOLYGON. Black dots indicate the buoys with temperature meters at 1200 m, which were used for the calculations. Reproduced with permission by Nauka.

for tidal internal waves with a wavelength of about 130–170 km.

The theoretical wavelength of a tidal internal wave has been calculated by integrating the equation for vertical velocity (1.11) and real depth profile of Brunt-Väisala frequency with zero boundary conditions at the surface and bottom of the ocean.

The wavelength for the first mode calculated this way for the MEGAPOLYGON region was approximately equal to 130 km. Near the Emperor Ridge the wavelength is somewhat greater, and it is equal to 167 km, while at a distance of 2000 km from the ridge it is equal to 156 km. The data of US moorings were analyzed in the northeastern part of the Pacific Ocean. A schematic of location of moorings is shown in figure 39.

The period of synchronous work of the moorings in the northwestern cluster lasted from August 10 to October 7, 1987, and for the southeastern cluster it lasted from September 22 to October 19. Intervals 150 h long (figure 38) were selected from these time series to estimate the tidal internal wave parameters. Two time independent intervals were distinguished in the southeastern cluster, and four in the northwestern one. Each of the estimates of spatiotemporal spectra calculated on the basis of independent data has a common maximum at a wavelength of the first mode of internal tide, which varies from 110 to 150 km, while the mean value is 130 km. The direction of wave propagation is also very stable and remains within 240–300 degrees, which corresponds to the propagation

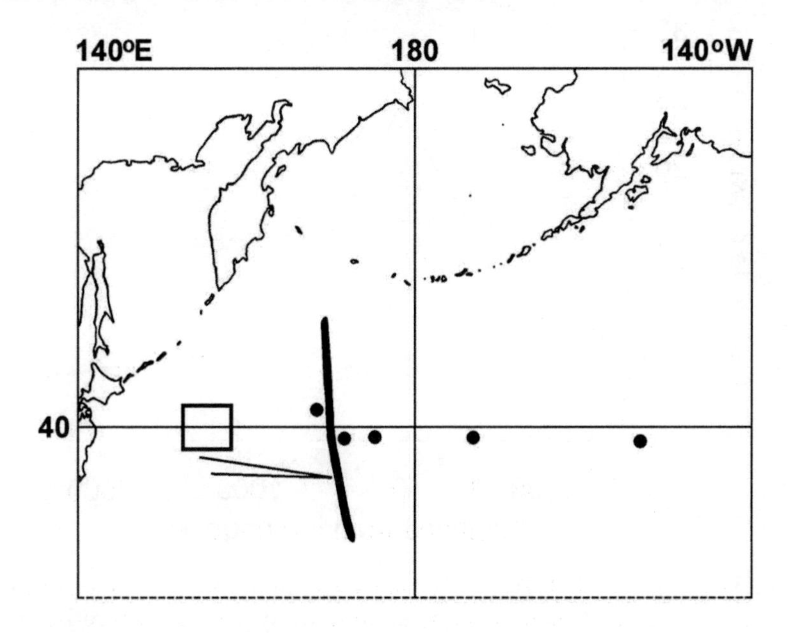

Figure 39. Schematic of location of moorings in the northwestern part of the Pacific Ocean. The Emperor Ridge is shown with a thick line. Black dots indicate US moorings. Two thin lines show the XBT-sections. The square indicates the position of the MEGAPOLYGON. Reproduced with permission by Nauka.

of waves to the west and northwest from the Emperor Ridge.

Let us investigate the variation of internal tide amplitudes, when waves propagate from the Emperor Ridge to the west. The amplitudes of internal waves were calculated on the basis of deviation of temperatures measured on moorings from their mean value with further division of this value by the mean vertical gradient of temperature. To distinguish the semidiurnal fluctuations we applied band filtering to the time series. We used the Matlab program with elliptic band filter of the second order, which provided a frequency band between $1/11\ \mathrm{h}^{-1}$ and $1/14\ \mathrm{h}^{-1}$ (Parks and Burrus, 1987).

The measurements at the MEGAPOLYGON were carried out at 1200 m depth, and the records from US buoys east of the ridge were obtained at 670–680 m depth. The dependence of internal tide amplitudes on the distance from the ridge is shown in figure 40. The solid line shows an experimental dependence.

6.2 Numerical modeling

Numerical modeling using the model described above was carried out to compare the results obtained from the analysis of measurements data on moored buoys with theoretical calculations. The amplitudes of waves and their decay in the course of propagation were analyzed.

Generation of internal waves over irregularities of the topography is determined by

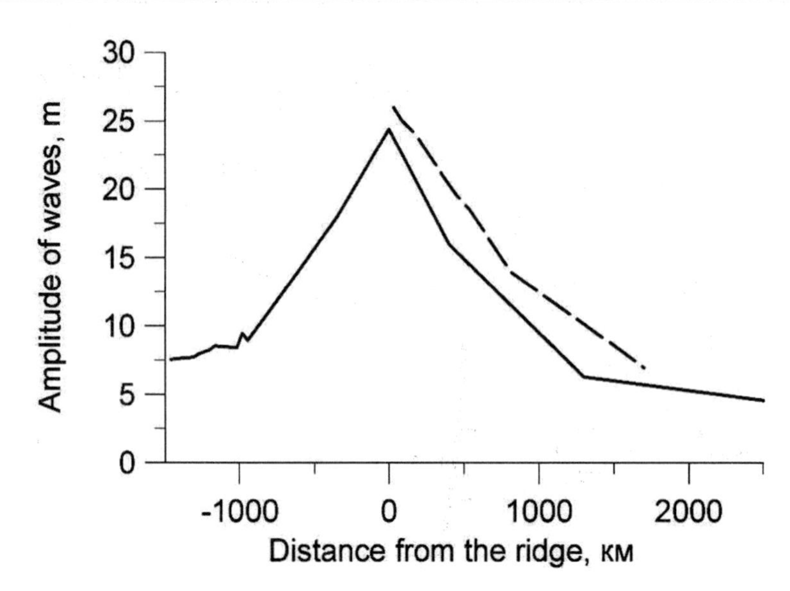

Figure 40. Dependence of the tidal internal wave amplitude on the distance from the Emperor Ridge. Solid line shows experimental data; dashed line corresponds to model calculation. Reproduced with permission by Nauka.

the specification of mass transport in the currents of a barotropic tide, by means of the boundary condition for the stream function at the bottom, in the form of a periodic function. The water transport in the barotropic tide current was specified on the basis of the measurements at moored buoy stations. It corresponded to the barotropic tide currents equal to 1.4 cm/s. The bottom topography was specified from the data of navigation charts. Parameters of stratification were specified on the basis of oceanographic atlases and hydrographic and CTD measurements during the MEGAPOLYGON experiment. The model calculations were carried out on a grid with a horizontal step of 1 km, 32 vertical levels and a time step of 7.5 s. The coefficients of horizontal and vertical turbulent exchange were equal to 200 and 0.01 m^2/s, respectively. The coefficient of horizontal turbulent exchange over the ridge was increased to 250 m^2/s. The size of the grid allowed us to study the behavior of the wave at a distance of 850 km at both sides of the ridge. The density field was calculated in the entire area of modeling. The amplitudes of vertical displacements caused by internal waves were determined at a depth of 1200 m on the basis of density deviation from the non-disturbed state at specific distances from the ridge. The decay of the model amplitudes is shown by the dashed line in figure 40. The value and character of the amplitude decay is close to that found experimentally. The wavelength determined from numerical modeling is equal to 112 km.

6.3 Analysis of data from sections made with expandable bathythermographs (XBT)

In a certain sense these data can be interpreted as towed measurements. The classification by Garrett-Munk (1975) refers to the spectral characteristics of the internal waves obtained from the data of such measurements as *towed spectra*. The unusual approach implies that, in general, the towed measurements are called those, which are obtained, when a vessel is towing a string of sensors or any other device. The present day devices limit the speed of towing by 5–6 knots, depth of measurement by 200 m, and the length of the tack by 100 km. These parameters are not acceptable for investigating internal tides, whose wavelength is usually greater than 100 km. Therefore, only short internal waves are investigated using the towed methods. Nevertheless, temperature measurements with expandable bathythermographs (XBT), which are carried out at full speed of the ship and reach the depths of 700-800 m, can be considered towed in the concept of Garrett-Munk the classification. The lengths of the tack exceed the wavelength several times, and these data can be used for analyzing internal waves, if the XBT launches are made frequently enough. The opposite direction of tack allows us to take into account the Doppler effect. We obtained the data of profiling by expandable bathythermographs from 35°48'N, 160°46'E to 33°54'N, 170°29'E and back to the point at 34°40'N, 161°32'E, which were measured in 1981. The length of the first tack is 1133 km, and the length of the second tack is 929 km. The tacks were made almost normal to the ridge, which allows us to consider that the first tack was directed almost against the direction of the tidal internal wave propagation, and the opposite tack was made approximately in the direction of the wave. The speed of the ship was about 7.2 m/s, and the approximate phase speed of the wave was equal to c = 3 m/s. Under the condition of this relation between the speeds, the Doppler shifted wavelength L_D is related to the real one L as

$$L_D = \frac{LU}{c \pm U}. \tag{6.1}$$

A plus sign in the denominator of relation (6.1) corresponds to opposite directions of the ship and wave, and minus corresponds to the same directions. The temperature profiles measured by XBTs reach a depth of 500–600 m, but most reliable records were obtained from the depths of about 300 m. The analog records of temperature profiles with depth were digitized and one-dimensional spatial spectra depending on the horizontal wavenumber were calculated on the basis of these data. The wavelength of an internal wave is approximately 7–9 times shorter than the length of the tack, hence, the statistical reliability of the spectra calculations is not very high. Nevertheless we can give an interpretation of the spectral peaks. The calculations for two tacks are shown in figure 41.

Let us use relation (6.1) to calculate the real wavelengths from the Doppler shifted values of wavenumbers corresponding to the peaks on the spectra. The wavelength taken from the spectra calculated from the tack directed to the ridge corresponding to a wavenumber of 0.014 cycles/km is equal to 72 km, which corresponds to a real wavelength of 92.7 km. For the tack directed from the ridge the peak at a value of wavenumber 0.0075 cycles/km corresponds to a Doppler wavelength of 137.3 km and a real wavelength equal to 96.2 km, respectively. These two values correlate well with each other, but both of

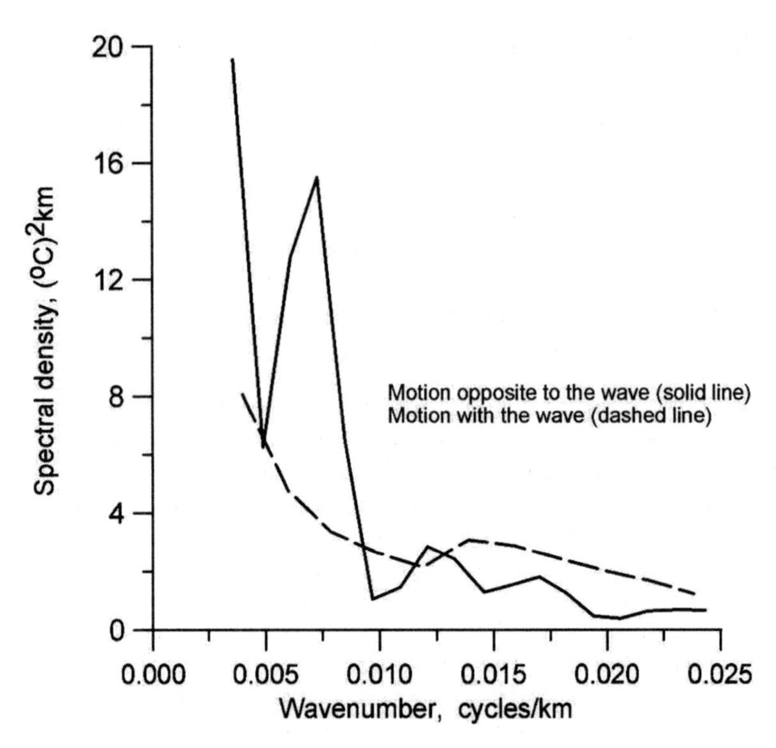

Figure 41. One-dimensional spatial temperature spectra at a depth of 300 m based on XBT-section in the direction to the Emperor Ridge (dash line) and back (solid line). Reproduced with permission by Nauka.

them are smaller than the estimates obtained from the measurements on the buoys and by integrating the equation for eigen values (1.11). A slight increase in wavelength is possible if we assume that our tacks were not strictly located in the direction of the wave propagation.

6.4 Analysis of CTD sections data

We want to show that vertical temperature profiles are strongly transformed due to the propagation of internal tides, which can be found from statistical analysis of vertical profiles of temperature and vertical temperature gradients. If we calculate spectral densities of temperature fluctuations, the level of spectral densities would be higher for those functions, where intensive internal waves are observed in contrast to those, where there are no strong internal waves. It is noteworthy that vertical profiles of temperature reflect the presence of homogeneous steps.

While interpreting the data of measurements in the light of the Garrett-Munk model, we cannot analyze the internal tides themselves. First, the model itself does not consider these internal waves. Second, we cannot get an estimate of spectral density for the vertical wavenumbers corresponding to tidal internal waves from the calculation of the spectra on the basis of vertical wavenumbers. The depth of the ocean itself is close to the inverse value of the vertical wavenumber for such waves. We can calculate the estimates for significantly larger wavenumbers. Thus, we can deal only with the secondary field of internal waves, generated as a result of the wave breaking of high amplitude tidal internal waves. Internal waves of this high-frequency and short wavelength scale are usually considered in the comparison of observations with the background Garrett-Munk model.

We shall analyze the fluctuations deeper than 500 m, where other processes associated with the close position to the ocean surface are not significant. In addition we exclude from the consideration the depths higher than 4500 m, where near bottom processes may play a certain role, besides the vertical density gradients are very low at those depths, so that there are almost no changes on vertical profiles.

The spectra of vertical profiles were calculated using Fourier transform of the correlation function. The maximum shift of the correlation function was equal to 480 m and provided 20 degrees of freedom for the calculations. Preliminary high filtering of the series eliminated all long wave fluctuations with vertical wavenumbers smaller than 0.001 cyc/m. The spectra were calculated from the vertical profiles of the temperature. In addition, we calculated the spectra of the vertical displacements, which were found as fluctuations of temperature normalized by the corresponding vertical temperature gradients. The spectra confirm significant differences of spectral functions of profiles measured near submarine ridges and at a distance from them.

Let us consider the spatial variability of spectral densities. The spatial variability of the values of spectral functions for the P1 section for the spectral point corresponding to 0.01 cyc/m is shown in figure 42. The upper curve illustrates the variability of the spectral densities of vertical displacements, and the lower one shows the depth of the ocean along the section. The plots clearly demonstrate the growth of the spectral densities in the region of sharp changes of the bottom topography.

This unusual method of analyzing the experimental data for investigating tidal internal waves has been suggested by Millard (1972). According to the Garrett-Munk classification spectral characteristics of internal waves based on the data of vertical profiling correspond to the concept of *dropped spectra* (Garrett and Munk, 1975). The measurements with CTD profilers can be compared with the theoretical *dropped* spectra depending on the vertical wavenumber. Tidal internal waves propagating in the ocean strain the vertical profile of temperature and salinity corresponding to the state of the ocean at rest. The complexity of this problem lies in the fact that it is not easy to isolate the signal caused by an internal wave, but the appeal of this approach is based on the abundance of these data of measurements.

To analyze the variability of tidal internal waves we have chosen a region in the northern part of the Pacific Ocean, with many moored measurements and two WOCE hydrographic sections made across the ocean from west to east. The P1 section was made along 40–46°N, while the P4 section followed 8–10°N. The sections were made in 1985 and 1989. The stations on sections were made with an interval of 30 nautical miles. The vertical spacing of measured data is 2 db. Both sections crossed such significant features of the bottom topography as submarine ridges. The P1 section crossed the Emperor Ridge (approximately by 170°E), and the P4 section crossed the Marshall Islands

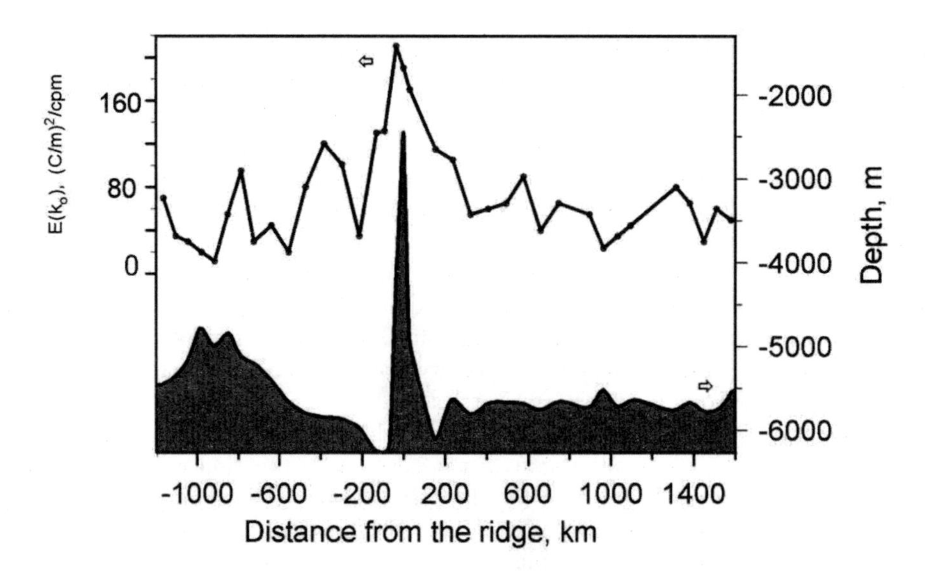

Figure 42. Variations of the values of spectral densities for the spectral estimate corresponding to the wavenumber equal to 0.01 cycl/m along the P1 section. The upper curve corresponds the spectrum of vertical displacements, the lower curve shows the depth of the ocean along the section. Reproduced with permission by Nauka.

(approximately by 170°E) and the Line Islands (approximately by 165°W). It is known that tidal internal waves are intensively generated near submarine ridges.

The analysis of the CTD-profiling in the northern part of the Pacific Ocean allowed us to show that the internal tides cause changes in the vertical structure of the hydrological fields caused by the breaking of long internal waves and energy transfer to the waves of the higher frequency range and formation of the fine vertical structure. These changes, in particular, fine structure steps and layers of high temperature and salinity gradients, can be registered by usual CTD profiling. The changes in the vertical structure are reflected in one-dimensional spatial spectra of temperature and temperature gradients, which are calculated from the data of vertical CTD profiling. In the regions, located far from the source, the spectral densities depending on the vertical wavenumber are close to the model spectra of Garrett-Munk, while those spectral densities measured near the ridge exceed the background level almost by one order of magnitude.

6.5 Data of drifters

The analysis is complemented by the results of drifters launched in the area. The drifters indicate a decay of semidiurnal orbits at the surface while the waves propagate away from the ridge. We processed the drifter data in the following way. First we filtered out the trend caused by the mean motion in the area. Second we filtered out the inertial motion with local inertial frequency. A graph of smoothed orbits of the drifters motion near the ridge (the longitude is 157°E), where tidal internal waves are intensive, shows

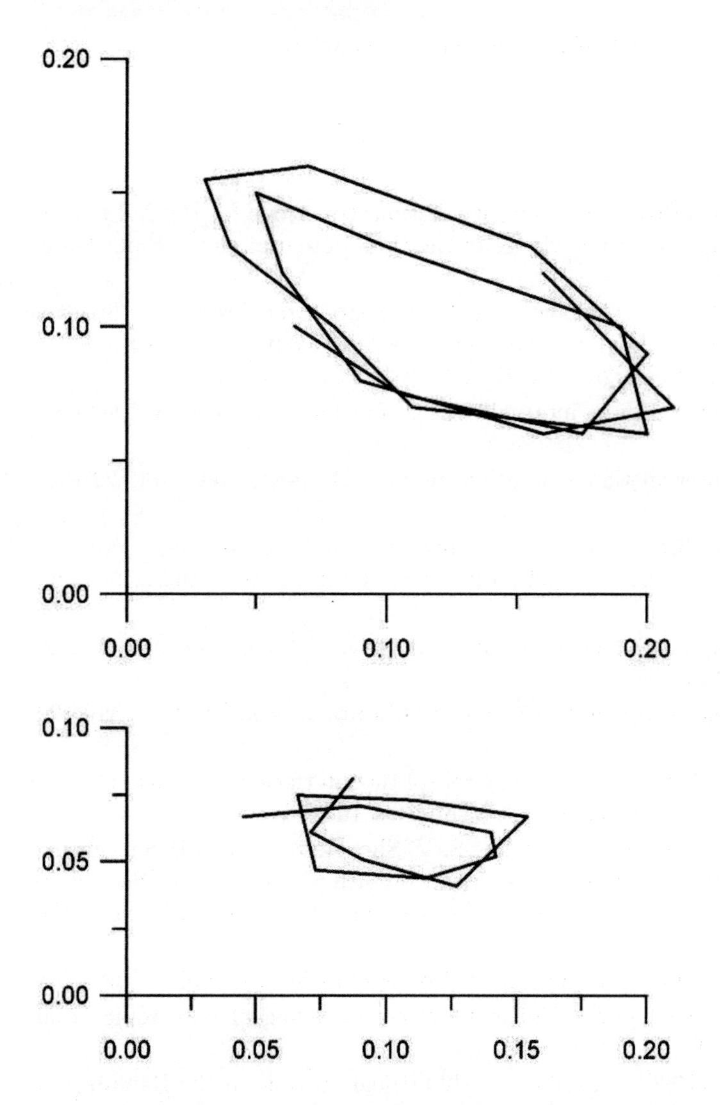

Figure 43. Orbits of drifters near the Emperor Ridge (above) and far from the ridge (below). The distances at the axes are given in nautical miles. Positions of drifters were observed with an interval of 2 hours.

that the displacement of drifters induced by internal waves have a diameter of about 0.15 nautical mile (1 nautical mile is 1852 m), whereas away from the ridge at a distance exceeding 1000 km from the ridge the diameter of the orbits is of the order of about 0.05 nautical mile (the longitude is 173°W) (figure 43). We should emphasize that the drifter data is very noisy, and it was not very easy to find the proper intervals in the records of the drifters positions to illustrate the effects associated with tidal internal waves.

This study was supported by the RFBR, project no. 03-05-64024.

Bibliography

Y. Accad and C. L. Pekeris (1978). Solution of the tidal equations for the M2 and S2 tides in the world oceans from a knowledge of the tidal potential alone, Phil. Trans. Roy. Soc. A290, 235-66.

L. Armi and D. M. Farmer (1988). The flow of Mediterranean water through the Strait of Gibraltar, D. M. Farmer and L. Armi, The flow of Atlantic water through the Strait of Gibraltar, Progress in Oceanogr. Vol. 21, 1-105.

P. G. Baines (1973). The generation of internal tides by flat-lump topography. Deep-Sea Res. Vol. 20, 179-205.

P. G. Baines (1982). On internal tide generation models. Deep-Sea Res. Vol. 29 (3a), 307-38.

N. F. Barber (1963). The directional resolving power of an array of wave detectors, In: Ocean wave spectra, Easton, Engelwood Cliffs, N.Y.: Prentice Hall, 1963, pp. 137-150.

T. H. Bell (1975). Topographically generated internal waves in the open ocean, J. Geophys. Res. Vol. 80, 320-27.

F. M. Boyce (1975). Internal waves in the Strait of Gibraltar. Deep Sea Res. Vol. 22, 597-610.

N. Bray, J. Ochoa and T. H. Kinder (1995). The role of the interface in exchange through the Strait of Gibraltar. J. Geophys. Res. C, Vol. 100, 10755-76.

V. I. Burenkov, S. V. Ershova, O. V. Kopelevich, S. V. Sheberstov and V. P. Shevchenko (2001). An estimate of the distribution of suspended matter in the Barents Sea waters on the basis of the SeaWiFS satellite ocean color scanner. Oceanology (Moscow), Vol. 41, (5), 622-8.

D. E. Cartwright (1977). Ocean tides, Reports of Progress in Physicas, Vol. 40, 665-708.

G. D. Egbert and S. Erofeeva (2002). Efficient inverse modeling of barotropic ocean tides. J. Atmos. Ocean Tech. Vol. 19, 183-204.

T. Furevik and A. Foldvik (1996). Stability at M2 critical latitude in the Barents Sea. J. Geophys. Res. C, Vol. 101, 8823-37.

C. Garrett and W. Munk (1972). Space time scales of internal waves. Geophys. Fluid Dyn. Vol. 3 (3), 225-64.

C. Garrett and W. Munk (1979). Internal waves in the ocean. Annu. Rev. Fluid Mech. Vol. 11, 339-69.

C. Garrett and W. Munk (1975). Space-time scales of internal waves: A progressive report. J. Geophys. Res. Vol. 80 (3), 291-8.

B. Gjevik, E. Nøst and T. Straume (1994). Model simulations of the tides in the Barents Sea. J. Geophys. Res. Vol. 99 (C2), 3337-50.

T. Hibia (1990). Generation mechanism of internal waves by a vertically sheared tidal flow over a sill. J. Geophys. Res. C, Vol. 95, 1757-64.

T. H. Kinder, G. Parrilla, N. Bray and D. A. Burns (1989). The hydrographic structure of the Strait of Gibraltar. Seminario Sobre La Oceanografia Fisica del Estrecho de Gibraltar, SECEG, Madrid, October 24-28, 1988, pp. 55-67.

W. Krauss (1966). Interne Wellen (Internal Waves). Berlin: Gebruder Borntraeger.

R. Millard (1972). Further comments on vertical temperature spectra in the MODE region, MODE Hot line News, no. 18, 1-6.

E. G. Morozov (1995). Semidiurnal internal wave global field. Deep-Sea Res. Vol. 42, 135-48.

E. G. Morozov, V. G. Neiman, S. V. Pisarev and S. Yu. Erofeeva (2003). Internal tidal waves in the Barents Sea. Doklady Earth Sciences Vol. 393, No. 8, 686-8.

E. G. Morozov, G. Parrilla-Barrera, M. G. Velarde and A. D. Scherbinin (2003). The Straits of Gibraltar and Kara Gates: A Comparison of Internal Tides. Oceanologica Acta, Vol. 26 (3), 231-41.

E. G. Morozov and S. V. Pisarev (2002). Internal tides at the Arctic latitudes (Numerical experiments). Okeanologiya Vol. 42 (2), 165-73.

E. G Morozov, K. Trulsen, M. G. Velarde and V. I. Vlasenko (2002). Internal tides in the Strait of Gibraltar. J. Phys. Oceanogr. Vol. 32, 3193-206.

W. H. Munk (1997). Once again: tidal friction. Progress in Oceanogr. Vol. 40, 7-35.

W. H. Munk and C. Wunsch (1998). Abyssal recipes II: energetics of tidal and wind mixing. Deep-Sea Res. Vol. 45, 1977-2010.

E. Nøst (1994). Calculating tidal currents profiles from vertically integrated models near the critical latitude in the Barents Sea. J. Geophys. Res. Vol. 99 (C4), 7885-901.

Oceanography from the Space Shuttle (1989): NASA Photograph S-17-34-081, ONR, p. 89.

T. W. Parks and C. S. Burrus (1987). Digital filter design. J. Wiley and Sons, Ch. 7, section 7.3.3.

A. R. Parson, R. H. Bourke, R. D. Muench, et al. (1996). The Barents Sea Polar Front in Summer, J. Geophys. Res. Vol. 101, no. 6, 14201-21.

R. D. Pingree and A. L. New (1991). Abyssal penetration and bottom reflection of internal tidal energy in the Bay of Biscay. J. Phys. Oceanogr. Vol. 21, 28-39.

R. Pinkel, W. Munk, P. Worchester, B. D. Cornuelle, D. Rudnick, J. Sherman, J. H. Filloux, B. D. Dushaw, B. M. Howe, T. B. Sanford, et al. (2000). Ocean Mixing Studies near the Hawaiian Ridge. EOS, Vol. 81 (46), 545 -53.

R. Robertson (2001). Internal tides and baroclinicity in the Southern Weddell Sea. 2. Effects of the critical latitude and stratification, J. Geophys. Res. C, Vol. 106, 27017-34.

F. J. Sandoval and G. L. Weatherly (2001). Evolution of the deep western boundary current of Antarctic Bottom Water in the Brazil Basin. J. Phys. Oceanogr. Vol. 31 (6), 1440-60.

G. M. Torgrimson and B. M. Hickey (1979). Barotropic and baroclinic tides over the continental slope and self off Oregon, J. Phys. Oceanogr. Vol. 9, 945-61.

V. I. Vlasenko (1991). A non-linear model of topographic generation of baroclinic tides of extended bottom irregularities. Morsk. Gidrofiz. J., No. 6, 9-16. (English translation: Sov. J. Phys. Oceanogr. Vol. 3, No 6, 417-424).

V. I. Vlasenko and K. Hutter (2002). Numerical experiments on the breaking of solitary internal waves over a slope-shelf topography. J. Phys. Oceanogr. Vol. 32, 1779-93.

J. C. Wesson and M. C. Gregg (1994). Mixing at Camarinal Sill in the Strait of Gibraltar, J. Phys. Oceanogr. Vol. 99, 9847-78.

World Ocean Atlas (1977). Vol. 1, Pacific Ocean, Vol. 2, Atlantic and Indian Oceans, Leningrad.